recent advances in phytochemistry

volume 30

Phytochemical Diversity and Redundancy in Ecological Interactions

RECENT ADVANCES IN PHYTOCHEMISTRY

Proceedings of the Phytochemical Society of North America
General Editor: John T. Romeo, University of South Florida, Tampa, Florida

Recent Volumes in the Series:

Volume 21 **The Phytochemical Effects of Environmental Compounds**
Proceedings of the Twenty-sixth Annual Meeting of the Phytochemical Society of North America, College Park, Maryland, July, 1986

Volume 22 **Opportunities for Phytochemistry in Plant Biotechnology**
Proceedings of the Twenty-seventh Annual Meeting of the Phytochemical Society of North America, Tampa, Florida, June, 1987

Volume 23 **Plant Nitrogen Metabolism**
Proceedings of the Twenty-eighth Annual Meeting of the Phytochemical Society of North America, Iowa City, Iowa, June, 1988

Volume 24 **Biochemistry of the Mevalonic Acid Pathway to Terpenoids**
Proceedings of the Twenty-ninth Annual Meeting of the Phytochemical Society of North America, Vancouver, British Columbia, Canada, June, 1989

Volume 25 **Modern Phytochemical Methods**
Proceedings of the Thirtieth Annual Meeting of the Phytochemical Society of North America, Quebec City, Quebec, Canada, August, 1990

Volume 26 **Phenolic Metabolism in Plants**
Proceedings of the Thirty-first Annual Meeting of the Phytochemical Society of North America, Fort Collins, Colorado, June, 1991

Volume 27 **Phytochemical Potential of Tropical Plants**
Proceedings of the Thirty-second Annual Meeting of the Phytochemical Society of North America, Miami Beach, Florida, August, 1992

Volume 28 **Genetic Engineering of Plant Secondary Metabolism**
Proceedings of the Thirty-third Annual Meeting of the Phytochemical Society of North America, Pacific Grove, California, June–July, 1993

Volume 29 **Phytochemistry of Medicinal Plants**
Proceedings of the Thirty-fourth Annual Meeting of the Phytochemical Society of North America, Mexico City, Mexico, August, 1994

Volume 30 **Phytochemical Diversity and Redundancy in Ecological Interactions**
Proceedings of the Thirty-fifth Annual Meeting of the Phytochemical Society of North America, Sault Ste. Marie, Ontario, Canada, August, 1995

A Continuation Order Plan is available for this series. A continuation order will bring delivery of each new volume immediately upon publication. Volumes are billed only upon actual shipment. For further information please contact the publisher.

recent advances in phytochemistry

volume 30

Phytochemical Diversity and Redundancy in Ecological Interactions

Edited by

John T. Romeo
University of South Florida
Tampa, Florida

James A. Saunders
USDA
Beltsville, Maryland

and

Pedro Barbosa
University of Maryland
College Park, Maryland

PLENUM PRESS • NEW YORK AND LONDON

Library of Congress Cataloging-in-Publication Data

On file

Cover photographs: Imago and larva of *Zygaena trifolii* which possesses a cyanogenic system similar to that of its host plant *Lotus*. It both accumulates linamarin and lotaustralin from the host and synthesizes the compounds *de novo*—redundancy?

Proceedings of the Thirty-Fifth Annual Meeting of the Phytochemical Society of North America on Phytochemical Diversity and Redundancy in Ecological Interactions, held August 12–16, 1995, in Sault Ste. Marie, Ontario, Canada

ISBN 0-306-45500-5

© 1996 Plenum Press, New York
A Division of Plenum Publishing Corporation
233 Spring Street, New York, N. Y. 10013

All rights reserved

10 9 8 7 6 5 4 3 2 1

No part of this book may be reproduced, stored in a retrieval system, or transmitted in any form or by any means, electronic, mechanical, photocopying, microfilming, recording, or otherwise, without written permission from the Publisher

Printed in the United States of America

PREFACE

Diversity within and among living organisms is both a biological imperative and a biological conundrum. Phenotypic and genotypic diversity is the critical currency of ecological interactions and the evolution of life. Thus, it is not unexpected to find vast phytochemical diversity among plants. However, among the most compelling questions which arise among those interested in ecological phytochemistry is the extent, nature, and reasons for the diversity of chemicals in plants.

The idea that natural products (secondary metabolites) are accidents of metabolism and have no biological function is an old one which has resurfaced recently under a new term "redundancy." Redundancy in the broader sense can be viewed as duplication of effort. The co-occurrence of several classes of phytochemicals in a given plant may be redundancy. Is there unnecessary duplication of chemical defense systems and if so, why? What selective forces have produced this result? On the other hand, why does the same compound often have multiple functions?

At a symposium of the Phytochemical Society of North America held in August 1995, in Sault Ste. Marie, Ontario, Canada, the topic "Phytochemical Redundancy in Ecological Interactions" was discussed. The chapters in this volume are based on that symposium. They both stimulate thought and provide some working hypotheses for future research. It is being increasingly recognized that functional diversity and multiplicity of function of natural products is the norm rather than the exception.

Berenbaum and Zangerl set the stage as they eloquently make the case that function in nature cannot be equated with biological kill in bioassays. With examples drawn from a single class of compounds, the furancoumarins, evidence for diverse functions, multiple functions, and the importance of synergism in mixtures is provided. Lindroth and Hwang, concentrating on a single genus of plants, the aspens, show how bioactivity of terpenoids and phenolics spans not only families, but kingdoms. Similar examples of multiplicity of function are provided by Renwick from *Pieris* butterfly/glucosinolate studies, and by Siegel and Bush from alkaloids in endophytic grasses.

Hammerschmidt and Schultz draw the parallel between insecticidal activity and defense against pathogens, emphasizing that different approaches to problems among plant/insect ecologists and plant pathologists have heretofore obscured the idea of the same or similar compounds having a range of biological function. Both groups may gain insight from studies of the other's discipline. Isman et al., in a discussion of compounds from the neem tree, emphasize the importance of nontoxic compounds which nonetheless can enhance toxicity. This is a relatively new concept which further complicates already complicated synergistic effects. It needs further attention. Cates' long-term study of the unique pine/sawfly interaction demonstrates convincingly that interactions mediated by chemicals are never simple, and that many relatively common natural products act in concert, possibly with common products of primary metabolism.

Nahrstedt's paper showing that an insect makes a toxic cyanogen which it can also accumulate from its host plant puts still a different twist on the debate. Is such duplication of effort redundancy in the narrow sense or an expanded functional role? In the same vein, the diversity of mechanisms which are responsible for systemic and non-systemic resistance to pathogens can be viewed as either "redundant" or adaptive to a variety of selective pressures. The papers by Constabel et al. and Uknes et al. are both enlightening and perplexing in this regard.

Finally, the ideas of complexity theory, made accessible in the paper by Jarvis and Miller, and the redundancy defense, here further refined by Firn and Jones, provoke a lively debate which ultimately may help ecologists and phytochemists alike refine their ideas about the ecological functions of natural products. While the papers in this volume go a long way towards solidifying the case for natural product functional diversity and multiplicity, they also emphasize a complexity of nature that is perhaps too convoluted and too varied to be explained by any single term, be it diversity, redundancy, or any new one on the horizon!

The senior editor expresses gratitude to Dawn McGowan for excellent technical expertise in the preparation of this volume. We have made an effort in all papers in this volume to emphasize the term "natural product" in lieu of the outdated term "secondary metabolite."

John T. Romeo, *University of South Florida*
James A. Saunders, *USDA, Beltsville*
Pedro Barbosa, *University of Maryland*

CONTENTS

1. Phytochemical Diversity: Adaptation or Random Variation? 1
    May R. Berenbaum and Arthur R. Zangerl

2. Diversity, Redundancy, and Multiplicity in Chemical Defense
        Systems of Aspen 25
    Richard L. Lindroth and Shaw-Yhi Hwang

3. Diversity and Dynamics of Crucifer Defenses against Adults and
        Larvae of Cabbage Butterflies 57
    J. A. A. Renwick

4. Defensive Chemicals in Grass-Fungal Endophyte Associations 81
    Malcolm R. Siegel and Lowell P. Bush

5. Multiple Defenses and Signals in Plant Defense against Pathogens
        and Herbivores 121
    Ray Hammerschmidt and Jack C. Schultz

6. Phytochemistry of the Meliaceae: So Many Terpenoids, So Few
        Insecticides 155
    Murray B. Isman, Hideyuki Matsuura, Shawna MacKinnon,
        Tony Durst, G. H. Neil Towers, and John T. Arnason

7. The Role of Mixtures and Variation in the Production of
        Terpenoids in Conifer-Insect-Pathogen Interactions 179
    Rex G. Cates

8. Relationships between the Defense Systems of Plants and Insects:
        The Cyanogenic System of the Moth *Zygaena trifolii* 217
    Adolf Nahrstedt

9. Polyphenol Oxidase as a Component of the Inducible Defense
   Response in Tomato against Herbivores 231
   C. Peter Constabel, Daniel R. Bergey, and Clarence A. Ryan

10. The Role of Benzoic Acid Derivatives in Systemic Acquired
    Resistance 253
    Scott Uknes, Shericca Morris, Bernard Vernooij, and John Ryals

11. Natural Products, Complexity, and Evolution 265
    Bruce B. Jarvis and J. David Miller

12. An Explanation of Secondary Product "Redundancy" 295
    Richard D. Firn and Clive G. Jones

Index 313

Chapter One

PHYTOCHEMICAL DIVERSITY
Adaptation or Random Variation?

May R. Berenbaum and Arthur R. Zangerl

Department of Entomology
320 Morrill Hall
University of Illinois
505 S. Goodwin, Urbana, Illinois 61801-3795

Introduction 1
Plant "Screening" vs Insecticide Screening 2
Parsnips as Paradigms—Why So Many Furanocoumarins? 4
    Historical Considerations 5
    Biosynthetic Considerations 6
    Activities of Furanocoumarins Assayed Alone 7
    Activities of Mixtures of Furanocoumarins 12
    Detoxification Systems and Their Contribution to the Evolution of
        Chemical Diversity 15
Does Common Sense Prevail? 16

INTRODUCTION

The spectacular diversification of plant secondary metabolism has prompted a century of speculation about its adaptive significance.[1-7] For much of the past hundred years it has been thought that natural selection effected by consumers is the driving force maintaining that diversity. Jones and Firn[8], however, recently have suggested that "the evolution of plant defence may...have proceeded independent of consumer adaptation." In other words, natural products may be maintained by plants not due to any selective advantage with respect to herbivory that accrues to genotypes that manufacture them but rather because "plants with a high absolute diversity of secondary metabolites have a greater probability of producing one or more active compounds at any time than plants with a low diversity."

According to this view, most secondary chemicals serve no role other than to contribute in a general way to the variability necessary to increase the probability of producing a few biologically active compounds for use when ecological circumstances requiring defense arise. Jones and Firn,[8] then, reject what they call the "common sense scenario"—that "well-defended plants should contain a moderate diversity of highly active compounds, and few if any inactive compounds other than essential precursors." This long-held latter view is predicated on the assumption that the production of natural products entails a significant fitness cost. In the absence of countervailing selection pressure from herbivores or pathogens, genotypes producing defense compounds would be selected against and defense compounds would become rare or absent from the population. A demonstrable lack of activity of a putative defense compound against an ecologically associated herbivore of a particular plant, then, is taken in this view as evidence of consumer adaptation.

Despite the importance that biological activity plays in their hypothesis, Jones and Firn[8] do not explictly define what they mean by the term. The legitimacy of the common-sense scenario, however, rests largely on the definition of "activity." In an attempt to address the question, "Are most secondary compounds highly active?", Jones and Firn[8] point to azadirachtin, pyrethroids, and nicotine as "highly active compounds," but temper this classification with the caveat that even these (and other such active compounds) are "rarely 100% lethal or inhibitory at naturally occurring concentrations." These authors then cite success rates in random screening programs at, among other places, the National Cancer Institute, Rohm and Haas, Dupont and Ciba-Geigy. Thus, the criterion for recognizing activity appears to be the laboratory bioassay designed to evaluate a single physiological or biochemical property. Reliance upon this criterion, however, overlooks the fact that plants "screen" for activity in ways that are completely different from the ways in which manufacturers of insecticides, anticancer drugs, or other commercially viable products do.

PLANT "SCREENING" VS INSECTICIDE SCREENING

One way that plant "screening" differs from drug discovery is in the nature of the target species against which biological activity is assessed. Most laboratory bioassays are conducted, not surprisingly, against laboratory bioassay species—the proverbial "guinea pigs", although in reality such "organisms" can be cells in culture as well as inbred laboratory strains of intact organisms. Bioassays of this sort are generally designed to detect only particular types of biological activities—cytotoxicity in human tumor cells, for example—and tend to have low rates of success (that is, the vast majority of compounds screened are inactive[8]). Success rates are likely to be especially low if the activity being sought is not ecologically relevant to the organism; there is no real reason to expect that plants should produce compounds that are toxic to human tumor cells, for example.

Screening studies that involve a range of taxonomically diverse and ecologically appropriate species, however, can yield very different results. For example, in an initial screeening, Thompson et al.[9] found antimicrobial activity in extracts of 28 of 40 marine sponges. From 8 of the most active species, 38 compounds were identified; 14 were isolated as pure compounds, and the others were characterized as simple mixtures of 2 to 4 compounds. These compounds and mixtures were then bioassayed for activity against 14 test species of higher organisms representing three kingdoms—three species of marine fungi, two algal species, a hydroid, two species of ectoproct, a limpet, a polychaete, an abalone, a sea star, a brine shrimp, and a goldfish. Only three substances showed no activity against any of the bioassay species; the remainder were active against at least one of the test organisms, although no single compound was active against all of the organisms tested.

Yet another way in which plant "screening" differs from laboratory "screening" of phytochemicals is that laboratory bioassays are generally administered under a single set of environmental conditions; environmental variation, if not ignored outright, is often deliberately controlled. This is despite the fact that nutritional factors,[10–11] light,[12–14] temperature,[15] and even humidity[16] are known to mediate the toxicity of phytochemicals. The practice of bioassaying chemicals under a limited range of (usually highly artificial) environmental conditions greatly reduces the likelihood of detecting activity.

In an ecological context, mortality is not necessarily as appropriate a criterion for phytochemical activity as it is in laboratory bioassays. A plant accrues the same selective advantage when a potential herbivore fails to find it or is repelled by an odor as it does when its potential herbivore dies. In fact, the plant that is avoided rather than eaten may be at a selective advantage if it loses no tissue in the process of driving away its enemies. Even in a situation in which a consumer that is neither killed nor repelled succeeds in destroying substantial amounts of tissue, a defense compound will be selectively advantageous if plants without the compound suffer greater losses and are consequently less fit.

Finally, plant screening differs from drug or insecticide screening in that the vast majority of laboratory bioassays are conducted on pure compounds (viz., "in the NCI anticancer screening programme, only 4.3% of plant species had any activity, and pure compounds with high activity resulted from 0.07% of species" Jones and Firn[8]). Assaying activity of pure compounds in isolation is not necessarily an ecologically justifiable approach to use when assessing biological efficacy. Natural selection perforce acts not on single compounds but rather on the entire chemical phenotype of a plant. As such, activity of any given compound must be evaluated in the context of this complete phenotype. Jones and Firn[8] themselves acknowledge that "A diversity of active compounds confers greatest resistance because mixtures of compounds enhance activity." In such mixtures, any individual compound need not have intrinsic toxicity in order to enhance the activity of a co-occurring toxin. The phenomenon of synergism, for example, is well known to insecticide manufacturers, who routinely combine nontoxic methylenedioxyphenyl-containing compounds, such as piperonyl butoxide, with insecticides to enhance their toxicity by interfering with cytochrome P450-mediated detoxification of the insecticide (Berenbaum and Neal[17] and references therein). Methylenedioxyphenyl-containing compounds are widespread in plants and may similarly potentiate naturally co-occurring toxins metabolized by cytochrome P450 monooxygenases.[18]

PARSNIPS AS PARADIGMS—WHY SO MANY FURANOCOUMARINS?

To demonstrate definitively that even a single compound is inactive, or more specifically selectively neutral, in a particular plant species it would really be necessary to assay the compound against all potential consumers, under a range of environmental factors experienced by the plant, and in mixture with cooccurring compounds. Such an assay would be a daunting undertaking, to say the least, and it can be safely said that no such set of assays has ever been (or ever will be) undertaken. It would be an even more daunting task to evaluate all chemicals in a single plant species for activity. In view of the technical challenges involved in testing the idea that phytochemical complexity is adaptive, it is indeed tempting to embrace the alternate hypothesis posed by Jones and Firn[8]—that the majority of phytochemicals serve no current function other than to act as raw material for the biosynthesis of defensive compounds whenever environmental stresses manifest themselves.

There is, however, sufficient evidence already available that, at least as far as one group of phytochemicals is concerned, the "common sense scenario" is the more reasonable one with respect to accounting for phytochemical diversity. Furanocoumarins are benz-2-pyrone derivatives characterized by the attachment of a furan ring at either the 6,7 position (in the case of linear furanocoumarins) or the 7,8 position (in the case of angular furanocoumarins). As is typical for plant natural products, the furanocoumarins are idiosyncratically distributed among plant families; reported to occur in only a dozen families or so, they are ubiquitous only in two families, the Rutaceae and Apiaceae.[19] As is the case for many groups of natural products, plants rarely contain only a single furanocoumarin; some species may contain a dozen or more different compounds.

One species that produces furanocoumarins in series is the wild parsnip, *Pastinaca sativa*, a biennial umbelliferous weed of roadsides, oldfields, and waste places. Above-ground parts of wild parsnip contain as many as nine furanocoumarins in readily measurable quantities. For the past fifty years, the furanocoumarins of parsnip have been subject to intense scrutiny for their ecological, agronomic, and pharmacological significance. As a result of this scrutiny, there is a substantial body of information available about furanocoumarin distribution, abundance, and biological activity that can be examined in order to evaluate competing hypotheses concerning the evolution and maintenance of phytochemical diversity in this plant species.

Historical Considerations

Before ecological hypotheses can be considered, it must be pointed out that the furanocoumarin diversity of *Pastinaca sativa* is at least in part a function of the intensity of scientific interest in the plant. Interest in the chemistry of wild parsnip arose initially not so much from the fact that the root of the cultivated form of the plant is considered edible by some; rather, the parsnip attracted attention because, when it establishes itself as a weed, it comes into close contact with humans, the result of which is a painful exanthematous rash. Establishing the connection between parsnip plants and phytophotodermatitis was a slow process. Stowers, in an 1897 talk presented before the London Dermatological Society, was the first to associate the parsnip plant with blistering eruptions (*dermatitis bullosa pratensis striata*), a suspicion shared by Hartmann and Briel.[21] Hirschberger and Fuchs[22] described a similar eruption in soldiers on camouflage maneuvers in Germany, identified parsnip as the plant responsible, eliciting symptoms by rubbing skin with parsnip leaves, and warned that such a rash might be mistaken for mustard gas poisoning in wartime. McKinlay[23] described nearly identical cases among soldiers unloading vegetables on Salisbury Plain and also ascribed them to parsnip. Kuske[24] showed conclusively that sunlight was necessary to elicit the skin reaction, and Jensen and Hansen,[25] at the Finsen Institute in Copenhagen, by using selective filters, demonstrated that only the UV spectrum was involved.

It was not until 1948, over fifty years after parsnip dermatitis made its first tentative appearance in the medical literature, that causative agents were identified. Fahmy and Abu-Shady[26] isolated three furanocoumarins, imperatorin, bergapten, and xanthotoxin, from fruits of wild parsnip—these compounds had earlier been isolated from bergamot oil and masterwort by Späth[27] and shown by Kuske[24] to cause dermatitis. Soine et al.[28] found imperatorin and bergapten in the fruits and, to prove these agents were responsible for parsnip dermatitis specifically, rubbed the pure compounds on their own skin. Several years later, Maksyutina and Kolesnikov[29] found xanthotoxol in the fruits as well. Isopimpinellin was reported in parsnips by Beyrich,[31] who was interested in determining whether the plant could serve as a useful source of xanthotoxin, at the time a promising new therapeutic agent for the treatment of skin diseases.[31] Improvements in analytical methods led to the discovery of angular furanocoumarins and dihydrofuranocoumarins in the plant, including sphondin,[32] isober-

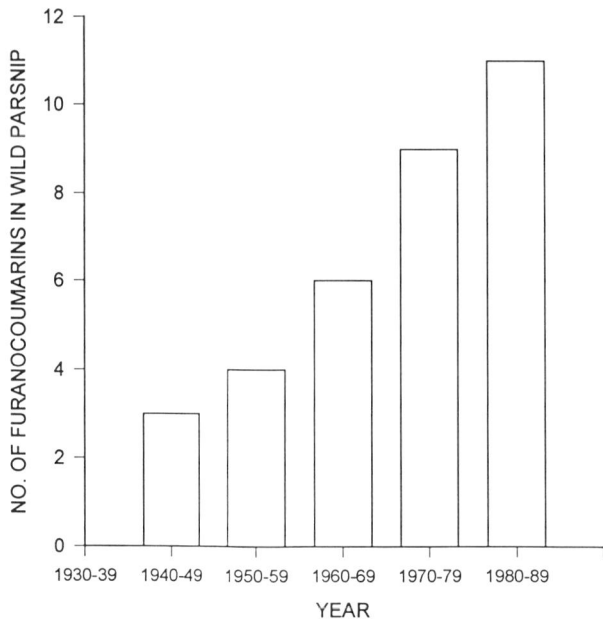

Figure 1. Number of furanocoumarins reported in *Pastinaca sativa* over time.

gapten, pimpinellin,[33] apterin,[34] and angelicin.[35] Psoralen was reported in roots[36] and in reproductive parts[37] a few years later.

Thus, the furanocoumarin diversity of *Pastinaca sativa* has been steadily increasing since 1948 (Fig. 1), assuredly not due to increasing selection pressures by consumers or fantastically elevated mutation rates but rather to increasing scientific interest in the phytochemicals and increasing sophistication and accuracy of chemical analytical methods.

Biosynthetic Considerations

Based upon biosynthetic studies in wild parsnip and other species, the furanocoumarins in wild parsnip are known to be derived from umbelliferone[18,38,39] (Fig. 2). In roughly increasing abundance, angelicin, psoralen, sphondin, isopimpinellin, bergapten, imperatorin, and xanthotoxin are all found in fruits of wild parsnip,[37] and all seven are sequestered together within oil tubes in the seeds.[40] Although two of the biosynthetic intermediates of these compounds, xanthotoxol and umbelliferone, have been reported in wild parsnip,[18] most intermediates are not known to occur, although their "absence" may result from accumulation in such low concentrations as to be undetectable. Thus, despite the potential for production and accumulation of 16 compounds within this pathway (including umbelliferone), only eleven have been reported, and only five (imperatorin, bergapten, isopimpinellin, xanthotoxin, and sphondin) are found with any consistency and regularity. Contrary to the assertion by Jones and Firn[8] that there is little penalty for accumulating inactive compounds that may serve as useful intermediates, this discrepancy between possible furanocoumarins and actual furanocoumarins suggests that not all compounds are sufficiently advantageous or inexpensive for a plant to produce on a continuous basis. The argument that these compounds are not found in parsnips because they are unstable precursors and are completely converted to other products cannot suffice because in this case, psoralen, angelicin, and xanthotoxin, which are precursors for other furanocoumarins, do accumulate in measurable quantities and because other precursors, such as marmesin, are accumulated by other species.[18]

Another contention by Jones and Firn[8] is that pathway diversification may come about by indiscriminate enzyme transformations. In their scenario, for example, a methylating enzyme may indiscriminately attach methyl groups at any of a number of positions within a single molecule, thereby producing a series of compounds that do not all possess activity. Such is not the case in certain well-characterized steps in furanocoumarin biosynthesis. In at least two steps, enzyme activity is highly site-specific. The enzyme that catalyzes the prenylation at the 6 position of umbelliferone to form demethylsuberosin, precursor to the linear furanocoumarins, does not prenylate the 8 position of umbelliferone,[41] a step necessary for production of osthenol, precursor of the angular furanocoumarins (Fig.2). Similarly, S-adenosylmethionine:xanthotoxol O-methyltransferase, which converts xanthotoxol to xanthotoxin, is distinct from the S-adenosylmethionine O-methyltransferase that converts bergaptol to bergapten.[42] Even the production of isopimpinellin (5,8-dimethoxypsoralen), which might theoretically occur by hydroxylation and subsequent methoxylation of either xanthotoxin or bergapten, appears to proceed only with xanthotoxin.[42] Inasmuch as specificity is the rule for those reactions about which something is known, indiscriminate enzyme activity does not account for much of the diversification of furanocoumarins.

Activities of Furanocoumarins Assayed Alone

Jones and Firn[8] argue that "molecular parsimony" may be an artifact of sampling, of the tendency for investigators to focus on chemical classes already known from particular plants to possess biological activity. They go on to say that "from our toxicological perspective, a 'single molecular type' does not produce a diversity of different types of biological activities." This singularity of biological activity, however, may just as likely be an artifact of bioassay as is the apparent chemical conservatism of some plant taxa. After their identification as the causative agents in photophytodermatitis, xanthotoxin and other furano-

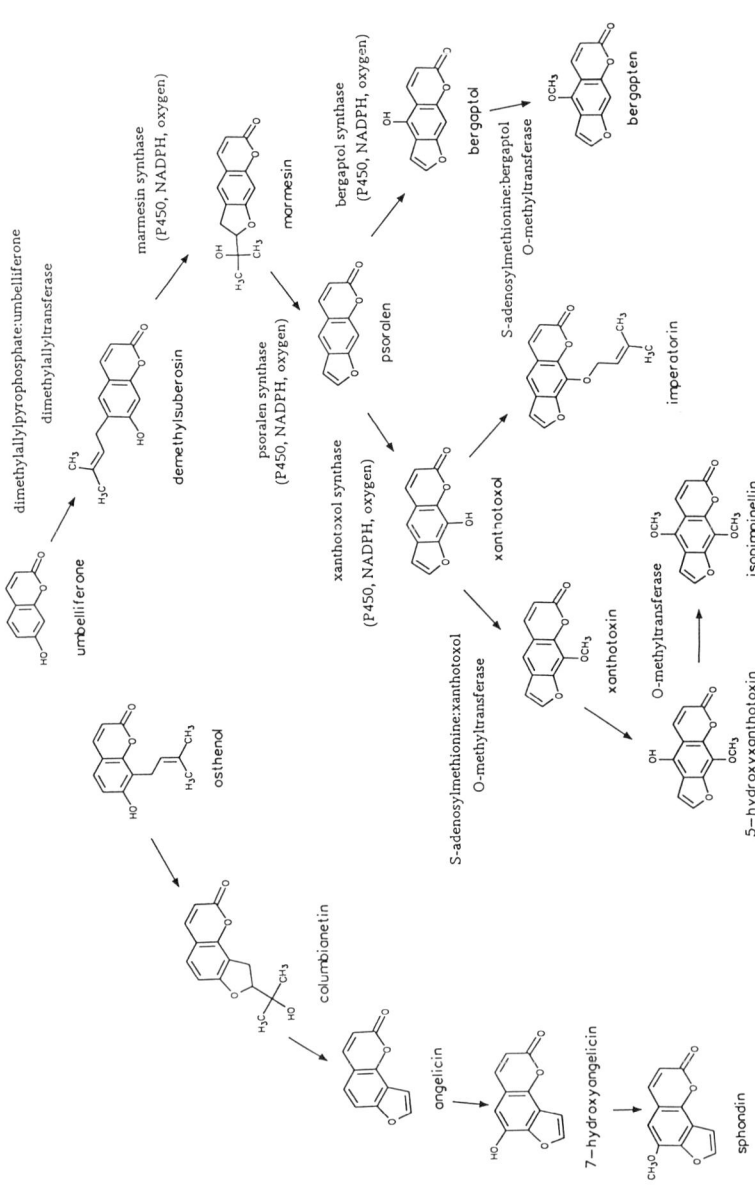

Figure 2. Biosynthesis of furanocoumarins (from refs. 38, 39, and 42).

coumarins were discovered to have remarkable therapeutic properties and quickly were developed as drugs for a wide variety of ailments. A factor contributing to their potential utility as therapeutic agents was their remarkably broad-spectrum toxicity; reports were published in relatively rapid succession documenting their toxicity against snails,[43] mice,[44] germinating plants,[45] fungi[46] and bacteria;[47] subsequent studies documented their efficacy against DNA viruses,[48] tumor cells,[49] sheep,[50] chickens, turkeys,[51] and bacteriophage[52] (cited in[19]).

Elucidating the mode of action of these potentially important therapeutic agents thus became a high priority. Initially, attention was focused on their ability to alter the structure of DNA.[53] The early demonstration of DNA-reactivity led to the assumption for years that DNA cycloaddition and crosslinkage are the dominant mechanisms by which furanocoumarins cause toxicity. In the past thirty years, literally hundreds of papers have been published detailing the nature of the interactions between DNA and furanocoumarins, and these compounds have become important structural probes for molecular studies. However, despite the vast wealth of information on these interactions, many manifestations of furanocoumarin toxicity are not entirely consistent with DNA damage. In recent years, investigators have demonstrated that, *in vitro*, furanocoumarins have other modes of action. They are capable of generating singlet oxygen and superoxide anion in the presence of ultraviolet light,[54,55] of binding irreversibly to proteins,[56] of denaturing or inhibiting enzymes,[57,58] and of forming covalent cycloadducts with unsaturated lipids.[59] A single "molecular type" thus possesses a multiplicity of biological activities. The extent to which each of these biological activities (affecting what are essentially universal target sites) contributes to effects *in vivo* is not known even for humans, who have been subjected to intense investigation for decades. It is entirely possible that a single furanocoumarin interacts with different targets or to differing degrees with the same target in different organisms.

This multiplicity of mechanisms may be responsible for the distinctive spectrum of biological activities that characterizes every furanocoumarin that has been investigated in detail. At least three dozen studies have compared the relative activities of at least two of the eleven furanocoumarins found in wild parsnip (Table 1). These studies are of two types—those conducted in the presence of activating wavelengths of ultraviolet light and those conducted in the absence of UV light. For both *in vitro* and *in vivo* studies, there is no consistent ordering of biological activities; the biological activity of any furanocoumarin relative to other furanocoumarins varies with the type of assay and the target organism, and no single furanocoumarin is consistently inactive. While psoralen tends to have highest activity, in comparison with other furanocoumarins tested, in tests conducted *in vitro* in the presence of UV radiation, particularly on unicellular organisms, it is often less active than other linear furanocoumarins in the absence of UV light or in tests *in vivo* on higher organisms. Angelicin, an

Table 1. Relative biological activity of furanocoumarins found in wild parsnip

UV-mediated effects			
in vitro effects			
DNA-binding in calf thymus	pso > xan > ber		89
DNA crosslinks in human cells	xan > ber > ang		90
DNA crosslinks in virus	iso > xan > ang		91
DNA crosslinks in mouse	xan > iso > ang		91
yeast ribosomal RNA binding	pso > xan > ber > xol*		89
HIV promoter activation	ber > xan > ang		92
inhibition of protein synthesis	pso > ang		93
death of tumor cells	pso > ang		93
formation of triplet state	pso > ang > xan > ber		55
formation of singlet oxygen	pso > xan > ber > ang*		55
formation of singlet oxygen	pso > xan > ber		94
formation of singlet oxygen	pso = ang > xan, ber		54
enzyme inactivation			
glutamate dehydrogenase	pso > ang > xan		57
6-phosphogluconate dehydrogenase	pso > xan = ang		57
lysozyme	pso > xan > ang		57
in vivo effects			
erythemal responses			
humans	pso > xan = ber > ang = isb iso*, imp*, xol*, bol*		95
albino guinea pigs	pso > xan > ber iso*, xol*		96
albino rabbits	pso > xan > ber		97
mutagenesis			
Bacillus subtilis	bol > xan = xol, ber*		98
Escherichia coli	pso > xan > ang		99
Allium root tips	ber > pso > ang > xol > xan		100
growth inhibition			
Candida albicans	ber > xan		101
Bacillus subtilis	pso = xan > ber > iso		47
Staphylococcus aureus	pso > xan > ber > iso		47
Streptococcus faecalis	pso > xan > ber > iso		47
Escherichia coli	pso > xan > ber*, iso*		47
Proteus vulgaris	pso > xan > ber > iso*		47
Pseudomonas aeruginosa	pso > xan*, ber*, iso*		47
cytolysis			
Paramecium caudatum	ber > xan		101
Tetrahymena pyriformis	ber > xan		
mortality			
Candida albicans	ber > xan		101
Bacillus subtilis	pso > xan > xol > bol		98
Escherichia coli	pso > xan > ber > iso		102
Spodoptera exigua	xan > pso > ber		63
Aedes aegypti	xan > sph		103

Table 1. *Continued*

UV-independent effects		
in vitro		
DNA crosslinks in virus	xan > iso > ang	91
DNA crosslinks in mouse	xan > iso > ang	91
inhibition of drug metabolism	iso > xan	104
tyrosinase induction	xan > ber	105
cytochrome P450 inhibition	xan > imp > ang > ber > pso > iso > xol	106
cytochrome P450 inhibition (mouse)	xan > osi > uno	107
frameshift mutations in *Escherichia coli*	pso > xan = ang	108
in vivo		
antifeedant activity		
Spodoptera litura	iso > ber > xan	109
Spodoptera litura	iso > ber > ang > xan > pso	61
Spodoptera litura	ber > xan	110
Spodoptera litura	iso > xan > pso > ang	60
Periplaneta americana	xan = iso > ber	61
Blattella germanica	ber > xan > iso	61
Leptinotarsa decemlineata	ber = iso > xan > imp	111
efficiency of food conversion		
Papilio polyxenes	ang > xan	64
growth inhibition		
Ancylostomidae	pso > ber*	112
Curvularia lunata	pso > ber=pim=imp	46
Aspergillus niger	pim > ber	46
Trichomonas vaginalis	ber>iso>xan	113
Lactuca sativa	pso = xan = ang, ber*, iso*, imp*, xol*, bol*	45
Heliothis virescens	xan = ber = ang = imp	114
Depressaria pastinacella	ber > xan*	83
Papilio polyxenes	ang > xan*	65
germination inhibition		
Lactuca sativa	xan > ang > pso, ber*, iso*, imp*, xol*, bol*	45
mortality		
Culex pipiens pallens	ber > xan > iso	61
Lebistes reticulatus	ber > imp > xan > iso	27
Biomphalaria mansoni	ber > iso > xan	43
toad	xan > imp > ber	115
rat	xan > imp > ber	115

* = no activity.

angular furanocoumarin, in contrast, almost invariably has lower activity than linear furanocoumarins in tests conducted in the presence of UV light, and is often more active than linear furanocoumarins in tests conducted in the absence of UV. Even for a single type of activity, no absolute hierarchy is apparent. The feeding deterrency of furanocoumarins, for example, varies with the taxon (isopimpinellin is more antifeedant to lepidopterans than to dictyopterans) and even, within a taxon, with the investigator (as an antifeedant against *Spodoptera litura*, Luthria et al.[60] found psoralen more effective than angelicin, and Yajima and Munakata[61] found angelicin more effective than psoralen). Rank-ordering the furanocoumarins found in wild parsnip to convey some sense of activity on an absolute scale, then, would be a meaningless, if not impossible, task.

Activities of Mixtures of Furanocoumarins

Biological activities of individual compounds are at best only an indicator of degree and quality of plant defense. For furanocoumarins, which almost never occur singly in plant tissues, ample evidence exists that the combined effect of a mixture of furanocoumarins does not necessarily amount to the summation of their individual effects. In early studies,[17,62] an extract of furanocoumarins of wild parsnip, bioassayed in artificial diet against *Helicoverpa zea*, was as toxic in the presence of UV as an equimolar amount of the most UV-photoactive component, xanthotoxin, and, in the absence of UV, the mixture was significantly more toxic than xanthotoxin. Diawara et al.[63] also report that mixtures of furanocoumarins have properties distinct from those of individual compounds. They investigated the toxicological effects of dietary xanthotoxin, bergapten, and psoralen (all of which occur in wild parsnip) on *Spodoptera exigua* (Lepidoptera: Noctuidae). When each compound was assayed individually in the presence of UV, xanthotoxin was most toxic, followed by psoralen and bergapten. In a subsequent experiment, the three compounds were assayed singly, in all pairwise combinations, and in a three-way combination at individual concentrations calculated to cause 25% mortality. The most lethal combination included all three furanocoumarins; however, the observed mortality in that case was lower than the expectation based upon a hypothesis of additivity. The toxicity produced by a combination of xanthotoxin and bergapten, isomers differing only in the placement of a methoxy group (Fig. 2), was additive. Finally, experiments with the oligophagous black swallowtail *Papilio polyxenes*, which feeds exclusively on furanocoumarin-containing plants including parsnip, provide evidence of nonadditive effects of combinations of furanocoumarins. In single compound assays, angelicin significantly reduces female pupal mass and adult fecundity compared to xanthotoxin.[64] The combination of xanthotoxin and angelicin, however, has a greater adverse affect on combined growth, consumption, and frass production than either compound administered singly in an amount equimolar to the mixture.[65]

Enhancement of activity of one toxin in the presence of its structural analogues has been called analogue synergism by McKey,[5] who postulated that such a phenomenon may be one possible explanation for the chemical diversity observed in plants. In the case of furanocoumarins, there are at least two possible mechanisms underlying analogue synergism; both relate to the functioning of cytochrome P450 monooxygenases. Cytochrome P450s are membrane-bound heme-based enzymes that are responsible for the detoxification of xenobiotics in a wide range of organisms.[66] Metabolism of furanocoumarins in combination may be significantly reduced over metabolism of furanocoumarins in isolation due to competition for binding sites on the P450. The black swallowtail, for example, is capable of rapid metabolism of xanthotoxin; angelicin, in contrast, is metabolized at a rate only about one-third that of xanthotoxin.[67,65] An equimolar mixture of xanthotoxin and angelicin is metabolized more slowly than the same amount of xanthotoxin.[65] The reduced rate of xanthotoxin metabolism, due to competition with angelicin, may allow unmetabolized xanthotoxin to accumulate and thus cause damage to the insect. This type of interference may explain patterns of metabolism observed when individual furanocoumarin-metabolizing isozymes of the black swallowtail (CYP6B1) are expressed in recombinant baculovirus-infected insect cell lines; while angelicin is not substantially metabolized by CYP6B1, rates of xanthotoxin metabolism are dramatically lower in its presence.[68] An evaluation of the heritabilities of metabolic capabilities in full-sib families of black swallowtails revealed significant additive genetic variation not only in the rates of metabolism of xanthotoxin and angelicin individually but also in the rate of metabolism of the two furanocoumarins in combination.[65] Significant family effects on metabolism of mixtures indicate that this trait is available and potentially responsive to selection.

Like the black swallowtail, the parsnip webworm (*Depressaria pastinacella*) feeds on wild parsnip and relies on cytochrome P450s to metabolize furanocoumarins in its hostplants.[69] Despite the fact that the furanocoumarin-metabolizing P450s of the parsnip webworm are not closely related to those of the black swallowtail, as evidenced by Northern analysis,[70] they share certain similar properties. When larvae of this species consume a mixture of the compounds in their hostplant, overall metabolism of the combined substrates is greatly reduced (Fig. 3). This decline in overall metabolism may result from competition for binding sites between rapidly metabolized furanocoumarins (such as xanthotoxin) and furanocoumarins that are turned over more slowly (in this case, bergapten and sphondin). As with the black swallowtail, an examination of familial variation in metabolic capabilities in webworms revealed additive genetic variance in metabolism of bergapten and xanthotoxin (linear furanocoumarins) alone as well as in metabolism of the compounds in mixtures (bergapten, xanthotoxin, and sphondin).[71] Again, the presence of additive genetic variance for this trait suggests that it is available for selection.

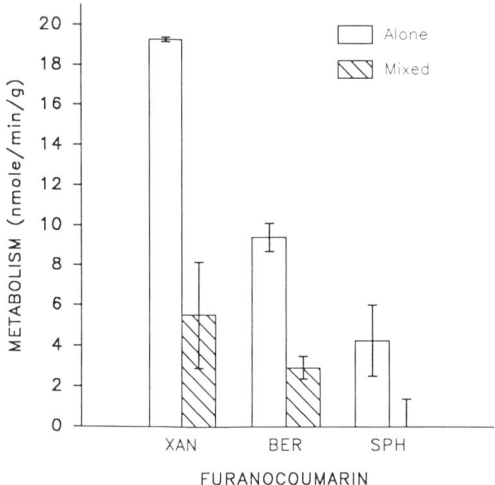

Figure 3. Rates (nmoles/min/g caterpillar) of *in vitro* metabolism of furanocoumarins individually and in mixtures by parsnip webworms (bars represent standard errors). Assays were conducted as described in;[71] the concentration of each furanocoumarin was 50 uM, so total concentrations in mixtures was 100 uM in assays with two furanocoumarins and 150 uM in assays with all 3 furanocoumarins. Rates of metabolism were compared by two-way ANOVA; significant main effects were found for bergapten and sphondin on xanthotoxin metabolism, for xanthotoxin and sphondin on bergapten metabolism, and for xanthotoxin on sphondin metabolism.

In addition to competing for binding sites on enzymes, a compound can interfere with metabolism of its structural analogue by binding irreversibly to enzymes and thus preempting binding sites. Bioactivation occurs when a compound is converted to a metabolite that is itself toxic. A special case of bioactivation is suicide substrate formation, in which a substrate is converted to a metabolite that destroys the activity of the enzyme that produced it.[72] Xanthotoxin has been shown to act as a suicide substrate of cytochrome P450s; cytochrome P450s that process hexobarbitol in mammalian tissue are destroyed *in vitro* when they are incubated with xanthotoxin.[58] The finding that inactivation is both O_2 and NADPH-dependent, characteristic requirements of P-450-mediated reactions, provides strong evidence that xanthotoxin must first be metabolized before inactivation occurs. Similar findings were reported for the effects of xanthotoxin on P450s that metabolize carbon tetrachloride[73] and coumarin[74] in mice, 7-ethoxycoumarin,[75] coumarin and testosterone in rats,[76] and aldrin in the tobacco hornworm *Manduca sexta*.[77] Bergapten and psoralen show similar activity against total cytochrome P450 activity in human liver microsomes.[78]

There is some *in vivo* evidence that furanocoumarins act as suicide substrates in the interaction between wild parsnip and its principal insect associate, the parsnip webworm. Laboratory-reared larvae that were fed umbels collected from field-grown plants that show no evidence of webworm damage did not grow as fast in laboratory bioassays as larvae that were fed umbels from plants harboring actively feeding caterpillars. Subsequent analysis of the furanocoumarin content of plants and the *in vitro* capacity of larvae to metabolize furanocoumarins revealed that the detoxification capacity of larvae fed umbels high in furanocoumarins (and thus avoided by larvae in the field) was significantly reduced.[79] This finding is consistent with irreversible inhibition of P450s by ingested furanocoumarins. When the detoxification capacity of the larvae is reduced by a suicide substrate (as, for example, by prior dietary exposure to furanocoumarin-rich plant material), unmetabolized furanocoumarins can accumulate and interfere with the growth of the insect.

Detoxification Systems and Their Contribution to the Evolution of Chemical Diversity

While enzymatic transformation is not the only mechanism of insect resistance to plant natural products, it is almost universally present in consumers of plant tissues.[66] Characteristics of metabolic detoxification may well have played a role in the evolutionary diversification of secondary compounds in general. Jones and Firn[8] point out that, due to the branching nature of most secondary metabolic pathways, "diversity of compounds within a class is always likely to exceed that between classes" and "a new mutuation is far more likely to add compounds to an existing pathway than to start a new pathway." There are two consequences of these biosynthetic tendencies. First, derivatives of active compounds are likely to be toxic themselves, due to the fact that they are structurally similar and thus likely to share active sites. Second, and more importantly in the context of detoxification, derivatives of active compounds, irrespective of whether they are themselves active, are more likely to interfere with or inhibit the enzymes that detoxify the active compounds, due to the fact that they share structural moieties that facilitate binding to the active site of the detoxifying enzyme.

Inhibition is here broadly defined as the reduction in the efficiency of a detoxification enzyme in the presence of an inhibitor; the inhibitor itself may or may not be metabolized by the target enzyme. A selective advantage can accrue to a plant which produces a compound that interferes with detoxification but is not itself metabolized even if the compound in and of itself is not toxic. Compounds that are co-metabolized may also be considered inhibitors, provided that they reduce overall metabolism of active compounds. Because enzymes that detoxify a particular compound are more likely to be inhibited by a similar compound, inhibition is a phenomenon that provides immediate benefits to plants for pathway diversification. A toxin to which consumers have evolved resistance by rapid detoxification can be instantly "reactivated" if a compound similar to the toxin, produced by the plant, inhibits metabolism of the toxin.

This reactivation can in turn generate selection pressure on the herbivore, favoring detoxification systems that can process the substrates in combination. The evidence for additive genetic variation in two parsnip-associated herbivores for metabolizing combinations of furanocoumarins that occur in parsnip foliage and flowers[71] suggests that such evolution has in fact taken place. Interspecific comparisons also provide suggestive evidence that resistance can evolve in response to combinations of chemicals. Neal and Berenbaum[80] demonstrated that *Papilio polyxenes*, an umbellifer specialist that routinely encounters both xanthotoxin and the methylenedioxyphenyl inhibitor myristicin in its foodplant, is better able to metabolize xanthotoxin or other substrates in the presence of the inhibitor than is its congener *P. troilus*, which never encounters these compounds in combination.

DOES COMMON SENSE PREVAIL?

Jones and Firn[8] predicted several patterns based on their theory: chief among these are the ideas that most natural products have little or no activity against any particular organism; that highly active compounds (compounds with broad efficacy) will be rare; and that high activity requires high receptor-site specificity. At least with respect to the distribution and abundance of furanocoumarins, most of these predictions do not appear to hold. Wild parsnip seems to be a "common sense" plant—all of its furanocoumarins appear to be toxic, one way or another, against both adapted and nonadapted species. The furanocoumarins that have been tested are active under certain conditions against several organisms (Table 1), and most can be considered highly active in the sense that they effect mortality in a broad cross section of taxa (although, granted, most of the bioassay species, with the exception of the insects, are not particularly relevant ecologically). These highly active furanocoumarins have as targets nucleic acids, proteins, and lipids (Table 1), and thus cannot be said to owe their high activity to high receptor site specificity.

The exceptional furanocoumarins from wild parsnip for which no biological activity has been reported (apterin, isobergapten) are not commercially available and thus may not have been widely tested; not coincidentally, the number of organisms reported to be sensitive to any particular furanocoumarin that is commercially available (e.g., psoralen, xanthotoxin, or bergapten) far exceeds the number of organisms reported to be sensitive to furanocoumarins that cannot be purchased and must be synthesized or borrowed for bioassay studies (e.g., xanthotoxol, sphondin, isopimpinellin) (Table 1).

Jones and Firn[8] also discuss implications of their theory with respect to the evolution of plant defense. They argue that: 1). there will be little or no relationship between plant chemotype and resistance (because so many of the chemicals are irrelevant to the plant's ecological milieu); 2) correlations between consumer abundance and chemical abundance are likely to be spurious; 3) compounds may be inactive for reasons other than consumer adaptation; and 4) the fact that artificial selection can bring about an increase in concentration of active chemicals cannot be taken to mean that natural selection can accomplish the same end.

While empirical work done over the past two decades with wild parsnip does not disprove the theory of Jones and Firn,[8] it is certainly inconsistent with

PHYTOCHEMICAL DIVERSITY 17

it in almost every conceivable way. At least in terms of the interaction between the plant and its principal herbivore, the parsnip webworm, *Depressaria pastinacella*, an oligophage restricted to feeding on the reproductive structures of the plant: 1) plant chemotype is consistently and predictably related to webworm densities;[81,82,79] 2) correlations between webworm abundance and concentrations of particular furanocoumarins associated with resistance are confirmed in laboratory bioassays;[83] 3) compounds that, in bioassay, have little or no effect on webworm growth and development are metabolized more efficiently than those associated with field resistance and larval growth in the laboratory;[71] and 4) in field populations of wild parsnips, furanocoumarin content of foliage and fruits is under genetic control, and webworm damage patterns are consistent with selection for increased concentrations of resistance-related furanocoumarins.[81,82,84]

It may well be that the association between wild parsnip and its insect herbivores is atypical of flowering plants, but there is no *a priori* reason to suppose that it is. On the contrary, it is more likely typical of the majority of old field forbs in that it is predictably associated with a fairly specialized fauna. The wild parsnip likely contains so many furanocoumarins because it has to contend with a highly variable biotic and abiotic environment: 1) like many, if not most, plants, wild parsnips must contend with a wide array of organisms, including microbial pathogens, insect and mammalian herbivores, and plant competitors; 2) these "enemies" are differentially adapted to the chemical defenses of parsnips; some may in fact use certain chemical compounds as host-recognition cues (e.g., Staedler et al.[85]); 3) these "enemies" are likely to be differentially important as agents of mortality in space and time; 4) as well, like many if not most plants, wild parsnips also interacts with mutualists such as pollinators, which may be sensitive to the effects of toxins in certain plant tissues (such as pollen and nectar); 5) because of the variability that exists in interactions between plants and their enemies, differential selection can lead to the evolution of a suite of chemicals within a species, at least in part due to synergistic interactions among chemicals; 6) because of the variability that exists in interactions between plants and other organisms, differential selection can lead to differentiation in chemistry between populations; and 7) because of the differential efficacy of these combinations of chemicals against individual "enemies", mechanisms allowing independent regulation (e.g., inducibility) of individual components should be operable, to allow for optimal biosynthesis of a suite of chemicals. Thus, at any given time and in any given population, the selection regime will be unique; it is not altogether unreasonable to expect that the selection response and resulting phytochemical profile will be unique as well.

There is a tendency in discussins of chemical defense to confound evolutionary origin with evolutionary maintenance. The evolutionary origin of chemical defenses is almost without exception a genetic event—mutation, recombination, or even insertion of a transposable element. It is futile to attempt to attribute the *origin* of a chemical defense to a biotic agent[86]—unless one invokes the notion of directed evolution (which has not been widely substantiated). Mutations are entirely random events. What can be attributed to biotic agents is evolutionary change in a trait—whether the trait is maintained and regulated, or eliminated altogether. Although some variability in furanocoumarin composition is undoubtedly nonadaptive—the result, for example, of transient polymorphisms or the residue of past selection by herbivores that are no longer relevant[87]—much of the diversity of furanocoumarins in the wild parsnip almost certainly results from both biotic and abiotic selection. The energetic and toxicological demands of furanocoumarin synthesis[81,88] and storage[40] argue against the idea that plants can maintain these compounds at little or no cost.

Jones and Firn[8] suggest a scenario that is somewhat antithetical to the doctrine of uniformitarianism: "the general metabolic traits that confer diversity may have been selected for very early in the evolution of plants...[and] The evolution of plant defence may...have proceeded independent of consumer adaptation, once these fundamental traits were in place." There is no compelling reason to assume that selection should work in only one phase of the process. Far more probable is a scenario in which mutation and recombination occur and those mutations that confer a selective advantage are favored over time. Such a scenario can more easily account for the quite remarkable fact that furanocoumarins have apparently evolved independently several times in the plant kingdom. Identical furanocoumarins are produced in multiple unrelated families (Table 2); random accumulation would more likely result in far greater structural diversity within the group than is actually seen. No current phylogenetic scheme can reconcile the present day distribution of the angular furanocoumarin angelicin, in Moraceae, Leguminosae, Rutaceae, Apiaceae, or the linear furanocoumarin bergapten, in Apiaceae, Amararanthaceae, Dipsacaceae, Rutaceae,

Table 2. Number of families containing furanocoumarins found in wild parsnip (from Murray et al.[19])

Furanocoumarin	No. of families	
Bergapten	8	Apiaceae, Amaranthaceae, Dipsacaceae, Moraceae, Pittosporaceae, Rutaceae, Samydaceae, Solanaceae
Imperatorin	4	Apiaceae, Goodeniaceae, Rosaceae, Rutaceae
Angelicin	4	Apiaceae, Leguminosae, Moraceae, Rutaceae
Psoralen	4	Apiaceae, Leguminosae, Moraceae, Rutaceae
Isopimpinellin	3	Amaranthaceae, Apiaceae, Compositae, Rutaceae
Xanthotoxin	3	Amaranthaceae, Apiaceae, Rutaceae
Pimpinellin	3	Apiaceae, Cyperaceae, Rutaceae
Sphondin	2	Apiaceae, Rutaceae
Xanthotoxol	2	Apiaceae, Rutaceae
Isobergapten	1	Apiaceae

Solanaceae, Moraceae, Pittosporaceae, and Samydaceae (representing five different subclasses!), with a common evolutionary origin and subsequent phylogenetic inertia. Selection for and retention of biologically active natural products can, however, more satisfyingly account for this distribution. Viewed from this perspective, it seems premature to pronounce phytochemical diversity as redundancy, or superfluity. In fact, the remarkable conservatism of structures within a biosynthetic class among the plant families producing these compounds prompts one to ask, "Why aren't there more natural products?" That question, however, must wait its turn for attention from phytochemists and ecologists, who have all they can handle at the moment inventorying the chemical diversity that does exist and understanding the ecological and physiological implications of that diversity in the life of the plant.

REFERENCES

1. STAHL, W. 1888. Pflanzen und Schecken. Jena. Z. Med. u. Naturw. 22: 557–684.
2. FRAENKEL, G.S. 1959. The raison d'etre of secondary plant substances. Science 129: 1466–1470.
3. EHRLICH, P. R., RAVEN, P.H. 1964. Butterflies and plants: a study in coevolution. Evol. 18: 586–608.
4. FEENY, P.P. 1976. Plant apparency and chemical defense. Rec. Adv. Phytochem. 10: 1–40.
5. MCKEY, D. 1979. The distribution of secondary compounds within plants. In: Herbivores Their Interaction with Plant Secondary Metabolites, (G. Rosenthal and D. Janzen, eds.). Academic Press, New York, pp. 56–133.
6. BERENBAUM, M.R. 1983. Coumarins and caterpillars: a case for coevolution. Evolution 37: 163–179.
7. BERNAYS, E.A., GRAHAM, M. 1988. On the evolution of host specificity of phytophagous arthropods. Ecol. 69: 886–892.
8. JONES, C.G., FIRN, R.D. 1991. On the evolution of plant secondary chemical diversity. Phil. Trans. Roy. Soc. London Ser. B. 333: 273–280.
9. THOMPSON, J.E., WALKER, R.P., FAULKNER, D.J. 1985. Screening and bioassays for biologically active substances from forty marine sponge species from San Diego, California, USA. Marine Biol. 88: 11–21.
10. HATHCOCK, J.N. 1982. Nutritional toxicology:definition and scope. In: Nutritional Toxicology (J. Hathcock ed.), Academic Press, New York, pp. 1–15.
11. SLANSKY, F. 1992. Allelochemical-nutrient interactions in herbivore nutritional ecology. In: Herbivores Their Interactions with Secondary Plant Metabolites, (G. Rosenthal and M. Berenbaum, eds.), Academic Press, San Diego, pp. 135–174.
12. DOWNUM, K.R. 1992. Tansley review no. 43: Light-activated plant defence. New Phytol. 122: 401–420.
13. ARNASON, J.T., PHILOGENE, B.J.R., TOWERS, G.H.N. 1992. Phototoxins in plant-insect interactions. In: Herbivores Their Interactions with Secondary Plant Metabolites (G. Rosenthal, M. Berenbaum, eds.), 2nd. ed. Academic Press, San Diego, pp. 317–342.
14. BERENBAUM, M.R. 1995. Phototoxicity of plant secondary metabolites: insect and mammalian perspectives. Arch. Insect Biochem Physiol. 29: 119–134.

15. WADLEIGH, R.W., KOEHLER, P.G., PREISLER, H.K., PATTERSON, R.S., ROBERTSON, J.L. 1991. Effect of temperature on the toxicities of ten pyrethroids to German cockroach (Dictyoptera: Blattellidae). J. Econ. Entomol. 84: 1433–1436.
16. REICHENBACH, N.G., COLLINS, W.J. 1984. Multiple logit analyses of the effects of temperature and humidity on the toxicity of propoxur to German cockroaches (Orthoptera:Blattellidae) and western spruce budworm larvae (Lepidoptera: Tortricidae). J. Econ. Entomol. 77: 31–35.
17. BERENBAUM, M.R., NEAL, J.J. 1987. Interactions among allelochemicals and insect resistance in crop plants. In: Allelochemicals: Role in Agriculture and Forestry (G.R. Waller, ed.). ACS Symp. Ser. 330: 416–430.
18. BERENBAUM, M., NEAL, J. 1985. Synergism between myristicin and xanthotoxin, a naturally co-occurring plant toxicant. J. Chem. Ecol. 11: 1349–1358.
19. MURRAY, R.D.H., MENDEZ, J., BROWN, S.A. 1982. The Natural Coumarins. Chichester: J. Wiley and Sons. p. 702.
20. KLABER, R. 1912. Phyto-photo-dermatitis. Brit. J. Derm. Syph. 54: 193–211.
21. HARTMANN, E., BRIEL, I. 1927. Uber gehauftes Auftreten einer bullosen Hauterkrankung in Strandbadern. Derm. Z. 50: 205–209.
22. HIRSCHBERGER, A., FUCHS, H. 1936. Munch. Med. Wschr. 83: 1965-(as cited in 20).
23. MCKINLAY, R., 1938. Vesicular dermatitis due to wild parsnip. J. Roy. Army M. Corps 71: 401–404.
24. KUSKE, H. 1938. Experimentelle Untersuchungen zur Photosensibilisierung der Haut durch pflanzliche Wirkstoffe. I. Lichtsensibilisierung durch Furocumarine als Ursache verschiedener phytogener Dermatosen. Arch. Derm. Syph. Wien 178: 112–123.
25. JENSEN, T., HANSEN, K.G. 1939. Active spectral range for phytogenic photodermatosis produced by *Pastinaca sativa*. Arch. Dermatol. Syph. 40: 566–577.
26. FAHMY, I.R., HIFNY-SABER, A., ABU-SHADY, H. 1956. A pharmacognostical study of the fruit of *Pastinaca sativa l.* cultivated in Egypt. J. Pharm. Pharmacol. 8: 653–660.
27. SPÄTH, E. 1936. Die naturlichen Cumarine und ihre Wirkung auf Fische. Monatsh. Chemie 69: 75–114.
28. SOINE, T.O., ABU-SHADY, H., DIGANGI, F.E. 1956. A note on the isolation of bergapten and imperatorin from the fruits of *Pastinaca sativa* L. J. AM. Pharm. Assoc. 45: 426–427.
29. MAKSYUTINA, N.P., KOLESNIKOV, D.G. 1959. Investigation of the furocoumarins of cultivated parsnips. J. Gen. Chem. USSR 29: 37970–3800 (Zhur. Obschei Khim. 29: 3836–40).
30. BEYRICH, T. 1966. Die Furocumarine von *Pastinaca sativa* L. Pharmazie 21: 365- 373.
31. BEYRICH, T. 1965. Xanthotoxin als Pharmakon. Pharmazie 21: 282–287.
32. MAKSYUTINA, N.P. 1965. Khim. Prir. Soed. 1: 133 (CA 63: 8123)
33. GUSAK, L.E., SAFINA, L.K. 1976. Coumarin content in fruit of Kazakhstan varieties of Pastinaceae. Tr. Inst. Bot. Akad. Nauk. Kaz. SSR 35: 145–50. (CA 86: 68378k).
34. FISCHER, F.C., BAERHEIM-SVENDSEN, A. 1976. Apterin, a common furanocoumarin glycoside in Umbelliferae. Phytochem. 15: 1079–1080.
35. BERENBAUM, M.R. 1981. Patterns of furanocoumarin production and insect herbivory in a population of wild parsnip (*Pastinaca sativa* L.). Oecologia 49: 236–244.
36. IVIE, G.W., HOLT, D.L., IVEY, M.C. 1981. Natural toxicants in human foods: psoralens in raw and cooked parsnip root. Science 213: 909–910.
37. NITAO, J.K., ZANGERL, A.R. 1987. Floral development and chemical defense allocation in wild parsnip (*Pastinaca sativa*). Ecol. 68: 521–529.
38. EBEL, J. 1986. Phytoalexin synthesis: the biochemical analysis of the induction process. Ann. Rev. Phytopathol. 24: 235–264.
39. HAMERSKI, D., SCHMITT, D., MATERN, U. 1990. Induction of two prenyltransferases for the accumulation of coumarin phytoalexins in elicitor-treated *Ammi majus* cell suspension cultures. Phytochem. 29: 1131–1135.

40. ZANGERL, A.R., BERENBAUM, M.R., LEVINE, E. 1989. Genetic control of seed chemistry and morphology in wild parsnip (*Pastinaca sativa*). J. Hered. 80: 404–407.
41. DHILLON, D.S., BROWN, S.A. 1976. Localization, purification, and characterization of dimethylallylpyrophosphate:umbelliferone dimethylallyltransferase from *Ruta graveolens*. Arch. Biochem. Biophys. 177: 74–83.
42. HAUFFE, K.D., HAHLBROCK, K., SCHEEL, D. 1986. Elicitor-stimulated furancoumarin biosynthesis in cultured parsley cells: S-adenosyl-L-methionine:bergaptol and S-adenosyl-L-methionine:xanthotoxol O-methyltransferases. Z. Naturforsch 41c: 228–239.
43. SCHONBERG, LATIH, N. 1954. Furochromones and coumarins. XI. The molluscicidal activity of bergapten, isopiminellin and xanthotoxin. J. Pharm. Pharmacol. 6: 6208.
44. EL-MOFTY, A.M. 1948. A preliminary clinical report on the treatment of leucodermia with *Ammi majus* Linn. J. Egypt. Med. Assoc. 31: 651- 665.
45. RODIGHIERO, G. 1954. Influenza di furocumarin naturali sulla germinazione dei semi e sullo sviluppo dei germogli e delle radici di lattuga. Giornale di Biochimica 3: 138–146.
46. CHAKRABORTY, D.P., DAS GUPTA, A., BOSE, R.K. 1957. On the antifungal action of some natural coumarins. Annals of Bioch. Exper. Medicine 17: 59–62.
47. FOWLKS, W.L., GRIFFITH, D.G., OGINSKY, E.L. 1958. Photosensitization of bacteria by furocoumarins and related compounds. Nature 181: 571–572.
48. MUSAJO, L., RODIGHIERO, G., COLUMBO, G., TORLONE, V., DALL'ACQUA, F. 1965. Photosensitizing furocoumarins: Interaction with DNA and photoinactivation of DNA containing viruses. Experientia, 23: 335–336.
49. MUSAJO, L., VISENTINI, P., BACCHINETTI, F., RAZZI, M.A. 1967. Photoinactivation of Ehrlich ascites tumor cells *in vitro* obtained with skin-photosensitizing furocoumarins. Experientia, 23: 335–336.
50. WILLIAMS, M.C. 1970. Xanthotoxin and bergapten in spring parsley. Weed Sci. 18: 479–480.
51. EGYED, M.N., WILLIAMS, M.C. 1977. Photosensitizing effects of *Cymopterus watsonii* and *Cymopterus longipes* in chickens and turkey poults. Avian Dis. 21: 566–575.
52. KITTLER, L., HRADECNÁ, Z., LÖBER, G. 1977. Photochemically induced finding of furocoumarins with lambda phage DNA *insitu*. Stud.Biophys. 6: 237–241.
53. MATHEWS, M.M. 1963. Comparative study of lethal photosensitization of Sarcina lutea by D-methoxypsoralen and by toluidine glue. J. Bacteriol. 85: 322–328.
54. JOSHI, P.C., PATHAK, M.A. 1983. Production of singlet oxygen and superoxide radicals by psoralens and their biological significance. Biochem. Biophys. Res. Comm. 112: 638–646.
55. GROSSWEINER, L.I. 1984. Mechanisms of photosensitization by furocoumarins. In: Photobiologic, Toxicologic, and Pharmacologic Aspects of Psoralens. National Cancer Institute Monogr. 66,pp. 47–54.
56. LASKIN, J.D., LEE, E., YURKOW, E.J., LASKIN, D.L., GALLO, M.A. 1985. A possible mechanism of psoralen phototoxicity not involving direct interaction with DNA. Proc. Nat. Acad. Sci. 82: 6158–6162.
57. VERONESI, F.M., SCHIAVON, O., BEVILACQUA, R., BORDIN, F., RODIGHIERO, G. 1982. Photoinactivation of enzymes by linear and angular furanocoumarins. Photochem Photobiol. 36: 25–30.
58. FOUIN-FORTUNET, H., TINEL, M., DESCATOIRE, V., LETTERON, P., LARREY, D., GENEVE, J., PESSAYRE, D. 1986. Inactivation of cytochrome P450 by the drug methoxsalen. J. Pharmacol. Exp. Ther. 236: 237–247.
59. CAFFIERI, S., DAGA, A., VEDALDI, D., DALL'ACQUA, F. 1988. Photoaddition of angelicin to linolenic acid methyl ester. J. Photochem. Photobiol. B: Biology 2: 515–521.
60. LUTHRIA, D.L., RAMAKRISHNAN, V., VERMA, G.S., PRABHU, B.R., BANERJI, A. 1989. Insect antifeedants from *Atalantia racemosa*. J. Ag. Food Chem 37: 1435–1437.
61. YAJIMA, T., MUNAKATA, K. 1979. Phloroglucinol-type furocoumarins, a group of potent naturally-occurring insect antifeedants. Agric. Biol. Chem. 43: 1701–1706.

62. BERENBAUM, M.R., NITAO, J.K., ZANGERL, ALR. 1991. Adaptive significance of furanocoumarin diversity in *Pastinaca sativa* (Apiaceae). J. Chem. Ecol. 17: 207–215.
63. DIAWARA, M.M., TRUMBLE, J.T., WHITE, K.K., CARSON, W.G., MARTINEZ, L.A. 1993. Toxicity of linear furanocoumarins to *Spodoptera exigua*: evidence for antagonistic interactions. J. Chem. Ecol. 19: 2473–2484.
64. BERENBAUM, M., FEENY, P. 1981. Toxicity of angular furanocoumarins to swallowtail butterflies: escalation in a coevolutionary arms race? Science 212: 927–929.
65. BERENBAUM, M.R., ZANGERL, A.R. 1993. Furanocoumarin metabolism in *Papilio polyxenes*: biochemistry, genetic variability, and ecological significance. Oecologia 95: 370–375.
66. BRATTSTEN, L.B. 1992. Metabolic defenses against plant allelochemicals. In: Herbivores Their Interactions with Secondary Plant Metabolites (G. Rosenthal, M. Berenbaum, eds.), 2nd ed. Academic Press, San Diego, pp. 176–242.
67. IVIE, G.W., BULL, D.L., BEIER, R.C., PRYOR, N.W. 1986. Comparative metabolism of [^{3}H]psoralen and [^{3}H]isopsoralen by black swallowtail (*Papilio polyxenes* Fabr.) caterpillars. J. Chem. Ecol. 12: 871–884.
68. MA, R., COHEN, M.B., BERENBAUM, M.R., SCHULER, M.A. 1994. Black swallowtail (*Papilio polyxenes*) alleles encode cytochrome P450s that selectively metabolize linear furanocoumarins. Archives Biochem. Biophys. 310: 332–340.
69. NITAO, J.K. 1989. Enzymatic adaptation in a specialist herbivore for feeding on furanocoumarin-containing plants. Ecol. 70: 629–635.
70. COHEN, M.B., SCHULER, M.A., BERENBAUM, M.R. 1992. A host-inducible cytochrome P450 from a host-specific caterpillar: molecular cloning and evolution. Proc. Natl. Acad. Sci 89: 10920–10924.
71. BERENBAUM, M.R., ZANGERL, A.R. 1992. Genetics of physiological and behavioral resistance to host furanocoumarins in the parsnip webworm. Evolution 46: 1373–1384.
72. WALSH, C.T. 1984. Suicide substrates, mechanism-based enzyme inactivators: recent developments. Ann. Rev. Biochem. 53: 493–535.
73. LABBE, G., DESCATOIRE, V., LETTERON, P., DEGOTT, C., TINEL, M., LARREY, D., CARRION-PAVLOV, Y., GENEVE, J., AMOUYAL, G., PESSAYRE, D. 1987. The drug methoxsalen, a suicide substrate for cytochrome P-450, decreases the metabolic activation, and prevents the hepatotoxicity, of carbon tetrachloride in mice. Biochem. Pharmacol. 36: 907–914.
74. MÄENPÄÄ, J., JUVONEN, R., RAUNIO, H., PELKONEN, O. 1994. Metabolic interactions of methoxsalen and coumarin in humans and mice. Biochem. Pharmacol. 48: 1363–1369.
75. WILKINSON, D.J., FRY, J.R. 1995. Rat liver cytochrome P450-mediated metabolic activation of methoxsalen and structurally related compounds and its relation to enzyme inhibition. J. Pharm. Pharmacol. 47: 79–84.
76. YAMAZAKI, H., MIMURA, M., SUGAHARA, C., SHIMADA, T. 1994. Catalytic roles of rat and human cytochrome P450 2A enzymes in testosterone 7-alpha- and coumarin-7-hydroxylase. Biochem. Pharmacol. 48: 1524–1527.
77. ZUMWALT, J.G., NEAL, J.J. 1993. Cytochromes P450 from Papilio polyxenes: Adaptations to host plant allelochemicals. Comp. Biochem. Physiol. 106C: 111.
78. TINEL, M., BELGHITI, J., DESCATOIRE, V., AMOUYAL, G., LETTERON, P., GENEVE, J., LARREY, D., PESSAYRE, D. 1987. Inactivation of human liver cytochrome P450 by the drug methoxsalen and other psoralen derivatives. Biochem. Pharmacol. 36: 951–955.
79. ZANGERL, A.R., BERENBAUM, M.R. 1993. Plant chemistry and insect adaptations to plant chemistry as determinants of hostplant utilization patterns. Ecology 74: 478–504.
80. NEAL, J.J., BERENBAUM, M.R. 1989. Decreased sensitivity of microsomal monooxygenases form *Papilio polyxenes* to inhibitors in their host plants. J. Chem. Ecol. 15: 439–446.
81. BERENBAUM, M.R., ZANGERL, A.R., NITAO, J.K. 1986. Constraints on chemical coevolution; Wild parsnips and the parsnip webworm. Evolution 40: 1215–1228.

82. ZANGERL, A.R., BERENBAUM, M.R. 1990. Furanocoumarin induction in wild parsnip: genetics and populational variation. Ecol. 17: 1933–1940.
83. BERENBAUM, M.R., ZANGERL, A.R., LEE., K. 1989. Chemical barriers to adaptation by a specialist herbivore. Oecologia 80: 501–506.
84. ZANGERL, A.R., RUTLEDGE, C.E. 1995. The probability of attack and patterns of constitutive and induced defense: a test of optimal defense theory. American Naturalist: 147: 599–608.
85. STAEDLER, E., BUSER, H.-R. 1984. Defense chemicals in leaf surface wax synergistically stimulate oviposition by a phytophagous insect. Experientia 40: 1157–1159.
86. SCHMITT, T.M., HAY, M.E., LINDQUIST, N. 1995. Constraints on chemically mediated coevolution: multiple functions for seaweed secondary metabolites. Ecol. 76: 107–123.
87. JANZEN, D.H., 1979. New horizons in the biology of plant defenses. In: Herbivores Their Interaction with Plant Secondary Metabolites (G. Rosenthal, D. Janzen, eds.), Academic Press, New York, pp. 331–350.
88. GERSHENZON, J. 1994. The cost of plant chemical defense against herbivory: a biochemical perspective. In: Insect-Plant Interactions (E.A. Bernays ed.). CRC Press, Boca Raton, pp. 105–173.
89. RODIGHIERO, G., CHANDRA, P., WACKER, A. 1970. Structural specificity for the photoinactivation of nucleic acids by furocoumarins. FEBS Letters 10: 29–32.
90. GRUENERT, D.C., ASHWOOD-SMITH, M., MITCHELL, R.H., CLEAVER, J.E. 1985. Induction of DNA-DNA cross-link formation in human cells by various psoralen derivatives. Cancer Research 45: 5394–5398.
91. ALTAMIRANO-DIMAS, M., HUDSON, J.B., TOWERS, G.H.N. 1986. Induction of cross-links in viral DNA by naturally occurring photosensitizers. Photochem. Photobiol. 44: 187–192.
92. ZMUDZKA, B.Z., STRICKLAND, A.G., MILLER, S.A., VALERIE, K., DALL'ACQUA, F., BEER, J.Z. 1993. Activation of the human immunodeficiency virus promoter by UVA radiation in combination with psotalens or angelicius. Photochem. Photobiol. 58: 226–232.
93. BORDIN, F., MARCIANA, S., BACCICHETTI, F.R., DALL'ACQUA, F., RODIGHIERO, G. 1975. Studies on the photosensitizing properties of angelicin, an angular furocoumarin forming only monofunctional adducts with the pyrimidine bases of DNA. Italian J. Biochem. 24: 258–287.
94. BLAN, Q.A., GROSSWEINER, L.I. 1987. Singlet oxygen generation by furocoumarins: effect of DNA and liposomes. Photochem. Photobiol. 45: 177–183.
95. MUSAJO, L., RODIGHIERO, G. 1962. The skin-photosensitizing furocoumarins. Experientia 18: 153–200.
96. PATHAK, M.A., FITZPATRICK, T.B. 1959. Relationship of molecular configuration to the activity of furocoumarins which increase the cutaneous responses following long wave ultraviolet radiation. J. Investigative Dermatology 32: 255–264.
97. POTAPENKO, A. YA., SUKHORUKOV, V.L., DAVIDOV, B.V. 1984. A comparison between skin-photosensitizing (334nm) activities of 8-methoxypsoralen and angelicin. Experientia 40: 264–265.
98. SONG, P-S, MANTULIN, W.W., MCINTURFF, D., FELKNER, I.C., HARTER, M.L. 1975. Photoreactivity of hydroxypsoralens and their photobiological effects in *Bacillus subtilis*. Photochem. Photobiol. 21: 317–325.
99. TUVESON, R.W., BERENBAUM, M.R., HEININGER, E. 1986. Inactivation and mutagenesis by phototoxins using *Escherichia coli* strains differing in sensitivity to near- and far- ultraviolet light. J. Chem. Ecol. 12:933–948.
100. MUSAJO, L. 1965. Interessanti proprieta delle furocumarine naturali. Il Farmaco Ed. Sci. 10: 539–558.
101. YOUNG, A.R., BARTH, J. 1982. Comparative studies on the photosensitizing potency of 5-methoxypsoralen and 8-methoxypsoralen as measured by cytolysis in *Paramecium caudatum*

and *Tetrahymena pyriformis*, and growth inhibition and survival in *Candida albicans*. Photochem. Photobiol. 35: 83–88.
102. ASHWOOD-SMITH, M.J., POULTON, G.A., BARKER, M., MILDENBERGER, M. 1980. 5-methoxypsoralen, and ingredient in several suntan preparations, has lethal, mutagenic and clastogenic properties. Nature 285: 407–409.
103. KAGAN, J.P., SZCZEPANSKI, BINDOKAS, V., WULFF,W.D., MCCALLUM, J.S. 1986. Delayed phototoxic effects of 8-methoxypsoralen, khellin, and sphondin in *Aedes aegypti*. J. Chem. Ecol. 12: 899–914.
104. WOO, W.S., SHIN, K.H., LEE, C.K. 1983. Effect of naturally occurring coumarins on the activity of drug metabolizing enzymes. Biochem. Pharmacol. 32: 1800–1803.
105. MARWAN, M.M., JIANG, J., CASTRUCCI, A. DELAURO, HADLEY, M.E. 1990. Psoralens stimulate mouse melanocyte and melanoma tyrosinase activity in the absence of ultraviolet radiation. Pigment Cell Res. 3: 214–221.
106. NEAL, J.J., WU, D. 1994. Inhibition of cytochromes P450 by furanocoumarins. Pest. Biochem. Physiol. 50: 43–50.
107. MÄENPÄÄ, J., SIGUSCH, H., RAUNIO, H., SYNGELMA, T., VOURELA, P., VOURELA, H., PELKONEN, O. 1993. Differential inhibition of coumarin 7-hydroxylase in mouse and human liver microsomes. Biochem. Pharmacol. 45: 1035–1042.
108. ASHWOOD-SMITH, M.J. 1978. Frameshift mutations in bacteria produced in the dark by several furocoumarins; absences of activity of 4,5',8-trimethylpsoralen. Mutation Research 58: 23–27.
109. YAJIMA, T., KATO, N., MUNAKATA, K. 1977. Isolation of insect anti-feeding principles in *Orixa japonica*. Agric. Biol. Chem. 41: 1263–1268.
110. ESCOUBAS, P., FUKUSHI, Y., LAJIDE, L., MIZUTANI, J. 1992. A new method for fast isolation of insect antifeedant compounds from complex mixtures. J. Chem. Ecol. 18: 1819–1832.
111. MUCKENSTURM, B., DUPLAY, D., ROBERT, P.C., SIMONIS, M.T., KIENLEN, J-C. 1981. Substances antiappetantes pour insectes phytophages presentes dans *Angelica silvestris* et *Heracleum sphondylium*. Bioch. Syst. Ecol. 9: 289–292.
112. POZETTI, G.L., GIAZZI, J.F., BERNADI, A.C., CABRERA, A., YASSUDA, H. 1976. Growth inhibition of larval stages of Ancylostomidae by furocoumarins obtained from Brosimum gaudichaudii Trecul.: bergapten and psoralen. Rev. Fac. Farm. Odontol. Araraquara 10: 221–223 (CA 89: 191745s).
113. VICHKONOVA, S.A., RUBINCHIK, M.A., ADGINA, V.V., IZOSIMOVA, S.B., MAKAROVA, L.V., SHIPULINA, L.D., GORYUNOVA, L.V. 1973. Antimicrobial and antiviral activity of some natural coumarins. Rast. Resur. 6: 370–379.
114. KLOCKE, J.A., BALANDRIN, M.F., BARNBY, M.A., BRYAN, R. 1989. Limonoids, phenolics, and furanocoumarins as insect antifeedants, repellants, and growth inhibitory compounds. In: Insecticides of Plant Origin (J.T. Arnason, R.J.R. Philogene, P. Morand eds.), ACS Symposium Series 387, Washington, D.C., pp. 136–149.
115. SHERIF, M.A.F., MADKOUR,M.K., HAKIM, R. 1957. Toxicity studies on the furocoumarin principles in Ammi majus. J. Egypt. Med. Assoc. 40: 463–466 (CA 55: 22588, 1961).

Chapter Two

DIVERSITY, REDUNDANCY, AND MULTIPLICITY IN CHEMICAL DEFENSE SYSTEMS OF ASPEN

Richard L. Lindroth and Shaw-Yhi Hwang

Department of Entomology
University of Wisconsin
Madison, Wisconsin 53706

Introduction	26
Taxonomy and Natural History	26
Secondary Chemistry	27
Phenolic Glycosides	27
Condensed Tannins	30
Coniferyl Benzoate	30
Miscellaneous Phenolic Compounds	30
Phytochemical Variation	30
Foliar Development and Seasonality	31
Induced Defenses	32
Resource Availability	33
Genetic Variation	35
Negative Genetic Correlations	35
Aspen Chemistry and Biotic Interactions	36
Aspen-Pathogen Interactions	37
Aspen-Insect Interactions	38
Aspen-Bird Interactions	47
Aspen-Mammal Interactions	47
Aspen Chemistry and Abiotic Interactions	49
Community and Ecosystem Effects	49
Conclusions	51

INTRODUCTION

Trembling aspen (*Populus tremuloides* Michx.) is the most widely distributed tree in North America, occurring over a variety of climatic, soil and topographical conditions.[1,2] Interactions between aspen and its biotic and abiotic environments play pivotal roles in the ecological dynamics of many early-successional ecosystems. These interactions, in turn, influence and are influenced by the chemical composition of aspen.

The focus of this paper is the role of phytochemistry, especially natural products, in mediating interactions between aspen and its environment. As will be illustrated, aspen produces a variety of biosynthetically related natural products that have multiple physiological and ecological functions. Moreover, both environmental and genetic factors contribute to substantial intraspecific variation in concentrations of these compounds. Such quantitative variation itself influences interactions between aspen and associated organisms.

TAXONOMY AND NATURAL HISTORY

Aspens are members of the plant family Salicaceae, genus *Populus*, section Leuce. Subsection Trepidae consists of the aspens, of which only two species are native to North America: trembling or quaking aspen (*P. tremuloides*) and bigtooth aspen (*P. grandidentata* Michx.).

The geographical range of trembling aspen covers 48° of latitude and 111° of longitude, extending throughout Canada and Alaska to the northern limit of forest, and throughout the northeastern, northcentral and Rocky Mountain regions of the United States.[2] The distribution of this aspen is determined primarily by water availability (areas where annual precipitation exceeds evapotranspiration) and secondarily by maximum and minimum temperatures during the growing season. Trembling aspen is a major component of three forest cover types, including Eastern Forest, Western Forest, and White Spruce—Aspen Forest (Society of American Foresters Types 16, 217 and 251, respectively).[2]

Trembling aspen grows on a variety of soils. It performs best on well-drained, basic, loamy soils, attaining heights of 23–25 m at age 50 years.[2] Growth is typically much lower in less optimal sites, such as coarse sandy soils in the Great Lakes area and granite-derived soils in the Rocky Mountains.

Trembling aspen is an early-successional species, one of the first woody plants to colonize a site after disturbance by logging, fire, or other factors. Reproduction occurs by both sexual and asexual means. The trees are dioecious; flowers are wind-pollinated and the very small seeds (~0.15 mg) are wind-dispersed. Asexual reproduction via suckering produces clones that vary in size from a few tenths of a hectare east of the Rocky Mountains to up to tens of hectares in the Rocky Mountains.[2] Indeed, an aspen clone covering over 40 ha in Utah may well be the largest individual organism on Earth![3] Aspen exhibits remarkable interclonal genetic variation with respect to traits as diverse as growth rate, leaf morphology, timing of leaf flushing and senescence, and resistance to insects, diseases, drought and pollution.[1,2]

In contrast to trembling aspen, bigtooth aspen has a fairly restricted distribution, occurring primarily in the Great Lakes region, southeastern Canada and northeastern United States.[2] This species is also early-successional, rapidly growing and clonal. The secondary chemistry of bigtooth aspen is similar to that of trembling aspen.[4] Because its role in mediating interactions, however, has seldom been studied, this chapter will deal mainly with trembling aspen, which henceforth will be referred to simply as "aspen."

SECONDARY CHEMISTRY

Aspen natural products with documented ecological roles are all products of the shikimic acid pathway (Fig. 1). These include a variety of phenolic glycosides, condensed tannins and coniferyl benzoate. Representation of other major groups of natural products, such as alkaloids and terpenoids, is poor to nonexistent.

Phenolic Glycosides

The distinguishing chemical feature of aspen is a suite of salicylate compounds generally known as phenolic glycosides. Scientific research on these compounds has a long and colorful history, associated with natural products pharmacology and paper chemistry. Two research groups in particular were responsible for isolation and characterization of a host of compounds in the 1960s: Thieme and co-workers in Germany (e.g., [5-7]) and Pearl and Darling in the United States (e.g., [8-10]).

Trembling aspen contains four phenolic glycosides: salicin, salicortin, tremuloidin and tremulacin (Fig. 2). The biosynthesis of these and related phenolic glycosides is poorly understood. Benzoic acid and 2-coumaric acid are the likely precursors for the aglycone moiety of salicin.[11] Salicin may then be benzoylated to form tremuloidin, esterified (with the 1-hydroxy-6-oxo-2-cyclo-hexene-1-carboxyloyl group) to form salicortin, or benzoylated and esterified to form tremulacin. Structural elaboration of the salicin precursor markedly alters the biological activity of phenolic glycosides. Compounds (e.g., salicortin and tremulacin) containing the cyclohexenone functional group have much greater activity than those lacking it.[12] Moreover, activity of tremulacin is greater than that of salicortin. These structural changes shift the relative polarity of the compounds (in the order tremulacin < tremuloidin < salicortin < salicin), which

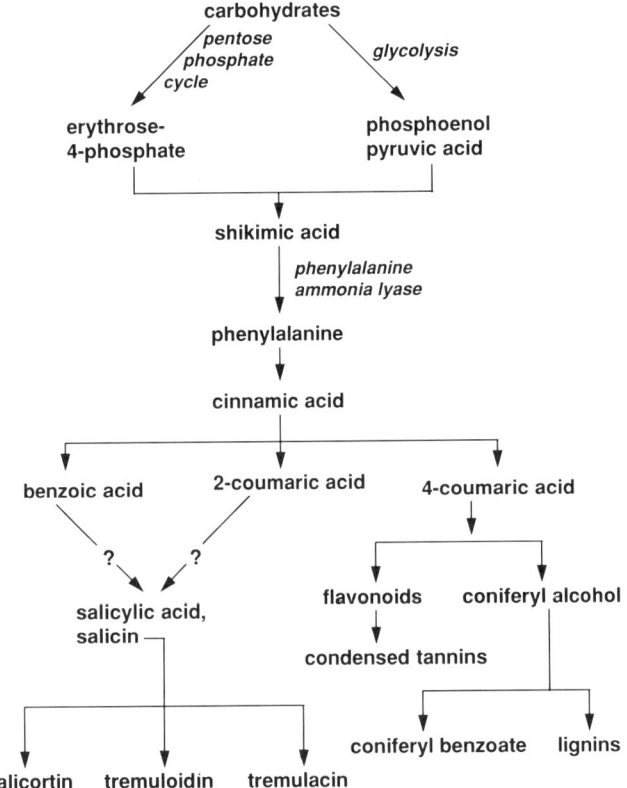

Figure 1. Biosynthetic pathways for production of simple and complex phenolic compounds in trembling aspen.

may in turn alter affinity for cellular membranes or receptor sites. Alternatively, as will be discussed, the structural changes may incorporate moieties that form reactive chemical species upon metabolism. For a fuller development of the biochemistry of phenolic glycosides, see the review by Pierpoint.[11]

Phenolic glycosides occur in both leaf and bark tissues of aspen.[13] Foliar concentrations of salicortin and tremulacin (1–7% dry weight each) are generally much higher than those of salicin and tremuloidin (<1% dry weight each).[4] Because the compounds are labile under tissue preservation and extraction schemes routinely employed prior to the mid-1980s, quantitative information on their occurrence and relative abundance should be interpreted with care.[14-17] Recent research suggests that appreciable levels of salicin reported in many early studies of the Salicaceae may have been artifacts of the hydrolysis of more complex phenolic glycosides such as salicortin.

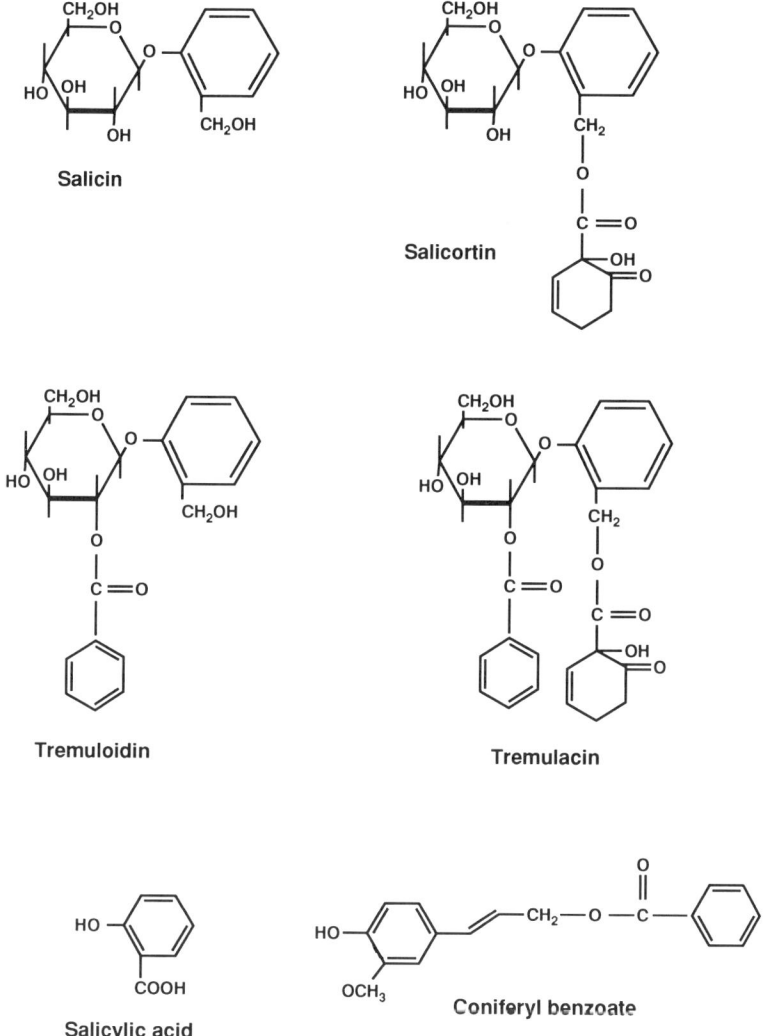

Figure 2. Simple phenolic compounds of trembling aspen.

No discussion of salycilate compounds would be complete without reference to the rapidly growing interest in salicylic acid (Fig. 2). Details of its biosynthesis are not fully known; two pathways have been described, proceeding through benzoic acid or 2-coumaric acid as precursors (Fig. 1).[11,18] A plant hormone, salicylic acid mediates physiological processes as diverse as flowering and the hypersensitive response around sites of pathogen infection.[11] Of particular interest recently is the role of salicylic acid as the chemical signal required for systemic induced resistance in plants. Systemic induced resistance is a defense response induced locally by pathogen or pest attack that then spreads systematically to provide immunity to the entire plant.[19] Concentrations of salicylic acid have been shown to increase throughout a plant after a localized pathogen attack, and application of salicylic acid or related salicylates induces expression of defense-related genes.[18-20] Recent studies (reviewed by Jones[19]) suggest that salicylic acid binds to catalase, leading to an increase in reactive oxygen species (e.g., hydrogen peroxide). The reactive oxygen compounds serve as secondary messengers, inducing defense-related genes that regulate production of various secondary metabolites. The roles of salicylic acid and related salicylates in defense responses of members of the Salicaceae have not been investigated, but should provide a fruitful area for future research.

Condensed Tannins

4-Coumaric acid is a precursor for production of flavan diols, which, when polymerized, give rise to condensed tannins (Fig. 1). Condensed tannins occur in both leaf and woody tissues of aspen. Concentrations in foliage are highly variable, ranging from 1 to 18% dry weight (Hwang and Lindroth, unpubl. data; Hemming and Lindroth, unpubl. data). Aspen does not produce hydrolyzable tannins.[21,22]

Coniferyl Benzoate

4-Coumaric acid also serves as a precursor for production of coniferyl benzoate, a phenylpropanoid ester (Figs. 1,2). This compound is produced exclusively in the flower buds of aspen. Concentrations average ~2.5 % but range from 0 to 7% dry weight.[23,24]

Miscellaneous Phenolic Compounds

Aspen produces a variety of other simple phenolic compounds. For example, aspen bud exudate contains benzoic and phenylpropenoic acids and their esters, and the flavonoids sakuranetin and isosakuranetin and their corresponding chalcones.[25] Because nothing is known, however, about the roles of these compounds in mediating ecological interactions, they will not be considered further.

PHYTOCHEMICAL VARIATION

A major emphasis of chemical ecology over several decades has been to identify the factors responsible for intraspecific variation in plant chemistry.

Such variation is wonderfully exemplified in trembling aspen. Recent research has begun to illustrate how a variety of developmental, environmental and genetic factors influence aspen chemistry.

Foliar Development and Seasonality

Populus species grow indeterminately, producing new leaves throughout a growing season. Nearly all leaves on a tree, however, are produced in the initial spring leaf flush. With the exception of the primary stem of small trees, aspen branches typically add only a few terminal leaves following initial leaf flush and stem elongation.

Several studies have monitored seasonal variation in macronutrient concentrations of aspen.[4,25–27] In general, levels of nitrogen, potassium, phosphorus and water decline with leaf age, especially following full leaf expansion (e.g., Fig. 3). Foliar levels of calcium and magnesium, however, increase. Seasonal variation in nutrient concentrations of twigs is small, and trends for nitrogen, potassium, phosphorus and calcium are opposite those for leaf tissues.[27]

Less is known about seasonal variation in aspen natural products. Lindroth et al.[4] monitored phenolic glycoside levels in individual *Populus* trees (separate

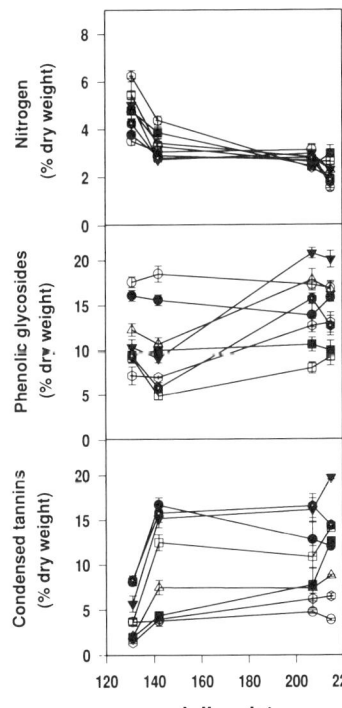

Figure 3. Seasonal variation in phytochemical composition of eight aspen clones grown in a common garden. Leaf samples were collected from each of six trees during early- and late-season studies of insect performance. Each symbol represents a clone; vertical lines indicate ± 1 SE.

clones). Tremendous among-tree variation obscured seasonal trends in trembling aspen, but seasonal declines occurred for tremulacin and salicortin in bigtooth aspen. More recently, Hwang and Lindroth evaluated the phytochemical composition of individuals from eight aspen clones propagated from root cuttings and grown in a common garden (Fig. 3). For the phenolic glycosides salicortin and tremulacin (summed together), concentrations generally declined during leaf expansion (early to mid-May in southern Wisconsin). Thereafter, levels remained constant in some clones but increased to a seasonal high in late summer in other clones. In contrast, condensed tannin concentrations increased most rapidly during leaf expansion, then remained constant or increased slightly for the remainder of the growing season (Fig. 3).

Within-tree variation in leaf age may also alter chemical composition, and thus affect interactions with organisms (e.g., insects) that selectively feed on particular leaf age classes. A preliminary assessment indicated that the most terminal leaves on aspen branches have higher levels of phenolic glycosides than do other leaves on the same branches.[4] Working with cottonwood (*P. deltoides*), Meyer and Montgomery [29] found that the most terminal leaves had higher levels of nitrogen, sugars, and total phenolics (probably phenolic glycosides), but lower levels of condensed tannins, than did older leaves.

Induced Defenses

That damage by pathogens and herbivores may elicit biochemical changes in remaining plant tissues is now well-established, although the mechanisms underlying such changes are not clear.[30] Several studies have addressed induced phytochemical changes in aspen. Mattson and Palmer[31] found that removal of 50% of the tissue of individual leaves led to an 18% increase in total phenolics in remaining tissue. Interestingly, removal of whole leaves at the petiole did not elicit a chemical response. Baldwin and Schultz[32] reported that not only do poplar (*P.* x *euroamericana*) ramets induce production of phenolics within 75 hours of damage, but that damaged plants elicit responses in undamaged plants inhabiting the same growth chamber. (The interpretation of interplant "communication" has been criticized, however, for lack of replication in the experimental design). In the only study to address specific phenolic glycosides, levels of salicortin and tremulacin were reported to increase by 28 and 18% respectively, whereas levels of salicin and tremuloidin did not change 24 hours after mechanical leaf damage in trembling aspen.[33] Chemical responses of individual trees varied greatly, from no increase to a 2-fold increase.

Cutting or heavy browsing of mature aspen trees by mammalian herbivores such as beaver and snowshoe hare typically results in the production of adventitious sprouts. Stump sprouts exhibit a juvenile growth form that differs from that of mature forms in terms of both morphology and chemical composition.[34,35] Although juvenile sprouts are better defended against subsequent browsing than are mature growth forms, the identity of the compounds conferring resistance remains unknown.[34,35] Moreover, the issue of whether juvenile reversion reflects an active induced response, or a passive response linked to developmental state and resource availability, is unresolved.[36]

Resource Availability

The role of resource availability as a modulator of plant chemical composition was formalized by Bryant et al.[37] as the carbon-nutrient balance (CNB) hypothesis. Briefly, the hypothesis asserts that plant allelochemical content is determined largely by resource availability. Plant growth requires carbon and mineral nutrients. Differences in the relative availability of these resources will alter plant carbohydrate stores, and consequently, accumulation of carbon-based compounds.

That requirements for growth, defense and reproduction compete within individual plants for limited resources is well-established.[38] Moreover, nutrient limitations typically reduce growth more than they do photosynthesis.[39] Thus, environmental conditions that increase carbon availability or decrease nutrient availability (e.g., high light, low soil fertility) alter internal plant reserves in favor of carbon. This in turn leads to an accumulation of C-based allelochemicals (e.g., phenolics) or storage compounds (e.g., starch).

Fast-growing species typically exhibit more plasticity in chemical response to changes in resource availability than do slow-growing species.[37,40] Thus, it is not surprising that *Populus* (and *Salix*[41]) species have been the subject of several empirical tests of the CNB hypothesis. Bryant et al.[21] found that when nitrogen fertilizer is applied to quaking aspen grown in N-deficient soil, levels of foliar nitrogen increase, whereas those of phenolic glycosides and tannins decrease. The same researchers, however, later found that the effects of light and nutrient availability on concentrations of several low molecular weight phenolics (including phenolic glycosides) in *Populus balsamifera* did not follow predictions of CNB theory.[42] They suggested that the theory best predicts accumulations of "static" secondary metabolites such as tannins, but not of "dynamic" secondary metabolites (those subjected to metabolic turnover), such as phenolic glycosides.

Our research group recently studied the direct and interactive effects of light and nutrient availability on foliar chemistry and growth of aspen saplings. The experimental design was a split-plot, with three fertilizer treatments (none, moderate, high) nested within two light treatments (30 and 85% shade). One-year-old seedlings were grown individually in 20 L pots for 1½ growing seasons (reaching heights of 1.5—3 m) before plants were harvested and leaves were chemically analyzed. Both light and nutrient availability strongly affected total plant growth.

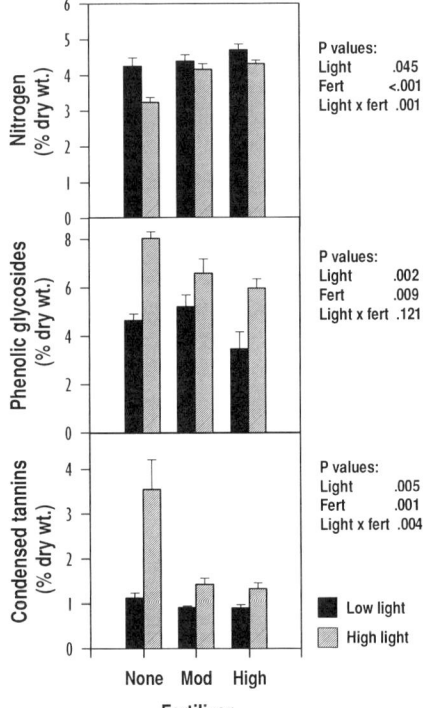

Figure 4. Effects of light and nutrient availability on aspen foliar chemistry. Low light = 85% shade; high light = 15% shade. Vertical lines indicate 1 SE. Phenolic glycosides measured as salicortin + tremulacin.

Leaf chemical profiles (Fig. 4) generally accorded well with predictions of CNB theory. Nitrogen concentrations were lower in high-light treatments than in low-light treatments and increased with increasing soil fertility, but the magnitude of differences among treatments was small. Phenolic glycoside concentrations averaged 54% higher in high-light treatments, and levels declined with increasing nutrient availability. Condensed tannin concentrations showed a strong nutrient x shade interaction response. Levels in high-light plants were 314% higher than in low-light plants under low nutrient availability, but only 51% higher under moderate to high nutrient availability.

Due to the growing awareness of the ecological consequences of increasing atmospheric CO_2 levels, researchers have begun to address the direct effects of CO_2 availability on plant chemistry. We found that CO_2-induced shifts in aspen chemistry are mostly consistent with predictions of the CNB hypothesis.[22,43] Under enriched CO_2 (levels anticipated in the next century), foliar nitrogen declines and starch increases. Levels of phenolic glycosides and condensed tannins tend to increase, but not always significantly so. Responses of some aspen constituents (e.g., starch) to CO_2 are in turn influenced by the availability of soil nutrients.[44]

Genetic Variation

A hallmark characteristic of aspen is striking variation among clones with respect to a variety of morphological, physiological and chemical characteristics. Such differences culminate in substantial variation in how clones respond to stress factors (pathogens, herbivores, drought, air pollution) in the environment. For example, during outbreaks of insects such as forest tent caterpillars (*Malacosoma disstria*) and gypsy moths (*Lymantria dispar*), defoliation rates may vary markedly among aspen clones (Lindroth, pers. obs.). Interclonal variation is most likely due to genetic factors, although other factors (e.g., endophytic symbionts) may also play a role.

High levels of genetic diversity are generally attributed to species with life history characteristics such as those exhibited by aspen (dioecious, wind-dispersed pollen and seeds, high fecundity and long generation time). Nevertheless, the infrequency of seedling establishment, propensity for asexual reproduction, and purported ancient age of extant clones (especially in the western interior of North America) have led some to question the degree of genetic diversity in localized aspen populations.[45] Recent studies have confirmed that it is comparable to or greater than that of other forest tree species.[3,45]

Surprisingly little is known about clonal variation in chemical composition, especially given the attention that has focused on other attributes. Jelinski and Fisher[46] assessed variation in nutrient content of dormant twigs collected from 24 clones. Concentrations of the macronutrients nitrogen, phosphorus, and potassium varied by only 1.6-, 1.6- and 1.4-fold, respectively, whereas those of calcium varied by 4.0-fold. Concentrations of structural constituents (neutral and acid detergent fiber, cellulose and lignin) varied less than those of macronutrients.

Our research group has been investigating phytochemical variation in aspen clones propagated from root cuttings. Trees were grown in a common garden for several years. As has been demonstrated by others,[46,47] variation in foliar nitrogen was minimal (1.3- to 1.8-fold, depending on season; Fig. 3). Variation in natural products, however, was much more pronounced. Phenolic glycoside concentrations differed by 2.4- to 3.2-fold and condensed tannin concentrations differed by 4.9- to 8.3-fold (Fig. 3). Variations of similar magnitude were documented for individual aspen trees from different clones in the field.[47] These results suggest that variation in secondary chemistry may be more important than variation in primary chemistry in explaining interclonal differences in aspen-organism interactions.

NEGATIVE GENETIC CORRELATIONS

Since the mid 1970s, several theories have emerged (or re-emerged), purporting to explain the physiological, ecological and evolutionary bases for intra- and interspecific variation in plant secondary chemistry *vis à vis* herbivory. These include the optimal defense hypothesis,[48] and several variants of resource allocation theory, including the carbon-nutrient balance hypothesis,[37] resource availability hypothesis[49] and growth-differentiation balance hypothesis.[50] A fundamental component of each is the notion of "trade-offs." Plants cannot simultaneously meet all physiological demands for resources, so trade-offs occur among growth, defense and reproduction.[38,50] The major hypotheses differ, however, in terms of whether the trade-offs occur because of the *physiological costs* of defense.

The optimal defense hypothesis proposes that defenses are costly because they divert resources from growth, and that herbivory is the major evolutionary factor affecting quantitative variation in secondary chemistry. The carbon-nutrient balance hypothesis, however, posits that when nutrient resources are limiting, growth is inhibited more than is photosynthesis, leading to an increase in carbon pools. Thus, secondary metabolism (especially C-based compounds) is supported by resources occurring in excess of those required by primary metabolism (e.g., growth), so defense incurs no cost. Consequently, herbivory does not play a central role in the evolution of quantitative variation in secondary chemistry. Finally, in subsuming the CNB hypothesis, the growth-differentiation balance hypothesis also maintains that synthesis of defensive metabolites is not costly because it does not compete with growth for resources. Herms and Mattson[50] do, however, acknowledge that there may be "a background level of secondary metabolism determined by selection" which would be evidenced by negative genetic correlations between growth and defense.

Despite being a widely held belief, little experimental evidence exists to support the notion that intraspecific, genetically-based trade-offs occur between growth and defense.[50,51] This is particularly true for tree species (but see [52]). One reason frequently cited for lack of such a relationship is the (oftentimes) small amount of resources required for defense relative to growth. The trade-off between growth and defense is more likely to occur in plants that allocate substantial amounts of resources to defense.

Thus, trade-offs are likely in aspen, in which secondary compounds comprise 10–35% of leaf dry weight. Indeed, in our recent study involving aspen clones grown in a common garden, differences in individual growth rates were pronounced. Future use of the trees precluded destructive harvests, so we measured above ground biomass by the index tree height x basal diameter. Tree growth and foliar defense were strongly and negatively correlated (Fig. 5). The proportion of variance explained in the correlation was 56%, a value higher than all but one of similar estimates in eight different studies of "trade-offs" reviewed by Vrieling and van Wijk.[53]

ASPEN CHEMISTRY AND BIOTIC INTERACTIONS

Qualitative variation in the secondary metabolite profile of aspen is rather limited, restricted mostly to products from the shikimic acid pathway. These

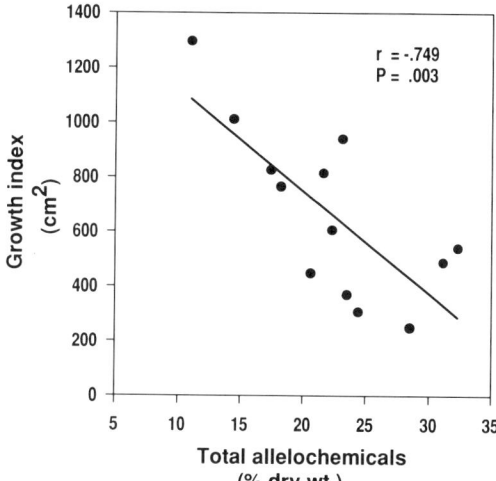

Figure 5. Negative genetic correlation between growth and defense in aspen clones. Each point represents the average response of eight trees/clone; "total allelochemicals" represents salicortin + tremulacin + condensed tannin.

compounds, however, play varied and multiple ecological roles, mediating trophic interactions between aspen and a host of associated organisms.

Aspen-Pathogen Interactions

Aspen is susceptible to a variety of pathogens, including viruses, bacteria, and, most importantly, fungi.[2,54] Although clonal variation in disease resistance is well-known, the contributions of secondary chemistry to such resistance remain poorly understood.

The role of natural products in disease resistance is best understood for *Hypoxylon* canker (*H. mammatum*). This stem canker is probably the most serious disease of aspen east of the Rocky Mountains, killing 1–2% of trees annually.[2] Most infections begin in scars or galls on young branches. The fungus rapidly invades bark tissues, girdling and killing the tree in a few years.[2] Drought-stressed aspen are particularly susceptible to *Hypoxylon* infection.

Clonal variation in resistance to *Hypoxylon* appears to be associated with phenolic inhibitors of ascospore germination. Catechol and various unidentified phenolic glycosides from aspen bark were the first compounds shown to inhibit *Hypoxylon*.[55,56] Unidentified phenolic glycoside phytoalexins were isolated from aspen twigs following *Hypoxylon* inoculation, and found to inhibit germination of *Hypoxylon* ascospores and *Alternaria* conidia, and growth of *Alternaria* mycelia.[57,58] Recently, Kruger and Manion[59] reported significant clonal variation in levels of salicin, salicortin and catechol in tissue culture plantlets, that concentrations of these compounds decrease under drought stress, and that the compounds inhibit *Hypoxylon* ascospore germination. Although much remains to be learned regarding the roles of aspen phenolic compounds against pathogens, that they do serve such a role is increasingly evident.

Aspen-Insect Interactions

Given its extensive range and particular life history characteristics (fast-growing, early successional), it is not surprising that aspen serves as host to a wide variety of insects. Included among these are lepidoptera, aphids, beetles and sawflies that chew, suck and bore on or in virtually all aboveground tissues.[1,2,54] As is true for most tree-feeding insect species,[60] the vast majority are of little consequence in terms of measurable impact on aspen fitness. Some, however, can defoliate trees on a scale rarely seen for other insect pests in North America. For example, annual defoliation by the forest tent caterpillar (*Malacosoma disstria*) and large aspen tortrix (*Choristoneura conflictana*) averages 935,000 and 246,000 ha, respectively, and outbreaks as large as 13.5 million and 2.6 million ha, respectively, have occurred.[61]

A number of recent studies, to be reviewed here, have assessed the role of specific natural products in mediating interactions between aspen and leaf-chewing insects. These studies have implicated phenolic glycosides, rather than condensed tannins, as the compounds exerting the most pronounced effects on performance of both aspen-adapted and nonadapted insects.

Sites and Modes of Allelochemical Toxicity. Phenolic glycosides. At toxic doses, phenolic glycosides cause formation of degenerative lesions in the midguts of insects.[12,62] The identity of the biologically active form of the salycilates, however, is unknown. Biochemical alteration of the compounds occurs in the digestive systems and/or tissues of herbivores, but the specific metabolic routes remain unclear.

Midgut β-glucosidases have been shown in several lepidopteran species to hydrolyze phenolic glycosides, releasing glucose and the phenolic aglycone.[63-65] More recent work using almond β-glucosidase revealed that the enzyme completely hydrolyzes glycosides with a free glucose moiety (salicin and salicortin) but is ineffective in hydrolyzing acyl-glycosides (tremuloidin and tremulacin).[66]

Esterases also have been implicated in playing important roles in the metabolism of phenolic glycosides. Whether ester hydrolysis precedes or follows glycoside hydrolysis is not known. Two opposing perspectives have been advanced regarding the role of esterases in the metabolism of ester-containing phenolic glycosides. Clausen et al.[33] proposed that foliar esterases released upon crushing of leaf cells during feeding cause the hydrolytic breakdown of salicortin and tremulacin and subsequent release of 6-hydroxy-2-cyclohexenone (6-HCH) (Fig. 6). 6-HCH or its byproduct catechol may be the constituents with biological activity against insects, as suggested by feeding studies with the large aspen tortrix (*Choristoneura conflictana*).[33] Recent research using artificial assay

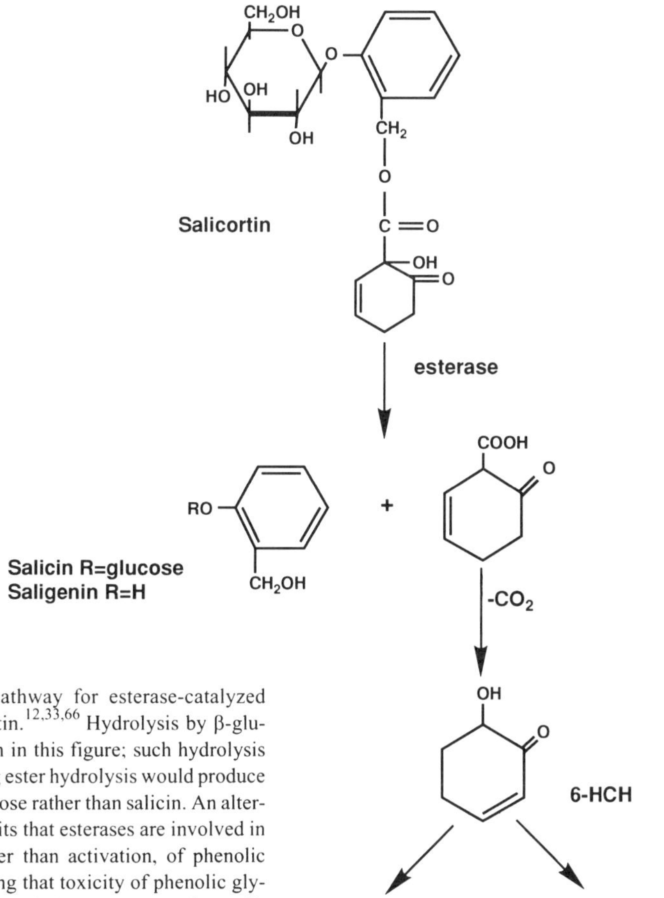

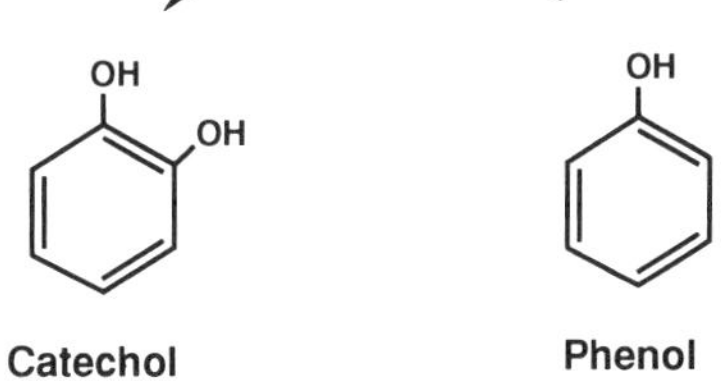

Figure 6. Proposed pathway for esterase-catalyzed metabolism of salicortin.[12,33,66] Hydrolysis by β-glucosidases is not shown in this figure; such hydrolysis preceding or following ester hydrolysis would produce saligenin and free glucose rather than salicin. An alternative perspective posits that esterases are involved in the detoxication, rather than activation, of phenolic glycosides. Recognizing that toxicity of phenolic glycosides containing the cyclohexenone ester (i.e., salicortin, tremulacin) far exceeds that of those lacking the substituent, and that toxicity of tremulacin (containing a second, potentially synergistic ester) is greater than that of salicortin, Lindroth et al.[12] proposed that esterases may be responsible for the metabolic detoxication of salicortin and tremulacin. Subsequent research showed that the toxicity of salicortin and/or tremulacin markedly increases when insect esterases are chemically inhibited[64,65,67] and that toxicity of the phenolic glycosides is retained in the absence of leaf esterases.[64,65,68] In addition, intra- and interspecific variation in resistance to phenolic glycosides is positively correlated with insect esterase activity,[68,69] and esterase activity can be induced in aspen-adapted insects by consumption of phenolic glycosides.[65,67]

systems and rabbit and porcine liver esterases further confirmed that the esterases can hydrolyze salicortin and tremulacin, and lead to production of 6-HCH and catechol.[66]

Although the two perspectives appear mutually exclusive, Clausen (pers. comm.) recently proposed a series of pathways reconciling the differences. Enzymatic decomposition of phenolic glycosides via β-glucosidases may produce a quinone methide with greater biological activity than 6-HCH. Because esterase-catalyzed hydrolysis does not form the quinone methide, the enzymes may serve a protective role in aspen-adapted insects.

Condensed Tannins. Condensed tannins appear to have little impact against aspen-adapted insects,[47] but may deter feeding by unadapted insects. The biochemical basis for activity of aspen tannins has not been investigated. Biological activity may be due to oxidative activation,[70,71] the importance of which is becoming increasingly recognized among tannin researchers. Oxidation of condensed tannins may be mediated by enzymes (e.g., polyphenol oxidase) or particular physicochemical conditions of the gut, although the latter (strongly acidic conditions) does not occur in insects.

Tiger Swallowtails (Papilio species). Plant chemistry has shaped the evolution of larval host preferences and diversification of North American tiger swallowtails (*Papilio glaucus* species complex).[12,72,73] Two very closely related species (formerly subspecies), *P. glaucus* and *P. canadensis*, exhibit strikingly different abilities to feed on members of the plant family Salicaceae. *Papilio canadensis*, which occurs throughout most of Alaska, Canada and the northern tier of states in the eastern United States, utilizes aspen as a primary host. In contrast, *P. glaucus*, which occurs throughout most of the eastern United States (overlapping narrowly with *P. canadensis*), cannot survive on aspen.

Lindroth et al.[121] reported that phenolic glycosides (particularly salicortin and tremulacin) are responsible for the differential abilities of *P. canadensis* and *P. glaucus* to utilize aspen as a host. Midgut esterases appear to confer resistance to the phenolics in *Papilio canadensis*, which has 2–3 times the activity measured in *P. glaucus*.[675,69] Scriber et al.[73] linked the recent evolutionary shift toward use of the Salicaceae by *P. canadensis* to isolation of an ancestral *Papilio* population in the northern Beringial refugiam during the Pleistocene glaciations of North America. Isolation for the period 36,000–9,500 years B.P. with members of the Salicaceae as the major woody hosts available would have facilitated such a shift.

Recent work with *P. canadensis* has shown that phenolic glycosides exact a cost in performance even in this aspen-adapted species (Thompson, Hwang and Lindroth, unpubl. data). Growth of larvae varies substantially among different aspen clones (Fig. 7). Moreover, larval weight is strongly and inversely correlated with foliar phenolic glycoside concentrations (Fig. 8). Reduced growth is due more to reductions in food processing efficiency than to reductions

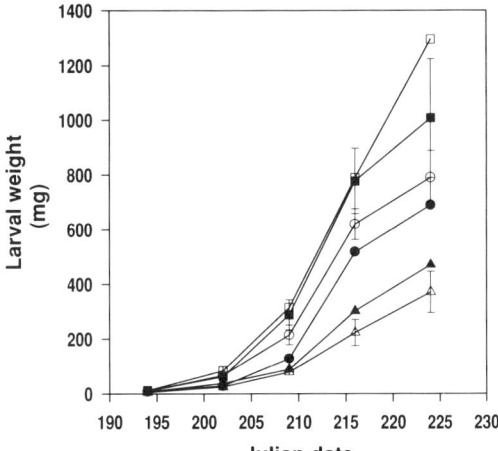

Figure 7. Growth of *Papilio canadensis* on six aspen clones. Each line represents the mean (fresh) weights of two larvae on each of six trees per clone, grown in a common garden. Vertical lines represent ± 1SE, and for clarity of presentation are shown for only half the clones.

in consumption, digestion or developmental rate, indicating toxicity and/or a "metabolic cost" to processing of phenolic glycosides.

Large Aspen Tortrix (Choristoneura conflictana). Trembling aspen is the principal host of the large aspen tortrix, which periodically causes extensive defoliation to aspen forests in Alaska and Canada.[74] Larval growth is marginally reduced by aspen condensed tannins.[21] Larval growth and pupal weights are, however, substantially reduced by phenolic glycosides (salicortin and tremulacin) and their decomposition products 6-HCH and catechol.[33]

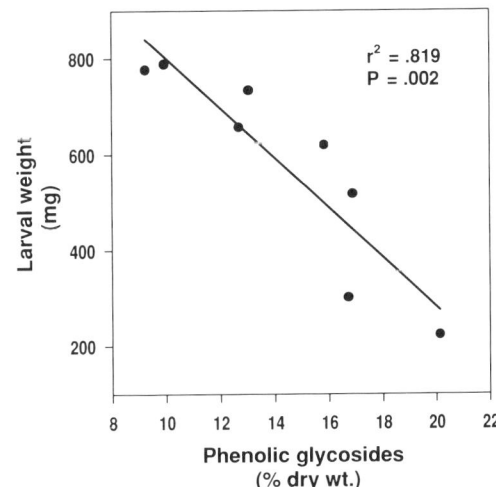

Figure 8. Relationship of fresh weight of *Papilio canadensis* larvae (~22 days after hatch) to phenolic glycoside (salicortin + tremulacin) content of aspen clones. Each point represents mean larval weight and phenolic glycoside content for six trees per clone.

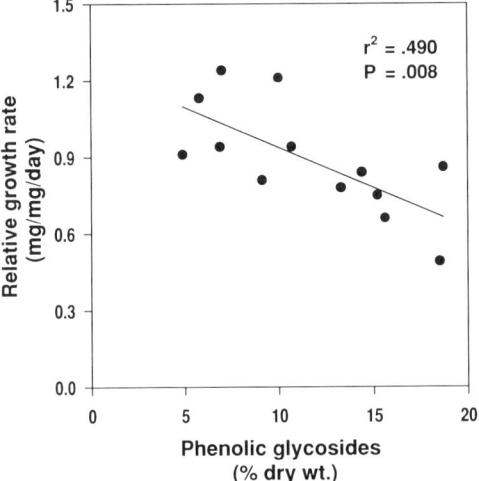

Figure 9. Relationship of growth of *Malacosoma disstria* (fourth instars) to phenolic glycoside (salicortin + tremulacin) content of aspen clones. Each point represents mean values for eight trees per clone, grown in a common garden.

Forest Tent Caterpillar (Malacosoma disstria). The forest tent caterpillar is another leaf-chewing lepidopteran that may cause severe and widespread defoliation to aspen forests. Outbreaks have been recorded since the late 1800s in the Great Lakes region, with a periodicity of about ten years.[75] Aspen trees and/or clones are not, however, uniformly susceptible to defoliation (Lindroth, pers. obs.).

Larval development times are prolonged and pupal weights are reduced for insects reared on aspen foliage containing low levels of foliar nitrogen or high levels of phenolic glycosides.[47] Condensed tannins do not affect insect performance. These results are corroborated by research that assessed responses of caterpillars to artificial diets varying in protein and phenolic glycoside content.[65] This work suggested that high levels of protein may enhance the capacity of larvae to detoxify phenolic glycosides. In general, forest tent caterpillars are influenced more by dietary nitrogen and less by dietary phenolic glycosides than are gypsy moths (*Lymantria dispar*).[47]

Recent studies conducted by our research group confirm that tent caterpillar survival, growth and development vary significantly among aspen clones (Hwang and Lindroth, unpubl. data). Interclonal variation in phenolic glycoside concentrations explains a significant amount of the variation in insect performance (Fig. 9).

Gypsy Moth (Lymantria dispar). The gypsy moth recently became established in the Great Lakes region. Its impact on deciduous forests has escalated exponentially; in 1992 defoliation exceeded 280,000 ha in Michigan alone.[76] Aspen is one of this insect's most favored hosts. As for the forest tent caterpillar,

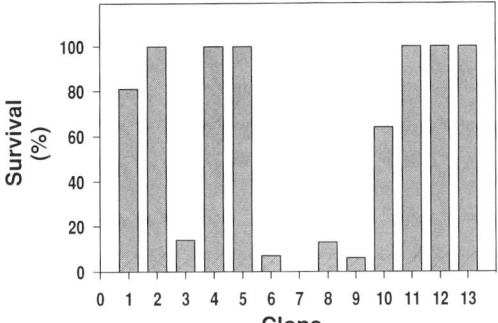

Figure 10. Survival rates of fourth instar *Lymantria dispar* (16 insects/clone) fed foliage from different aspen clones grown in a common garden.

however, rates of defoliation by gypsy moths vary among trees/clones (Lindroth, pers. obs.).

Larval development times and pupal weights are not significantly affected by natural variation in foliar nitrogen or condensed tannin concentrations.[47] These performance parameters are strongly affected, however, by phenolic glycosides. In one study in which larvae were reared on foliage from seven different trees in the field, female pupal weights varied by 3-fold, and variation in phenolic glycoside concentrations explained 98% of the variation in pupal weights.[47] Several laboratory studies with purified phenolic glycosides (salicortin and tremulacin) have shown that at moderate to high (but naturally-occurring) concentrations, the compounds reduce survival, growth and food processing efficiencies of gypsy moth larvae.[47,64,68]

Another study showed that gypsy moths exhibit only minor changes in performance in response to resource-mediated shifts in aspen chemistry (Fig. 4). Results suggest that even though foliar chemical composition responded in a manner consistent with the CNB hypothesis, overall concentrations of phenolic glycosides were too low (approx. 1.5–2.5% fresh wt.) for treatments to effect significant changes in larval performance. Low levels of both phenolic glycosides and condensed tannins were attributed to the specific genotype of aspen used in the study.

Indeed, genetic (clonal) variation in aspen chemistry can markedly affect gypsy moth performance. Survival, development, growth, and food processing efficiencies vary significantly for larvae fed aspen foliage from different clones (Fig. 10). Phenolic glycosides account for most of the interclonal variation in insect performance (Fig. 11).

Big Poplar Sphinx Moth (Pachysphinx modesta). The big poplar sphinx moth is one of the largest North American insects to specialize on aspen. The insects are solitary feeders, and rarely cause significant defoliation of trees. Compared to the better-known forest pest species, little is know about the

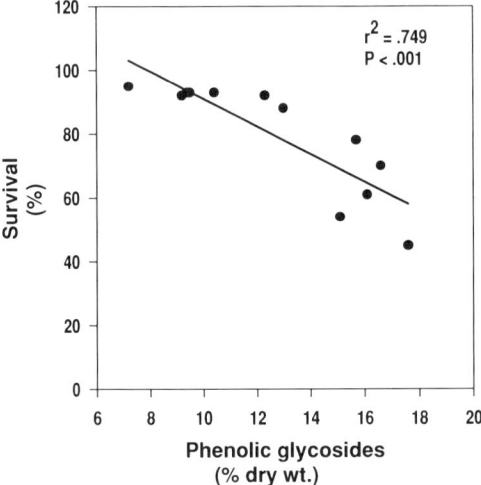

Figure 11. Relationship of fourth instar *Lymantria dispar* survival to phenolic glycoside (salicortin + tremulacin) content of aspen clones. Each point represents mean values for eight trees per clone.

nutritional or chemical ecology of poplar sphinx moths. Preliminary work suggests that performance of even this aspen specialist is altered by the secondary chemistry of its host. Larvae exhibited considerable variation in growth rates when reared on trees from different aspen clones (Fig. 12). Interclonal variation in phenolic glycoside content again explains most of the variation in insect growth (Fig. 13). Corroborating feeding studies with isolated phenolic glycosides have yet to be conducted.

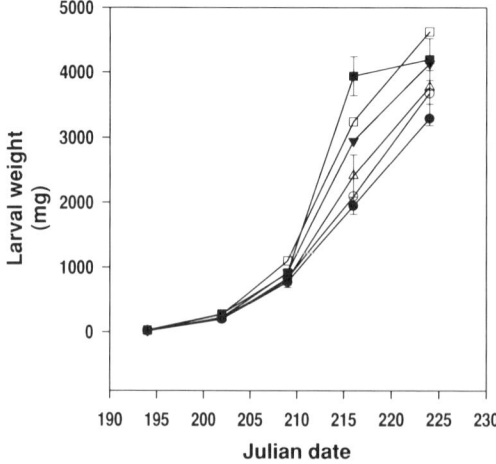

Figure 12. Growth of *Pachysphinx modesta* on six aspen clones. Each line represents the mean (fresh) weights of two larvae on each of six trees per clone, grown in a common garden. Vertical lines represent ± 1SE, and are shown for only half the clones.

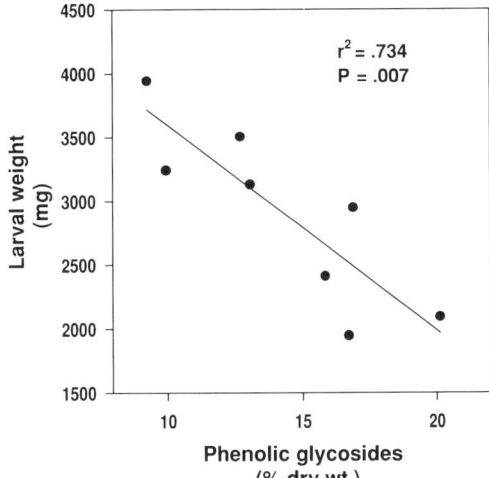

Figure 13. Relationship of weight of *Pachysphinx modesta* larvae (~22 days after hatch) to phenolic glycoside (salicortin + tremulacin) content of aspen clones. Each point represents mean (fresh) larval weight and phenolic glycoside content for six trees per clone.

Chrysomelid Beetles. The diversity of beetle species feeding on North American poplars is probably higher than for any other type of tree.[77] Those known to defoliate aspen include the aspen leaf beetle (*Chrysomela crotchi*), American aspen beetle (*Gonioctena americana*), the cottonwood leaf beetle (*C. scripta*) and the gray willow leaf beetle (*Pyrrhalta decora*).[2,77] Larvae feed on the lower surface of leaves, whereas adults feed on entire leaves.

Little is known about the chemical ecology of beetles feeding on aspen trees. In contrast, however, considerable research has been conducted on chrysomelids or related beetles feeding on hybrid poplar and willow (*Salix* species). A number of chrysomelid species have specialized defense glands, wherein β-glucosidases convert phenolic glycosides to salicylaldehyde.[78,79] Salicylaldehyde is used for defense against generalist predators (e.g., ants), whereas the free glucose is likely metabolized as an energy source. Thus in these insects, the protective role of phenolic glycosides has been altered to serve the insects themselves, and the beetles selectively feed on species or clones with high levels of phenolic constituents.[80,81] Various willow-feeding species are not equally adapted to phenolic glycosides, however, and the performance of some can be negatively affected by high concentrations or unique combinations of the compounds.[81,82] Although use of phenolic glycosides for defensive purposes by beetles feeding on aspen has not been documented, it is likely to occur.

Tritrophic Interactions. Almost no attention has been given to the role of aspen chemistry in tritrophic interactions. The work that has been done has focused on the effects of aspen phenolics on insect pathogens. Hwang et al.[83] showed that efficacy of the bacterial insecticide *Bacillus thuringiensis* subsp.

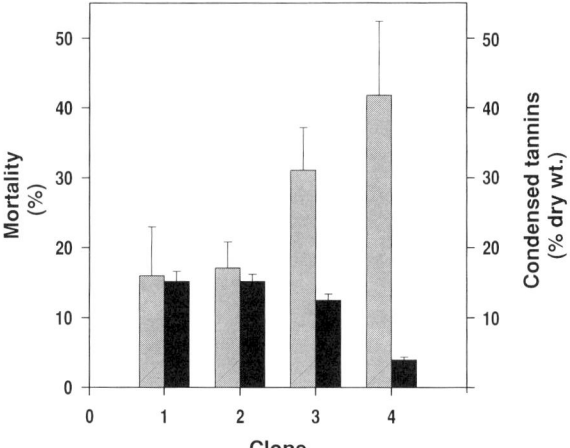

Figure 14. Variation in *Lymantria dispar* NPV mortality (hatched bars) and condensed tannin concentrations (solid bars) in four aspen clones. Assays conducted on 24 insects per tree, four trees per clone. Vertical lines indicate 1 SE.

kurstaki varied 100-fold for gypsy moth larvae reared on different aspen trees. They suggested that condensed tannins reduce toxicity of the Bt endotoxin, but that phenolic glycosides enhance its toxicity. Histopathological examination showed that phenolic glycosides and Bt damage midgut epithelial cells, and that the phenolic compounds inhibit cell regeneration following damage from subacute doses of Bt (Shields, Hwang and Lindroth, unpubl. data).

Aspen secondary chemistry also influences pathogenicity of the gypsy moth nucleopolyhedrosis virus (NPV). This virus is the primary biological agent contributing to population declines following outbreaks in the field. Preliminary work revealed that larval susceptibility to the pathogen is inversely proportional to foliar condensed tannin concentrations (Fig. 14). Similar results have been reported in other studies including different tree species; the mechanism underlying the tannin-NPV interaction is not known.[84,85]

Conclusion and Caveat. Clearly, aspen chemistry has played a central role in the evolution of host range in aspen-feeding insects, and continues to play a pivotal role in the dynamics of interactions between aspen and a variety of ecologically and economically important insects. This does not mean, however, that secondary chemistry strongly influences ecological relationships between aspen and all aspen-feeding insects. For example, research by Auerbach and colleagues[86,87] (and pers. comm.) has shown that performance of the aspen blotch leaf-miner (*Phyllonorycter salicifoliella*) varies markedly among aspen species, clones and branches, but that host defensive chemistry is not a major factor in differential performance. Rather, synchronization of egg hatch with physical characteristics (pubescence and toughness) of expanding leaves appears to be of critical importance.

Aspen-Bird Interactions

The buds and catkins of aspen are eaten by various songbirds to a limited extent, but provide an important food resource for several species of grouse during winter and spring. This is particularly true for ruffed grouse (*Bonasa umbellus*), for which staminate flower buds may comprise up to two-thirds of the winter diet, and extended catkins from these buds may comprise nearly the entire diet during a brief springtime period.[23]

Winter use of aspen buds by ruffed grouse varies from year to year; periods of low use coincide with decreasing or low population densities.[88] Within a winter season, use of aspen is restricted to particular trees or clones.[88,89] A series of studies by Jakubas and colleagues[23,24,88–90] has shown that differential use of aspen by grouse is largely due to variation in levels of coniferyl benzoate (Fig. 2), and not to variation in tannins or "total phenolics." In the field, ruffed grouse normally ingest aspen buds with an average concentration of 1% coniferyl benzoate, and will consume buds with concentrations up to 1.8% (dry weight). Both values lie below the level of 2.5% that reflects the average concentration of coniferyl benzoate in buds available in the field. Ingestion of coniferyl benzoate by ruffed grouse causes nitrogen loss, acidosis from acidic detoxication products, and reduction of metabolizable energy.[23,24]

Aspen-Mammal Interactions

Nonhuman Herbivores. Leaves, twigs and bark of aspen and other members of the Salicaceae provide an important food resource to a variety of rodents, lagomorphs and cervids, some of which are known to exhibit preferences among tissues, trees and clones. Our understanding of the role of plant chemistry in such interactions, however, is rudimentary, and relies primarily on correlative studies of food preferences and host chemistry.

The earliest studies of the role of Salicaceae defensive chemistry in interactions with mammals were conducted with opossums (*Trichosurus vulpecula*) in New Zealand. The herbivores were found to avoid *Populus* and *Salix* species that contain high concentrations of salicin and related compounds.[91,92]

Winter browsing of aspen and willow by hares also appears to be linked to plant chemistry. Adventitious shoots of aspen contain higher levels of phenolic and terpene resins than do twigs of mature trees, and these compounds (identity unknown) render the shoots unpalatable to snowshoe hares (*Lepus ameri-*

canus).[34] Such reversion of shoots to a chemically protected juvenile growth form may play a role in the 10-year hare population cycle. Recently, Reichardt et al.[93] reported that 6-HCH and salicaldehyde, but not tannins, protect internodes of juvenile balsam poplar (*Populus balsamifera*) against grazing by hares. Such compounds may serve a similar role in trembling aspen. In related work, Tahvanainen et al.[94] found that mountain hares (*Lepus timidus*) show distinct preferences for some willow species over others, and for mature shoots over juvenile shoots. Consumption rates were inversely correlated with phenolic glycoside contents, and supplemental feeding studies confirmed that the compounds deter browsing.

Beaver (*Castor canadensis*) can be the dominant herbivore of aspen in riparian habitats. Stands of aspen previously cut by beaver consist of root sprouts with the juvenile growth form. Work by Basey et al.[35] in the intermountain west showed that beaver avoid feeding on juvenile form aspen when mature growth-form branches or sprouts are available. Food preferences are not related to phenolic glycoside levels, but to levels of an unknown phenolic compound with high concentrations in juvenile tissues. Moreover, levels of the compound are inversely proportional to tree size in sites occupied by beaver for many years.[95] Thus, in these sites beaver preferentially cut larger trees, a behavior contrary to their normal behavior of selecting small trees over large trees.

Aspen is a principal browse species for a variety of cervids, including deer (*Odocoileus virginianus*, *O. hemionus*), elk (*Cervus canadensis*) and moose (*Alces alces*). Virtually nothing is known, however, about the influence of aspen secondary chemistry on the foraging of these species. Given that both tannin and nontannin phenolics affect diet selection by ruminant browsers,[96] such effects are likely.

Humans.

> *"There is a bark of an English tree, which I have found by experience to be a powerful astringent, and very efficacious in curing aguish and intermitting disorders."*
> —Written April 25, 1763, in a letter from Rev. Mr. Edmund Stone in Chipping-Norton (Oxfordshire) to the Right Honourable George, Earl of Macclesfield, President of the Royal Society.[97]

The central theme of this series of papers, *viz.*, multiplicity of chemical roles, is probably best exemplified for salicylates in the realm of human medicine.

Concoctions derived from poplars and willows have long been used as folk remedies to treat problems as diverse as fever, earaches, warts and dandruff.[98]

By the early 1800s, the active constituents were recognized as salicylates. Since then, use of various salicylates (e.g., acetylsalicylic acid, a.k.a "aspirin") and related drugs (e.g., acetominophen, ibuprofen) in medical practice has grown to a level vastly exceeding that of any other pharmacological agent. Americans ingest 16,000 tons of aspirin tablets per year.[97]

Salycilates are commonly and widely used because of their therapeutic value in treating a wide range of human maladies. Salycilates reduce pain and inflammation, relieve fever, inhibit blood platelet aggregation and function, alter uric acid excretion and eliminate warts. The compounds also contribute to formation of peptic ulcers, and inhibit bone metabolism and proteoglycan synthesis. This tremendous range of effects is due to several modes of action at the biochemical level. Salicylates have been known for several decades to inhibit synthesis of prostaglandins, cellular hormones involved in localized pain and inflammation responses. Recent work by Weissman[97] and colleagues has identified an additional mode of action: inhibition of neutrophil activation and aggregation. Neutrophils are the most abundant cells associated with acute inflammation, and exacerbate damage by releasing proteases and other reactive compounds.

ASPEN CHEMISTRY AND ABIOTIC INTERACTIONS

Virtually no published information exists on the role of aspen secondary chemistry with respect to protection from the physical environment. Such a role is likely, however, with respect to ultraviolet (UV) radiation. Like the better-known flavonoids, salicylate phenolic glycosides absorb UV radiation (particularly in the 200–320 nm band). If these phenolic glycosides are effective in attenuating incident UV radiation, one would expect relatively high concentrations to occur in aspen inhabiting high UV environments. Preliminary evidence supports the prediction. We propagated four clones of aspen collected as root cuttings from an alpine environment in Colorado (approx. 2,500 m), and grew them in a common garden with clones collected in Wisconsin. Foliage of trees from Colorado had 60% higher levels of phenolic glycosides, and less variation among clones, than did foliage of trees from Wisconsin.

COMMUNITY AND ECOSYSTEM EFFECTS

Shifts in chemistry induced by genetic changes, resource availability, pathogens, herbivory or abiotic factors may alter the dynamics of organismal interactions. These in turn will have consequences for community development and organization, and ecosystem structure and function. Such ecological effects are beyond the scope of this paper, but we will mention several examples for the sake of illustration.

Two decades ago Mattson and Addy[99] proposed that phytophagous insects may regulate primary production of forest ecosystems via their impact on nutrient cycling. Their ideas expanded on (among other things) observations of aspen forest response to defoliation by forest tent caterpillars. In short, they argued that although heavy defoliation may reduce tree growth in the short term, it may accelerate growth in the long term due to increases in nutrient cycling rates. The research described in this paper suggests, however, that aspen forests are far from chemically uniform. One would expect then, that the combination of chemical variation, insect outbreaks, and positive and negative feedbacks would result in a spatial mosaic of changes in nutrient cycling rates and primary production in an aspen-dominated forest. This may be especially true in western North America where aspen clones are extensive in size.

Plant chemistry and attendant effects on herbivory and decomposition are now recognized to play a central role in the succession of North American boreal forest communities.[100] Aspen, balsam poplar and birch are the dominant woody species common to early successional boreal forests. Tissues of these species generally have high concentrations of nitrogen but relatively low concentrations of phenolics, terpenoids and lignin. As a consequence, litter is rapidly decomposed and soil nitrogen availability is high. In contrast, late successional species (balsam fir and spruce) produce tissues low in nitrogen but high in phenolics, terpenoids and lignin. Slowly decomposing litter of these species depresses soil nitrogen availability. In short, soil nitrogen availability controls net primary production, but the types of carbon compounds produced (e.g., starch and cellulose vs. phenolics and lignin) affect both herbivory and soil nitrogen availability. Nutrient cycles are linked by positive and negative feedback loops among plants, herbivores and decomposers.

Beaver and moose, which feed on the same tree species, are keystone herbivores in boreal forests. Their foraging behaviors differ, however, resulting in marked differences in their impact on community and ecosystem dynamics. Beaver cut large aspen and other early-successional trees near their ponds, opening the canopy and facilitating regeneration of the shade-intolerant species. The effects of aspen chemistry on beaver foraging behavior are not known for these systems. If, however, they are similar to those in the intermountain west, differing levels of phenolic compounds may deter use of juvenile aspen, promote use of large aspen, and thereby contribute to successional delays by changing mid-successional stands to early-successional stands.

In contrast to beaver, moose prefer to feed on young aspen and other early successional species. Long-term and intensive browsing can kill trees. Selective feeding on aspen and avoidance of unpalatable, late-successional species therefore accelerate community succession. Thus, differential browsing by beaver and moose on different size classes in an aspen population may lead to a divergence in community succession and ecosystem dynamics.[100]

CONCLUSIONS

Aspen secondary chemistry is characterized by a fairly limited variety of products from the shikimic acid pathway, including phenolic glycosides, phenylpropanoids and condensed tannins. What is lacking in qualitative variation, however, is more than compensated for by tremendous quantitative variation and multiplicity of function. Aspen phenolics provide protection against pathogens, insects, birds, mammals and possibly UV radiation. Biological activity varies significantly even among close structural analogues. Moreover, some constituents (e.g., salicortin, tremulacin) have broad-spectrum activity against a variety of organisms, while others (e.g., coniferyl benzoate) have a narrower spectrum of activity. The biochemical modes of action are diverse, as illustrated by the phenomenal array of effects that salicylates exhibit in human pharmacology. Tissue concentrations of the compounds are influenced by tree developmental status, prior herbivory, resource availability and genetic variation. Quantitative variation, in turn, strongly affects the dynamics of interactions between aspen and associated organisms. Aspen's broad geographic distribution, its abundance in early-successional, ecosystems and its associaton with a great variety of pathogens and herbivores combine to make its suite of phenolic natural products one of the most important groups of compounds mediating ecological interacitons in North America.

ACKNOWLEDGMENTS

We thank Mark Scriber, John Bryant, Tom Clausen and Paul Reichardt for introduction to, and advice regarding, the fascinating realm of aspen chemoecology. Much of the work reported here was conducted as part of the thesis research of graduate students Jocelyn Hemming, Cindy Thompson, Karl Kinney, and Sherry Roth. We are indebted to them and to our technical assistants Annie Weisbrod and Steve Jung. Thanks to Nancy Lindroth for preparation of the figures. Funding for the research described here was provided by the National Science Foundation, U.S. Department of Agriculture, and University of Wisconsin-Madison.

REFERENCES

1. DICKMANN, D.I., STUART, K.W. 1983. The Culture of Poplars in Eastern North America. Department of Forestry, Michigan State University, East Lansing, MI. p. 168.
2. PERALA, D.A. 1990. *Populus tremuloides* Michx. Quaking Aspen. In: Silvics of North America. Volume 2, Hardwoods. (R.M. Burns, B.H. Honkala, eds.), Forest Service, United States Department of Agriculture, Washington, D.C., pp. 555–569.

3. MITTON, J.B., GRANT, M.C. 1996. Genetic variation and the natural history of quaking aspen. Bioscience 46: 25–31.
4. LINDROTH, R.L., HSIA, M.T.S., SCRIBER, J.M. 1987. Seasonal patterns in the phytochemistry of three *Populus* species. Biochem. Syst. Ecol. 15: 681–686.
5. THIEME, H. 1965. Die phenolglykoside der Salicaceen. Planta Medica 13: 431–438.
6. THIEME, H. 1967. Uber die phenolglykoside der gattung *Populus*. Planta Medica 15: 35–40.
7. THIEME, H., BENECKE, R. 1970. Die phenolglykoside der Salicaceen. 7. Mitteilung: Uber die glykosidfuhrung einheimischer bzw. mitteleuropa kultivierter *Populus*-arten. Pharmazie 25: 780–789.
8. PEARL, I.A., DARLING, S.F. 1969. Investigation of the hot water extractives of *Populus balsamifera* bark. Phytochemistry 8: 2393–2396.
9. PEARL, I.A., DARLING, S.F. 1968. Mass spectrometry as an aid for determining structure of natural glucosides. Phytochemistry 7: 831–837.
10. PEARL, I.A., DARLING, S.F. 1971. The structures of salicortin and tremulacin. Phytochemistry 10: 3161–3166.
11. PIERPOINT, W.S. 1994. Salicylic acid and its derivatives in plants: medicines, metabolites and messenger molecules. Adv. Bot. Res. 20: 163–235.
12. LINDROTH, R.L., SCRIBER, J.M., HSIA, M.T.S. 1988. Chemical ecology of the tiger swallowtail: mediation of host use by phenolic glycosides. Ecology 69: 814–822.
13. PALO, R.T. 1984. Distribution of birch (*Betula* spp.), willow (*Salix* spp.), and poplar (*Populus* spp.) secondary metabolites and their potential role as chemical defense against herbivores. J. Chem. Ecol. 10: 499–520.
14. LINDROTH, R.L., PAJUTEE, M.S. 1987. Chemical analysis of phenolic glycosides: art, facts, and artifacts. Oecologia 76: 144–148.
15. JULKUNEN-TIITTO, R., TAHVANAINEN, J. 1989. The effect of the sample preparation method of extractable phenolics of Salicaceae species. Planta Medica 55: 55–58.
16. ORIANS, C.M. 1996. Preserving leaves for tannin and phenolic glycoside analyses: a comparison of methods using three willow taxa. J. Chem. Ecol. 21: 1235–1243.
17. LINDROTH, R.L., KOSS, P.A. 1996. Preservation of Salicaceae leaves for phytochemical analyses: further assessment. J. Chem. Ecol. (in press).
18. BENNETT, R.N., WALLSGROVE, R.M. 1994. Tansley review no. 72: Secondary metabolites in plant defence mechanisms. New Phytol. 127: 617–633.
19. JONES, A.M. 1994. Surprising signals in plant cells. Science 263: 183–184.
20. GAFFNEY, T., FRIEDRICH, L., VERNOOIJ, B., NEGROTTO, D., NYE, G., UKNES, S., WARD, E., KESSMANN, H., RYALS, J. 1993. Requirement of salicylic acid for the induction of systemic acquired resistance. Science 261: 754–756.
21. BRYANT, J.P., CLAUSEN, T.P., REICHARDT, P.B., MCCARTHY, M.C., WERNER, R.A. 1987. Effect of nitrogen fertilization upon the secondary chemistry and nutritional value of quaking aspen (*Populus tremuloides* Michx.) leaves for the large aspen tortrix (*Choristoneura conflictana* (Walker)). Oecologia 73: 513–517.
22. LINDROTH, R.L., KINNEY, K.K., PLATZ, C.L. 1993. Responses of deciduous trees to elevated atmospheric CO_2: productivity, phytochemistry and insect performance. Ecology 74: 763–777.
23. JAKUBAS, W.J., KARASOV, W.H., GUGLIELMO, C.G. 1993. Ruffed grouse tolerance and biotransformation of the plant secondary metabolite coniferyl benzoate. The Condor 95: 625–640.
24. JAKUBAS, W.J., KARASOV, W.H., GUGLIELMO, C.G. 1993. Coniferyl benzoate in quaking aspen (*Populus tremuloides*): its effect on energy and nitrogen digestion and retention in ruffed grouse (*Bonasa umbellus*). Physiol. Zool. 66: 580–601.
25. ENGLISH, S., GREENAWAY, W., WHATLEY, F.R. 1991. Bud exudate composition of *Populus tremuloides*. Can. J. Bot. 69: 2291–2295.

26. TEW, R. 1970. Seasonal variation in the nutrient content of aspen foliage. J. Wildl. Manage. 34: 475–478.
27. JAMES, T.D.W., SMITH, D.W. 1978. Seasonal changes in the major ash constituents of leaves and some woody components of trembling aspen and red osier dogwood. Can. J. Bot. 56: 1798–1803.
28. HUNTER, A.F., LECHOWICZ, M.J. 1992. Foilage quality changes during canopy development of some northern hardwood trees. Oecologia 89: 316–323.
29. MEYER, G.A., MONTGOMERY, M.E. 1987. Relationships between leaf age and the food quality of cottonwood foliage for the gypsy moth, *Lymantria dispar*. Oecologia 72: 527–532.
30. TALLAMY, D.W., RAUPP, M.J. (eds.) 1991. Phytochemical Induction by Herbivores. John Wiley and Sons, New York. p. 431.
31. MATTSON, W.J., PALMER, S.R. 1988. Changes in foliar minerals and phenolics in trembling aspen, *Populus tremuloides*, in response to artificial defoliation. In: Mechanisms of Woody Plant Defenses Against Insects. Search for Pattern. (W.J. Mattson, J. Levieux, C. Bernard-Dagen, eds.), Springer-Verlag, New York, pp. 157–169.
32. BALDWIN, I.T., SCHULTZ, J.C. 1983. Rapid changes in tree leaf chemistry induced by damage: evidence for communication between plants. Science 221: 277–279.
33. CLAUSEN, T.P., REICHARDT, P.B., BRYANT, J.P., WERNER, R.A., POST, K., FRISBY, K. 1989. A chemical model for short-term induction in quaking aspen (*Populus tremuloides*) foliage against herbivores. J. Chem. Ecol. 15: 2335–2346.
34. BRYANT, J.P. 1981. Phytochemical deterrence of snowshoe hare browsing by adventitious shoots of four Alaskan trees. Science 213: 889–890.
35. BASEY, J.M., JENKINS, S.H., MILLER, G.C. 1990. Food selection by beavers in relation to inducible defenses of Populus tremuloides. Oikos 59: 57–62.
36. BRYANT, J.P., DANELL, K., PROVENZA, F., REICHARDT, P.B., CLAUSEN, T.A., WERNER, R.A. 1991. Effects of mammal browsing on the chemistry of deciduous woody plants. In: Phytochemical Induction by Herbivores, (D.W. Tallamy, M.J. Raupp, eds.), John Wiley and Sons, New York, pp. 135–154.
37. BRYANT, J.P., CHAPIN III, F.S., KLEIN, D.R. 1983. Carbon/nutrient balance of boreal plants in relation to vertebrate herbivory. Oikos 40: 357–368.
38. BAZZAZ, F.A., CHIARIELLO, N.R., COLEY, P.D., PITELKA, L.F. 1987. Allocating resources to reproduction and defense. Bioscience 37: 58–67.
39. CHAPIN III, F.S. 1980. The mineral nutrition of wild plants. Ann. Rev. Ecol. Syst. 11: 233–260.
40. BRYANT, J.P., CHAPIN III, F.S., REICHARDT, P.B., CLAUSEN, T.P. 1987. Response of winter chemical defense in Alaska paper birch and green alder to manipulation of plant carbon/nutrient balance. Oecologia 72: 510–514.
41. PRICE, P.W., WARING, G.L., JULKUNEN-TIITTO, R., TAHVANAINEN, J., MOONEY, H. A., CRAIG, T.P. 1989. Carbon-nutrient balance hypothesis in within-species phytochemical variation of *Salix lasiolepis*. J. Chem. Ecol. 15: 1117–1131.
42. REICHARDT, P.B., CHAPIN III, F.S., BRYANT, J.P., MATTES, B.R., CLAUSEN, T.P. 1991. Carbon/nutrient balance as a predictor of plant defense in Alaskan balsam poplar: potential importance of metabolite turnover. Oecologia 88: 401–406.
43. ROTH, S.K., LINDROTH, R.L. 1995. Elevated atmospheric CO_2: effects on phytochemistry, insect performance and insect-parasitoid interactions. Global Change Ecology 1: 173–182.
44. KINNEY, K.K., LINDROTH, R.L. JUNG, S.M., NORDHEIM, E.V. 1996. Effects of atmospheric CO_2 and soil NO_3^- availability on deciduous trees: phytochemistry and insect performance. Ecology (in press).
45. JELINSKI, D.E., CHELIAK, W.M. 1992. Genetic diversity and spatial subdivision of *Populus tremuloides* (Salicaceae) in a heterogeneous landscape. Am. J. Bot. 79: 728–736.
46. JELINSKI, D.E., FISHER, L.J. 1991. Spatial variability in the nutrient composition of *Populus tremuloides*: clone-to-clone differences and implications for cervids. Oecologia 88: 116–124.

47. HEMMING, J.D.C., LINDROTH, R.L. 1995. Intraspecific variation in aspen phytochemistry: effects on performance of gypsy moths and forest tent caterpillars. Oecologia 103: 79–88.
48. RHOADES, D.F. 1979. Evolution of plant chemical defense against herbivores, In: Herbivores: Their Interaction with Secondary Plant Metabolites, (G.A. Rosenthal, D.H. Janzen, eds.), Academic Press, New York, pp. 3–54.
49. COLEY, P.D. 1986. Costs and benefits of defense by tannins in a neotropical tree. Oecologia 70: 238–241.
50. HERMS, D.A., MATTSON, W.J. 1992. The dilemma of plants: to grow or defend. Quart. Rev. Biol. 67: 283–335.
51. MOLE, S. 1994. Trade-offs and constraints in plant-herbivore defense theory: a life-history perspective. Oikos 71: 3–12.
52. COLEY, P.D., BRYANT, J.P., CHAPIN III, F.S. 1985. Resource availability and plant antiherbivore defense. Science 230: 895–899.
53. VRIELING, K., VAN WIJK, C.A.M. 1994. Estimating costs and benefits of the pyrrolizidine alkaloids of *Senecio jacobaea* under natural conditions. Oikos 70: 449–454.
54. OSTRY, M.E., WILSON, L.F., MCNABB Jr., H.S., MOORE, L.M. 1988. A Guide to Insect, Disease, and Animal Pests of Poplars. United States Department of Agriculture; Forest Service, Washington, DC. p. 118.
55. HUBBES, M. 1966. Inhibition of *Hypoxylon pruinatum* (Klotzsche)Cke. by aspen bark meal and the nature of active extractives. Can. J. Bot. 44: 365–386.
56. HUBBES, M. 1962. Inhibition of *Hypoxylon pruinatum* by pyrocatechol isolated from bark of aspen. Science 136: 156.
57. FLORES, G., HUBBES, M. 1979. Phytoalexin production by aspen (*Populus tremuloides* Michx.) in response to infection by *Hypoxylon mammatum* (Wahl.) Mill and *Alternaria* spp. Eur. J. For. Pathol. 9: 280–288.
58. FLORES, G., HUBBES, M. 1980. The nature and role of phytoalexin produced by aspen (*Populus tremuloides* Michx.). Eur. J. For. Pathol. 10: 95–103.
59. KRUGER, B.M., MANION, P.D. 1994. Antifungal compounds in aspen: effect of water stress. Can. J. Bot. 72: 454–460.
60. MASON, R.R. 1987. Nonoutbreak species of forest lepidoptera. In: Insect Outbreaks, (P. Barbosa, J.C. Schultz, eds.), Academic Press, New York, pp. 31–57.
61. MATTSON, W.J., HERMS, D.A., WITTER, J.A., ALLEN, D.C. 1991. Woody plant grazing systems: North American outbreak folivores and their host plants. In: Forest Insect Guilds: Patterns of Interaction with Host Trees, (Y.N. Baranchikov, W.J. Mattson, F.P. Hain, T.L. Payne, eds.), Gen. Tech. Rep. NE-153, United States Department of Agriculture Forest Service, Northeastern Forest Experiment Station, Radnor, PA, pp. 53–84.
62. LINDROTH, R.L., PETERSON, S.S. 1988. Effects of plant phenols on performance of southern armyworm larvae. Oecologia 75: 185–189.
63. LINDROTH, R.L. 1988. Hydrolysis of phenolic glycosides by midgut β-glucosidases in *Papilio glaucus* subspecies. Insect Biochem. 8: 789–792.
64. LINDROTH, R.L., HEMMING, J.D.C. 1990. Responses of the gypsy moth (Lepidoptera: Lymantriidae) to tremulacin, an aspen phenolic glycoside. Environ. Entomol. 19: 842–847.
65. LINDROTH, R.L., BLOOMER, M.S. 1991. Biochemical ecology of the forest tent caterpillar: responses to dietary protein and phenolic glycosides. Oecologia 86: 408–413.
66. JULKUNEN-TIITTO, R., MEIER, B. 1992. The enzymatic decomposition of salicin and its derivatives obtained from Salicaceae species. J. Nat. Prod. 55: 1204–1212.
67. LINDROTH, R.L. 1989. Biochemical detoxication: mechanism of differential tiger swallowtail tolerance to phenolic glycosides. Oecologia 81: 219–224.
68. LINDROTH, R.L., WEISBROD, A.V. 1991. Genetic variation in response of the gypsy moth to aspen phenolic glycosides. Biochem. Syst. Ecol. 19: 97–103.

69. SCRIBER, J.M., LINDROTH, R.L., NITAO, J. 1989. Differential toxicity of a phenolic glycoside from quaking aspen to *Papilio glaucus* butterfly subspecies, hybrids and backcrosses. Oecologia 81: 186–191.
70. APPEL, H.M. 1993. Phenolics in ecological interactions: the importance of oxidation. J. Chem. Ecol. 19: 1521–1552.
71. BARBEHENN, R.V., MARTIN, M.M. 1994. Tannin sensitivity in larvae of *Malacosoma disstria* (Lepidoptera): roles of the peritrophic envelope and midgut oxidation. J. Chem. Ecol. 20: 1985–2001.
72. FEENY, P. 1991. Chemical constraints on the evolution of swallowtail butterflies. In: Plant-Animal Interactions: Evolutionary Ecology in Tropical and Temperate Regions, (P.W. Price, T.M. Lewinsohn, G.W. Fernandes, W.W. Benson, eds.), John Wiley and Sons, New York, pp. 315–340.
73. SCRIBER, J.M., LEDERHOUSE, R.C., HAGEN, R.H. 1991. Foodplants and evolution within *Papilio glaucus* and *Papilio troilus* species groups (Lepidoptera: Papilionidae). In: Plant-Animal Interactions: Evolutionary Ecology in Tropical and Temperate Region, (P.W. Price, T.M. Lewinsohn, G.W. Fernandes, W.W. Benson, eds.), John Wiley and Sons, New York, pp. 341–373.
74. BECKWITH, R.C. 1973. The large aspen tortrix. USDA Forest Service, Forest Pest Leaflet 139: 1–5.
75. DUNCAN, D.P., HODSON, A.C. 1958. Influence of the forest tent caterpillar upon the aspen forests of Minnesota. For. Sci. 4: 71–93.
76. Gypsy Moth News. 1993. Gypsy moth defoliation 1992. In: Gypsy Moth News 31, (D.B. Twardus, ed.), United States Department of Agriculture Forest Service, Morgantown, West Virginia.
77. ROSE, A.H., LINDQUIST, O.H. 1982. Insects of Eastern Hardwood Trees. Department of the Environment, Canadian Forestry Service, Ottawa. p. 304.
78. PASTEELS, J.M., ROWELL-RAHIER, M., BRAEKMAN, J.C., DUPONT, A. 1983. Salicin from host plant as precursor of salicylaldehyde in defensive secretion of chrysomeline larvae. Physiol. Entomol. 8: 307–314.
79. PASTEELS, J.M., ROWELL-RAHIER, M., RAUPP, M.J. 1988. Plant-derived defense in chrysomelid beetles. In: Novel Aspects of Insect-Plant Interactions, (P. Barbosa, D. Letourneau, eds.), John Wiley and Sons, pp. 235–272.
80. SMILEY, J. 1978. Plant chemistry and the evolution of host specificity: new evidence from *Heliconius* and *Passiflora*. Science 201: 745–747.
81. TAHVANAINEN, J., HELLE, E., JULKUNEN-TIITTO, R., LAVOLA, A. 1985. Phenolic compounds of willow bark as deterrents against feeding by mountain hare. Oecologia 65: 319–323.
82. KELLY, M.T., CURRY, J.P. 1991. The influence of phenolic compounds on the suitability of three *Salix* species as hosts for the willow beetle *Phratora vulgatissima*. Ent. exp. appl. 61: 25–32.
83. HWANG, S.Y., LINDROTH, R.L. 1995. Effects of aspen leaf quality on gypsy moth (Lepidoptera: Lymantriidae) susceptibility to *Bacillus thuringiensis*. J. Econ. Entomol. 88: 278–282.
84. KEATING, S.T., HUNTER, M.D., SCHULTZ, J.C. 1990. Leaf phenolic inhibition of gypsy moth nuclear polyhedrosis virus. J. Chem. Ecol. 16: 1445–1457.
85. HUNTER, M.D., SCHULTZ, J.C. 1993. Induced plant defenses breached? Phytochemical induction protects an herbivore from disease. Oecologia 94: 195–203.
86. AUERBACH, M., ALBERTS, J.D. 1992. Occurrence and performance of the aspen blotch miner, *Phyllonorycter salicifoliella*, on three host-tree species. Oecologia 89: 1–9.
87. AUERBACH, M. 1991. Relative impact of interactions within and between trophic levels during an insect outbreak. Ecology 72: 1599–1608.

88. JAKUBAS, W.J., GULLION, G.W. 1991. Use of quaking aspen flower buds by ruffed grouse: its relationship to grouse densities and bud chemical composition. The Condor 93: 473–485.
89. JAKUBAS, W.J., GULLION, G.W., CLAUSEN, T.P. 1989. Ruffed grouse feeding behavior and its relationship to the secondary metabolites of quaking aspen flower buds. J. Chem. Ecol. 15: 1899–1917.
90. JAKUBAS, W.J., GULLION, G.W. 1990. Coniferyl benzoate in quaking aspen. J. Chem. Ecol. 16: 1077–1087.
91. MARKHAM, K.R. 1971. A chemotaxonomic approach to the food selection of opossum resistant willows and poplars for use in soil conservation. New Zeal. J. Sci. 14: 179–186.
92. EDWARDS, W.R.N. 1978. Effect of salicin content on palatability of *Populus* foliage to opossum (*Trichosurus vulpecula*). New Zeal. J. Sci. 21: 103–106.
93. REICHARDT, P.B., BRYANT, J.P., MATTES, B.R., CLAUSEN, T.P., CHAPIN III, F.S., MEYER, M. 1990. Winter chemical defense of Alaskan balsam poplar against snowshoe hares. J. Chem. Ecol. 16: 1941–1959.
94. TAHVANAINEN, J., JULKUNEN-TIITTO, R., KETTUNEN, J. 1985. Phenolic glycosides govern the food selection pattern of willow feeding leaf beetles. Oecologia 67: 52–56.
95. BASEY, J.M., JENKINS, S.H., BUSHER, P.E. 1988. Optimal central-place foraging by beavers: tree-size selection in relation to defensive chemicals of quaking aspen. Oecologia 76: 278–282.
96. MCARTHUR, C., ROBBINS, C.T., HAGERMAN, A.E., HANLEY, T.A. 1993. Diet selection by a ruminant generalist browser in relation to plant chemistry. Can. J. Zool. 71: 2236–2243.
97. WEISSMANN, G. 1991. Aspirin. Sci. Amer. 264: 84–90.
98. CULPEPER, N. 1826. Culpeper's Complete Herbal and English Physician. J. Gleave and Son, Deansgate, Manchester, England. p. 241.
99. MATTSON, W.J., ADDY, N.D. 1975. Phytophagous insects as regulators of forest primary production. Science 190: 515–522.
100. PASTOR, J., NAIMAN, R.J. 1992. Selective foraging and ecosystem processes in boreal forests. Am. Nat. 139: 690–705.

Chapter Three

DIVERSITY AND DYNAMICS OF CRUCIFER DEFENSES AGAINST ADULTS AND LARVAE OF CABBAGE BUTTERFLIES

J. A. A. Renwick

Boyce Thompson Institute
Tower Road
Ithaca, New York 14853

Introduction 57
Chemical Defenses of Cruciferae 59
The *Pieris*-Crucifer Model System 60
Deterrents in *Erysimum cheiranthoides* 62
Deterrents in *Iberis amara* 64
Deterrents in Host Plants of *Pieris rapae* 66
Differential Responses of *Pieris* Species 67
Balancing Stimulants and Deterrents 68
Effects of Experience on Behavioral Responses 72
Conclusions 74

INTRODUCTION

Plants have evolved the ability to protect themselves from invading organisms in many diverse ways that involve both physical and chemical strategies. Chemical mechanisms of defense are of particular interest to plant breeders in their quest for varieties of crop plants that are resistant to insects and disease. Such resistance to insects is found in many different forms that have been conveniently classified according to mode of action as antixenosis, tolerance or antibiosis.[1,2] Antixenosis results in avoidance by, or nonpreference of potential insect pests, whereas antibiosis implies that an invading insect will be adversely affected, either by ingestion of, or contact with plant constituents.[3] From a

Phytochemical Diversity and Redundancy in Ecological Interactions,
edited by Romeo et al., Plenum Press, New York, 1996

phytocentric point of view, the chemical options available to a plant could be simply described as "repel, deter, or kill".

However, the defensive chemistry of plants is not nearly as simple as these earlier views suggested. Recent research has revealed the dynamic nature of plant defenses, involving complex interactions with invading organisms as well as the influence of abiotic forces (Fig.1). The biosynthesis and/or release of secondary metabolites may be triggered by mechanical damage, microorganisms, herbivory, or a variety of other environmental factors.[4] The active compounds may repel flying insects, deter oviposition or feeding, interfere with development, or cause rapid death, of the invaders. Furthermore, the production of allelochemicals and toxins may be regulated to a large degree by nutritional status, light quality and other growth conditions. In addition to the direct effects of defensive chemicals, plants may also benefit indirectly from the emission of volatiles which serve to recruit natural enemies — the so-called "cry for help" syndrome.[5]

Although rapid advances towards understanding the dynamics of plant chemical defenses are now being made, several questions still remain to be answered. The identities of all the compounds involved and their exact roles in the defensive scenario need to be determined. The efficiency of resource allocation by the plant to deal with different invaders or to combat the different life stages of an insect has yet to be addressed. Despite the fact that the chain of events involved in the production of defensive chemicals is now becoming much

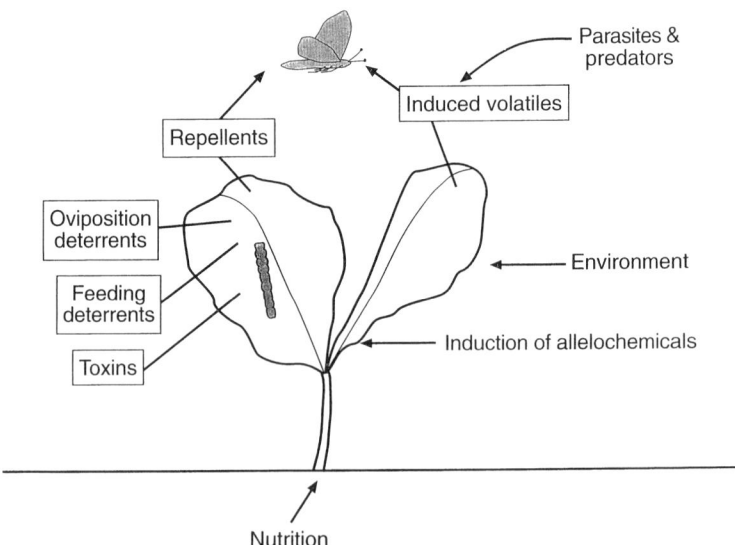

Figure 1. Schematic representation of the dynamics of a plant's chemical defense against insects.

clearer, many details of the factors that influence this process are still missing. The effectiveness of defensive chemicals against their presumed victims is not always obvious, and variable responses of insect species, or even individuals within a species, to plant chemicals need to be explained.

To answer some of these outstanding questions, a model plant-insect system, in which the most important chemicals are known and the insect behavior is well documented, might provide some valuable insight. Here we examine a model system involving closely related crucifer specialists to document and analyze factors that bear on the effectiveness of defensive chemistry in a specific group of plants. Comparative analyses of *Pieris rapae* and *P. napi oleracea* adult and larval sensitivity to deterrents from a range of plants confirm that we cannot generalize about the activity of any compound against adults and larvae of a species, or against related species. Although this example using related cabbage butterflies may not be typical of insect-plant relationships, it provides a clear indication of the multiplicity of plant and insect variables involved.

CHEMICAL DEFENSES OF CRUCIFERAE

Members of the family Cruciferae (= Brassicaceae) are chemically linked by the almost universal presence of glucosinolates, a class of sulfur-containing glycosides, also known as mustard oil glycosides or thioglycosides (Fig. 2). These compounds are considered to be the first line of defense of crucifers against insects and other organisms, either directly or as their hydrolysis products.[6–8] When the tissue of a crucifer is disrupted, as for example by insect feeding, the glucosinolates are hydrolyzed by a thioglucosidase (myrosinase) to

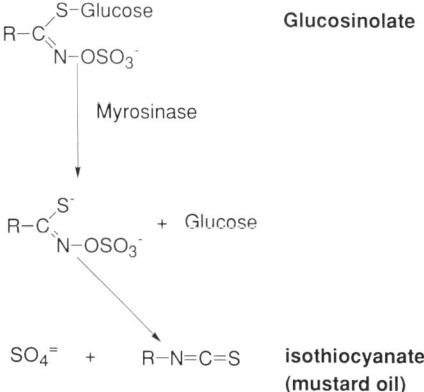

Figure 2. General structure of glucosinolates and hydrolysis leading to production of mustard oil.

yield an unstable aglycone that rearranges to form a variety of products, including isothiocyanates (mustard oils), thiocyanates, nitriles, cyanoepithioalkanes and oxazolidine-2-thiones. The isothiocyanates are toxic to many insects[6] and also have both bacteriostatic and fungistatic properties.[8]

As might be expected, many insects have adapted to this first line of defense in crucifers and actually make use of glucosinolates and their volatile hydrolysis products to recognize or locate suitable host plants.[9] However, not all crucifers are acceptable to these insects, and secondary defense of the avoided species has been associated with the presence of other classes of compounds.[6] For example, species of *Cheiranthus* and *Erysimum* are known to contain cardenolides, *Iberis* contains cucurbitacins, and both *Lunaria* and *Capsella* produce alkaloids.[10] Most of these plants are avoided by crucifer specialists or they support very poor growth. In a study of pierid herbivory on Moroccan species within the Capparales, Chew[8] found that *Alyssum* spp. and *Matthiola parviflora* were also rejected by larvae. But the compounds responsible for rejection have not been investigated.

Until recently, the role of reputed secondary defense compounds in crucifers has not been clearly demonstrated. Some support for implicating cardenolides and cucurbitacins present in *Erysimum* and *Iberis* was provided by Nielsen[11,12] for crucifer-feeding flea beetles, which avoided host leaf material that was treated with standard compounds. Less convincing evidence was obtained by Usher and Feeny[13] who added known compounds to artificial diet and found that *Pieris rapae* larvae fed and developed almost normally. Such results would suggest the need for bioassays under as natural conditions as possible to isolate active constituents that could explain avoidance or rejection of specific plants.

THE *PIERIS*-CRUCIFER MODEL SYSTEM

The pierid butterflies, in particular *Pieris brassicae* and *P. rapae*, have long been used as model insects to demonstrate chemical associations between insects and their host plants and to examine the sensory basis for behavioral responses to plant constituents. Their popularity has stemmed in part from the early observations of Verschaffelt,[14] who first demonstrated the association between the host range of *P. brassicae* and the presence of glucosinolates. Subsequent studies have further emphasized the link between the presence of glucosinolates and host plant use by a variety of crucifer specialists. The involvement of glucosinolates and mustard oils in attracting the insects or stimulating oviposition and/or feeding by flea beetles, leaf beetles, root flies, aphids, cabbage butterflies and the diamondback moth have now been amply demonstrated (reviewed by Städler[9]). However, despite the close association of pierids with glucosinolate-containing plants, the actual role of specific glucosinolates has not been clear.

The behavioral events leading to oviposition by gravid *Pieris* butterflies have recently been described in some detail.[15] Landing on a potential host is primarily guided by the spectral reflectance of the leaf surface,[16, 17] and this is quickly followed by contact evaluation of the leaf surface by drumming with the foretarsi.[18] Behavioral observations as well as ablation experiments on *P. brassicae* and other butterflies have been used to show that chemosensory information at the leaf surface is perceived by specialized tarsal sensilla.[19, 20] Trichoid tarsal sensilla of *P. brassicae*, *P. rapae* and subspecies of *P. napi* have been shown to respond to glucosinolates and to cardenolides.[21, 22]

Although the involvement of glucosinolates in stimulating oviposition had been previously suggested, early experiments with sinigrin cast some doubt on the role of these compounds.[8, 23] It was not until the stimulants in cabbage were systematically isolated, with behavioral assays to monitor the process, that glucobrassicin was identified as the most effective stimulant in cabbage.[24, 25] Other glucosinolates, glucoiberin and sinigrin were also present in the cabbage extracts, but were much less active. Traynier and Truscott[26] had also found that glucobrassicin was much more active than sinigrin in stimulating oviposition by *P. rapae* on an artificial substrate. Subsequent studies have shown that *P. rapae* and *P. napi oleracea* have different sensitivities to specific glucosinolates, and that definite structure-activity relationships exist.[27, 28] However, the superior activity of glucobrassicin for *P. rapae* raised questions about the stimulation of oviposition on other host plants that lack this compound. Stimulatory fractions were isolated from a variety of crucifers as well as non-crucifer hosts, and in each case, the presence of a prominent glucosinolate could explain the stimulatory activity.[29] However, *P. rapae* showed a distinct preference for aromatic glucosinolates, whereas *P. napi oleracea* preferred aliphatic members of the group.[27]

The recognition of food plants by larvae of *Pieris* species is known to involve gustatory chemoreception of plant constituents, including glucosinolates.[30, 31] Tests of individual compounds on larvae of *P. brassicae* showed that sinigrin was active as a stimulant,[30] but when the activities of different glucosinolates were compared, the methyl and benzyl glucosinolates were the most effective feeding stimulants.[31] However, the selection of test compounds in these studies was based on availability and their effects on other insects. No systematic isolation of feeding stimulants for *Pieris* has yet been reported. The discovery that flavonol glycosides in horseradish foliage stimulate feeding by a specialized flea beetle[32, 33] suggests additional compounds could be involved for other crucifer specialists.

Since the major stimulants responsible for host recognition by ovipositing *P. rapae* adults are now known, it is logical to ask why these butterflies avoid or reject certain crucifers. If a second line of chemical defense is produced by these plants to discourage oviposition, are larvae also affected by the same compounds? We know that butterflies can make oviposition mistakes, laying their eggs on plants which do not support larval growth and development.[34] The presence of oviposition stimulants in unsuitable plants can explain these mistakes, but the chemical basis for lack of larval success is known or suspected in only a few cases.[6] *P. rapae* has been a good model insect for answering some of these questions. Studies of the chemical factors mediating oviposition have been extended to include examination of the chemical signals that influence feeding. In addition, correlations between behavioral responses and electrophysiological responses of the taste receptors of butterflies have been found.[21, 28] Thus, it is now possible to compare plant defenses against adults and larvae of one species and to begin to find a chemical and sensory basis for differences in host ranges of closely related species.

DETERRENTS IN *ERYSIMUM CHEIRANTHOIDES*

The rejection of *Erysimum asperum* by *Pieris* species in a montane area of Colorado was first pointed out by Chew,[35] and this observation prompted speculation about the chemical basis for lack of herbivory on this plant.[36] Subsequent studies have shown that the presence of oviposition deterrents in polar extracts of *Erysimum cheiranthoides* and other plants that were unacceptable to *P. rapae* could explain the observed avoidance behavior of this species.[37] During the bioassay-guided chemical fractionation of *E. cheiranthoides* extracts, it was found that the foliage contains both stimulants and deterrents for oviposition.[23] The deterrent was subsequently isolated from n-butanol extracts, and the activity was associated with a group of three cardenolides,[38] which were identified as erysimoside, erychroside and erycordin.[39] Erysimoside and erychroside were strongly deterrent to *P. rapae*, but erycordin was inactive (Fig. 3). This was the first indication that specific structural requirements were necessary for deterrent activity. Comparison of five related cardenolides then indicated that a strophanthidin nucleus attached to a 2,6-dideoxy sugar was important, and the additional presence of a substituent on the sugar was essential[39] (see Fig. 3).

Rejection of *E. cheiranthoides* by larvae of *P. rapae* was investigated in a similar manner as for adults. Bioassays in which extracts were applied to cabbage leaf discs in both choice and no-choice situations clearly indicated the presence of feeding deterrents.[40] As for the oviposition deterrents, activity was associated with butanol-extractable constituents obtained from aqueous extracts. When separated by HPLC, the fraction containing oviposition deterrents was also deterrent to larval feeding, but the most active material in larval assays was eluted in a later fraction. These results suggested that the defense of *E. cheiranthoides* against *P. rapae* involves two groups of compounds acting on the different developmental stages of the insect.[40] Further work on the chemistry of the feeding deterrents resulted in the identification of four additional cardenolides, three of which were highly active in assays of the pure compounds.[41] These

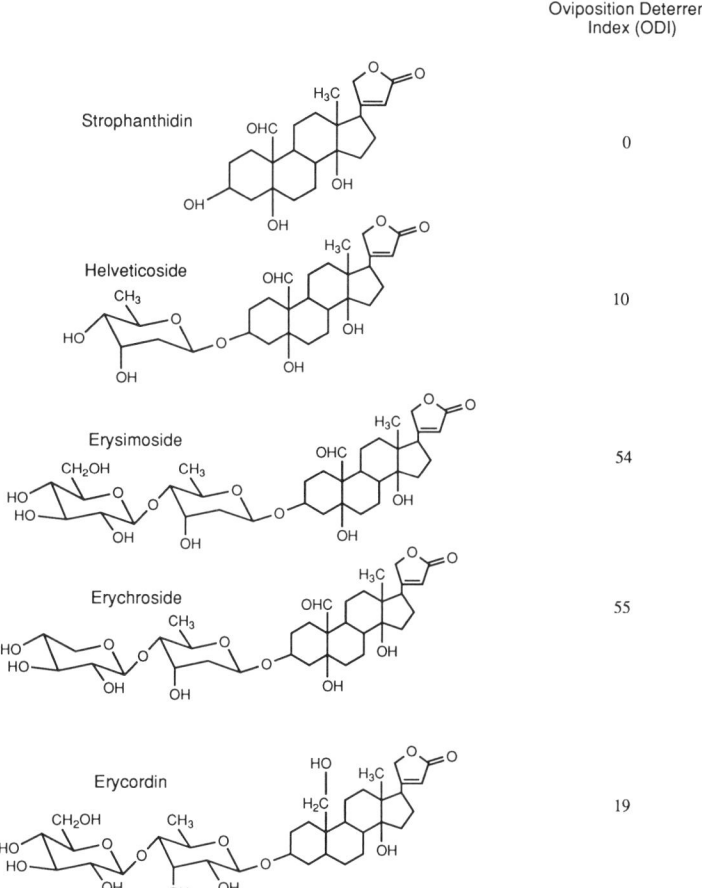

Figure 3. Relative activities of strophanthidin and related cardenolides isolated from *Erysimum cheiranthoides* as oviposition deterrents for *Pieris rapae*.

were all derivatives of digitoxigenin, viz. digitoxigenin 3-O-β-D-glucoside, glucodigigulomethyloside and glucodigifucoside (Fig. 4). An inactive cardenolide was also identified as cheirotoxin.[41] Comparative assays of a range of cardenolides indicated that high activity was associated with glycosides of digitoxigenin, although some of the strophanthidin-based cardenolides were also quite strongly deterrent.

Comparison of results from oviposition and feeding studies suggest that a limited number of specific cardenolides can explain the rejection of *E. cheiranthoides* by adults of *P. rapae*, but that a larger array of cardenolides comes into

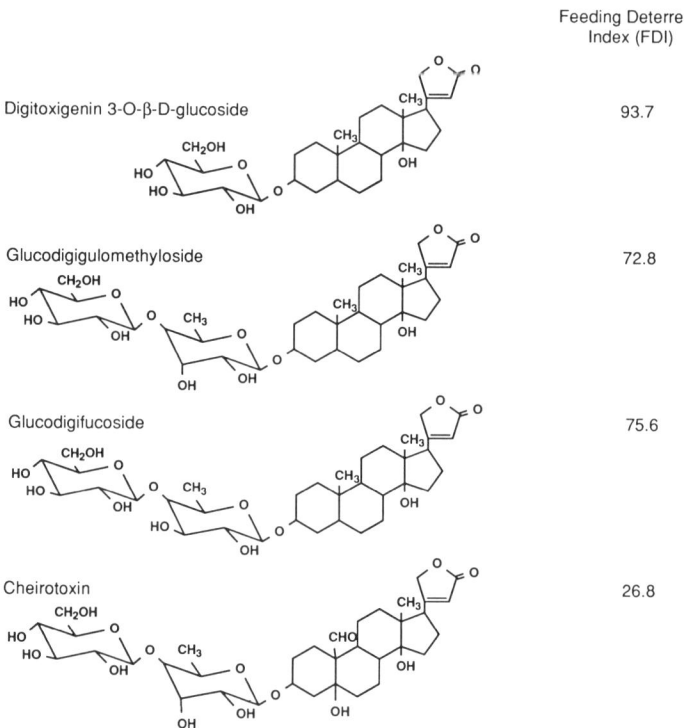

Figure 4. Relative activities of cardenolides isolated from *Erysimum cheiranthoides* as feeding deterrents for larvae of *Pieris rapae*.

play in defending the plant against feeding larvae. Two distinct classes of cardenolides are involved, and precise structural requirements exist for each biological activity.

DETERRENTS IN *IBERIS AMARA*

The existence of secondary defense chemistry in *Iberis amara* has been suggested by the fact that this crucifer is not generally used as a host for *Pieris rapae* (F.S.Chew, personal communication) and is not acceptable to crucifer-feeding flea beetles.[42] Cucurbitacins in the foliage were found to act as feeding deterrents to *Phyllotreta nemorum*,[12] but until recently no explanation for avoidance by *P. rapae* butterflies was available. In a series of experiments with a range of glucosinolate-containing plants, Huang & Renwick[43] showed that

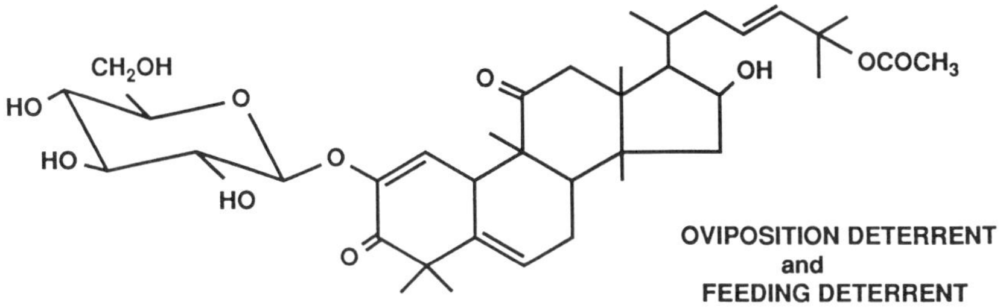

2-O-β-D-Glucosyl cucurbitacin E

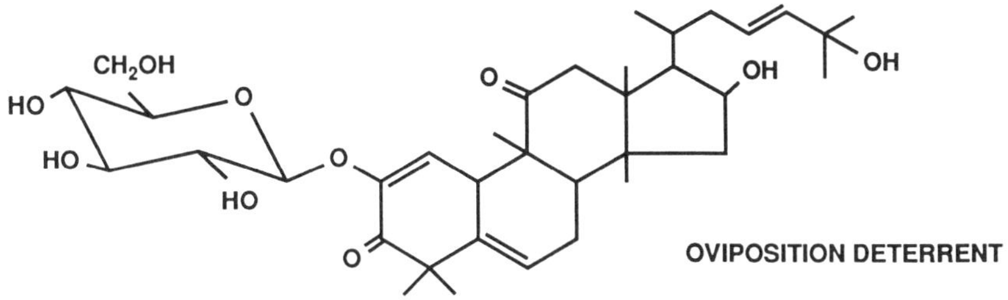

2-O-β-D-Glucosyl cucurbitacin I

Figure 5. Cucurbitacin glycosides identified as oviposition deterrents and a feeding deterrent to *Pieris rapae* from *Iberis amara*.

in a choice between *I. amara* and cabbage plants, *P. rapae* oviposited almost exclusively on cabbage. Extraction of foliage and tests of solvent fractions revealed the presence of deterrents in the butanol fraction obtained by partitioning between water and butanol. Subsequent chromatographic separations led to the isolation of two active compounds that were identified as 2-O-β-D-glucosyl cucurbitacin E and 2-O-β-D-glucosyl cucurbitacin I[44] (Fig. 5). Both glycosides were significantly deterrent when tested at natural concentrations in oviposition assays.[45]

Independent examination of antifeedant activity of *I. amara* extracts against larvae of *P. rapae* led to the isolation of an active fraction that contained the same two cucurbitacin glycosides. However, bioassays of the purified compounds showed that only the 2-O-β-D-glucosyl cucurbitacin E was active as a feeding deterrent.[44]

These results of comparative experiments suggest that *I. amara* is doubly defended against ovipositing adults of *P. rapae*, but that it relies on only one of these compounds for its defense against feeding larvae.

DETERRENTS IN HOST PLANTS OF *PIERIS RAPAE*

Although the presence of deterrents is almost always associated with non-host plants, some deterrent activity can also be found in hosts of *P. rapae*. When the insects display a preference for one plant species over another, this may be explained in part by the relative concentrations of stimulant and deterrent in each of the two plants. Huang and Renwick[43] found that *P. rapae* butterflies preferred cabbage over most other glucosinolate-containing plants for oviposition, and this could be attributed both to the presence of deterrents in the less preferred plants as well as high stimulatory activity in the cabbage. Particularly strong oviposition deterrency was demonstrated by butanol extracts of garden nasturtium,*Tropaeolum majus* (Tropaeolaceae). Although this plant is not a crucifer, it contains high concentrations of a glucosinolate (benzyl glucosinolate) and is a good host for *P. rapae*. Further studies have shown that when *P. rapae* larvae are transferred as second or later instars from cabbage to nasturtium, they refuse to feed on the new host. Bioassays of nasturtium foliage extracts have demonstrated the presence of feeding deterrents to account for this rejection behavior,[46] and the most conspicuous compound in deterrent fractions has been identified as chlorogenic acid (Fig. 6). Although other active constituents have yet to be identified, chlorogenic acid accounts for about half of the deterrent activity.[47] However, when tested in oviposition bioassays, pure chlorogenic acid was inactive (unpublished results), and the identities of the oviposition deterrents remain to be determined.

In the case of nasturtium, therefore, more than one compound is responsible for deterring feeding by *P. rapae*, but the major active constituent has no effect on ovipositing females. We do not yet know whether the oviposition deterrents have any effect on larval feeding or if unidentified feeding deterrents will affect oviposition behavior.

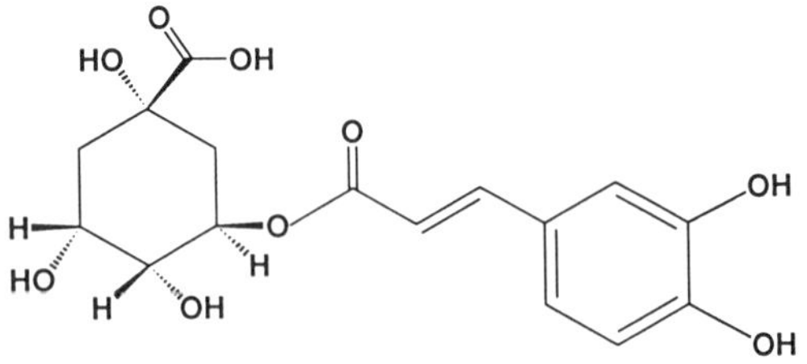

Chlorogenic Acid

(3-Caffeoylquinic acid)

Figure 6. Structure of chlorogenic acid, which was identified as a feeding deterrent to *Pieris rapae* from *Tropaeolum majus*.

A preference for one plant over another does not always indicate the presence or absence of deterrents. *P. rapae* strongly prefers cabbage over *Alyssum saxatile* in choice tests.[43] However, no deterrent effect was obtained with butanol extracts of *A. saxatile*. Furthermore, aqueous extracts of *A. saxatile* were slightly more stimulatory than the corresponding cabbage extracts. So the avoidance of this plant for oviposition is still a mystery, but might involve physical factors or lipophilic deterrents. In contrast, *Lunaria annua* was also avoided in the presence of cabbage, and the butanol extract of this crucifer is deterrent to the butterflies. Similarly, both *Isatis tinctoria* and *Barbarea vulgaris* contain deterrents to *P. rapae* adults,[48,49] and *I. tinctoria* is only slightly less preferred in choice tests with cabbage. However, *B. vulgaris*, which is a common host of *P. rapae*, is just as acceptable as cabbage for oviposition, despite the fact that it contains deterrents. This may be at least partly explained by the strong stimulatory activity of the major glucosinolate, glucobarbarin, present in the foliage of this plant.[49]

Garlic mustard, *Alliaria petiolata* is another crucifer that is now widely used as a host by *P. rapae* (F.S. Chew, pesonal communication). However, in oviposition choice tests, cabbage was strongly preferred. Deterrent assays of the butanol extract from *A. petiolata* revealed only weak activity, and the glucosinolate fraction from *A. petiolata*, which consisted primarily of sinigrin, was less stimulatory than the corresponding cabbage fraction.[50] Thus the preference for cabbage could be explained by the combined effects of weak deterrents in garlic mustard and stronger stimulation by the cabbage glucosinolates.

DIFFERENTIAL RESPONSES OF *PIERIS* SPECIES

Most studies of host selection behavior of *Pieris* butterflies and correlations with plant chemistry have centered on the European species, *P. rapae* and *P. brassicae*.[14, 31, 51] The responses of these two species to different plants, extracts and pure chemicals are remarkably similar, despite their different strategies for egg distribution.[52] Both species are strongly stimulated to oviposit by glucobrassicin in *Brassica oleracea*,[24, 25] and both are deterred by strophanthidin-based cardenolides.[22, 39] Olfactory responses of the two species to a range of chemicals also were almost identical,[53] and electrophysiological studies have shown similarities in sensory responses of the larvae that correlate well with behavioral responses.[54]

Comparisons of *P. rapae*, which was introduced from Europe into North America, with the indigenous *P. napi oleracea* are revealing many differences as well as similarities in behavioral responses to plant chemistry. Although these two species co-occur in some areas of New England, their host ranges and preferences are somewhat different.[55] Favorite hosts of *P. rapae* include cabbage and other varieties of *Brassica oleracea*, whereas these are seldom used by *P.*

napi oleracea in nature. However, the host ranges of the various *P. napi* subspecies in North America include many plants that do not appear to be acceptable to *P. rapae* (F.S. Chew, personal communication).

The choice of host plants by *P. rapae* and *P. napi oleracea* have been studied in some detail to find chemical and/or sensory explanations for differences. Butterflies were given a choice between cabbage and each of nine crucifers, one Capparidaceae and one Tropaeolaceae for oviposition.[43] *P. rapae* preferred cabbage over most of the test plants, but *P. napi oleracea* strongly preferred those plants that were avoided by *P. rapae*. Bioassays of fractions from foliage extracts were used to demonstrate the involvement of stimulants and deterrents in this differential behavior. Subsequent comparisons of responses to other crucifers and their constituents have provided a more complete picture of the chemical effects responsible[50,56] (Table 1). These two *Pieris* species have apparently evolved different sensitivities to the chemical stimuli that mediate oviposition. In general, it appears that *P. rapae* is more strongly stimulated by aromatic glucosinolates, whereas *P. napi oleracea* is more strongly stimulated by aliphatic members of the group, and especially by the glucosinolates that have a sulfinyl or sulfonyl side chain.[51] On the other hand, *P. napi oleracea* is less sensitive to the deterrents that have been identified so far, although bioassays with extracts have indicated that strong deterrents for this species have yet to be identified from other plants. Electrophysiological studies are expected to determine whether differences in sensitivity of these two species to allelochemicals are due to differences at the peripheral receptor level or in the central nervous system.[28]

Recent studies of *Barbarea vulgaris* and *Alliaria petiolata*, both of which were introduced from Europe, have shown that the indigenous *P. napi oleracea* butterflies readily accept the plants for oviposition, but that their offspring do not usually survive.[49,50] Acceptability for oviposition is readily explained by the high concentrations of glucosinolates present in these crucifers, but the poor performance of hatching larvae is not yet clear. Preliminary evidence points to the presence of feeding deterrents in *A. petiolata*, but it remains to be determined whether these or other compounds also have toxic effects.

BALANCING STIMULANTS AND DETERRENTS

We now know that the effectiveness of chemical deterrents in defending a plant against insect attack may be influenced by numerous physiological and environmental factors that mediate both the production of the chemicals and the responses of the insects. The dynamic nature of plant-insect relationships was addressed by Dethier,[57] who suggested the existence of a balance between positive and negative stimuli. He further proposed that the ultimate behavioral response of an insect will depend on internal physiological factors such as motivation and degree of satiety. This idea was used by Miller and Strickler[58] as

Table 1. Sensitivities of *Pieris rapae* and *P. napi oleracea* to stimulants and deterrents affecting oviposition on various plants, related to their preference for these plants

Plant	Allelochemicals		Responses					
	Major Stimulants	Deterrents	*P. rapae*			*P. napi oleracea*		
			Pref.[1]	Stim.[2]	Det.[3]	Pref.	Stim.	Det.
Erysimum cheiranthoides	Glucocheirolin glucoiberin	Cardenolides	−98	+	− − −	+40	+++	
Alyssum saxatile	Glucoalyssin		−64	+++		+70	+++	− −
Iberis amara	Sinigrin glucoiberin	Cucurbitacin glycosides	−88	++	− − −	+42	++	− −
Cleome spinosa	Glucocapparin		−4	++		+43	+++	− −
Lunaria annua	?	?	−87		− −	−73	++	− − −
Brassica juncea	Sinigrin		+22	+++		+47	+++	
Isatis tinctoria	Indole glucosinolates*	?	−20	+++	− −	−56		− −
Tropaeolum majus	Glucotropaeolin*	?	−12	+	− − −	+12		− −
Barbarea vulgaris	Glucobarbarin*	?	+11	++	−	+61	+++	− −
Alliaria petiolata	Sinigrin	?	−42	+	−	−6	+++	− −

Based on data from Refs. 43, 49 and 50.   * = aromatic (other stimulants are aliphatic) glucosinolates

1. Preferences are based on a choice between test plant and cabbage. Positive values indicate a preference for the test plant.

2. Stimulant activity is based on comparison with a standard cabbage extract. (+) = slightly more stimulatory than cabbage extract; (+++) = strongly stimulatory.

3. Deterrent activity is based on response to butanol extracts of test plants applied to cabbage. (−) = slightly deterrent; (− − −) = strongly deterrent.

a basis for their "rolling fulcrum model" to take into account the internal excitatory factors that could tip the balance in favor of rejection or acceptance of a plant. However, until recently, no suitable insect-plant system was available to test the model. Progress in elucidating the chemistry and behavioral nuances of *Pieris*--crucifer relationships has now provided us with sufficient information to examine some crucial components of this model.[59]

The evidence that we have so far suggests that the "decision" of any one insect species whether or not to oviposit is dependent to a large extent on the relative concentrations of stimulants and deterrents perceived at the surface of the foliage. Thus the biosynthesis of the relevant chemical stimuli and movement of the products to the leaf cuticle are of primary importance. Any factor that selectively influences the production of either stimulant or deterrent could directly affect the direction in which the balance is tipped. As mentioned before, production of allelochemicals may be induced by abiotic as well as biotic stress. Concentrations of specific chemicals may also depend on age or stage of development of the plant, and can vary from leaf to leaf. Nutritional regimes are also likely to affect the relative rates of production of nitrogen-rich, sulfur-rich or carbon-rich compounds in a plant.

Induced changes in the chemistry of crucifers have been studied in only a few cases, and all of these have focused on glucosinolates. Koritsas et al.[60] reported the increased accumulation of indole glucosinolates as a result of flea beetle damage or mechanical wounding in oilseed rape. Similar changes were observed in the cotyledons of oilseed rape and mustard after mechanical wounding or feeding by *Phyllotreta cruciferae*,[61] and this effect could be duplicated by application of jasmonic acid or methyl jasmonate.[62] Feeding by larvae of the turnip root fly was found to cause an increase in total glucosinolate content and an increase in the proportion of indole glucosinolates in both oilseed and forage rape.[63] However, in this case, artificial damage caused a reduction in total glucosinolate content of the roots. In general, therefore, when induction of glucosinolates occurs, the production of indoles appears to be strongly favored. At this time, induction of secondary defense compounds in crucifers has not been reported.

Plant nutrition has been shown to affect herbivory in several cases,[64] and nitrogen fertilization resulted in increased levels of oviposition by *Pieris rapae* on cabbage and mustard.[65,66] Such behavioral changes are undoubtedly related to plant chemistry, but the importance of studying the balance between positive and negative cues for the insects must be emphasized. The demonstrated involvement of both stimulants and deterrents in acceptance or rejection of *Erysimum cheiranthoides* by *Pieris* butterflies has provided a convenient model for studying the effects of nutrition on this balance. When seedlings of *E. cheiranthoides* were provided with three levels of nitrogen for a period of 15 days, the ratio of cardenolides to glucosinolates in plants at elevated C/N ratios followed the carbon/nutrient balance hypothesis. But a high nitrogen supply increased

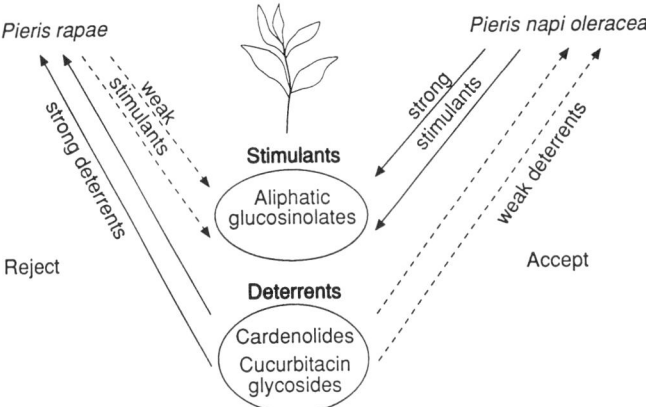

Figure 7. Schematic of responses of *Pieris rapae* and *P. napi oleracea* to stimulants and deterrents that mediate acceptance or rejection of *Erysimum cheiranthoides* and *Iberis amara* for oviposition.

biomass production to the extent that concentrations of these natural products were unchanged or reduced.[67] Concentrations of glucosinolates at the leaf surface were positively related to total tissue levels. Cardenolide levels at the leaf surface, however, did not reflect the total content in the leaves of nitrogen-deficient plants. These plants had the highest concentrations in whole leaves, but the lowest concentrations at the surface. This means that the ratio of glucosinolates to cardenolides at leaf surfaces of nitrogen-deficient plants might favor oviposition by *Pieris* on these plants, but larval feeding would be strongly inhibited by the high cardenolide content that would be encoutered upon penetration of the cuticle.

The sensory system of each insect is obviously a key component contributing to the balance of inputs that mediate host acceptance. Differences in sensitivity of related species to allelochemicals have provided us with the opportunity to closely examine the interaction between the various stimuli to explain different levels of acceptability. Acceptance of *Erysimum cheiranthoides* and *Iberis amara* by *P. napi oleracea* hinges on the insensitivity of this species to the cardenolides and cucurbitacin glycosides that are deterrent to *P. rapae* (Fig. 7). In addition, sensitivity to the aliphatic glucosinolates in these plants is much higher for *P. napi oleracea* than for *P. rapae*. This combination of differential sensory responses results in a dramatic difference in the way that the two butterflies react to these plants.[45,48,59]

In cases where a plant is a marginal host for one or more butterfly species, the balance is likely to be much more delicate. When *P. napi oleracea* and *P. rapae* were offered a choice between cabbage and *Barbarea vulgaris* for oviposition, *P. rapae* showed no preference, but *P. napi oleracea* preferred *B. vulgaris*.

This difference was explained by the fact that *P. napi oleracea* was more strongly stimulated by the glucosinolates, particularly (2R)-glucobarbarin. The response of both species to deterrents in butanol extracts of the plant were quite similar.[56] Similar differences in acceptance of *Alliaria petiolata* by the two *Pieris* species have been attributed to different sensitivities to the stimulants (primarily sinigrin) and deterrents. In contrast to the relative sensitivities to previously isolated deterrents, *P. napi oleracea* appears to be more sensitive than *P. rapae* to the deterrents present in *A. petiolata*.[50]

It seems reasonable to assume that larval feeding will also be dependent on a balance of positive and negative cues. Different sensitivities of larvae of the different *Pieris* species to feeding deterrents are already apparent from their acceptance or rejection of specific plants. However, systematic studies on the sensitivity of larvae to feeding stimulants have yet to be conducted. Glucosinolates appear to be involved, but comparisons of activity are lacking.[30,31] We have already shown that adults and larvae do not always respond to the same deterrents, so it will be interesting to determine the extent to which responses to stimulants either coincide or diverge.

EFFECTS OF EXPERIENCE ON BEHAVIORAL RESPONSES

The dynamic nature of plant defensive chemistry has been pointed out, and we now know that the sensitivity of insects to allelochemicals varies among species, subspecies, populations, and even individuals. However, previous experience of an insect has emerged as an additional factor that can affect the outcome of encounters with potential host plants. We usually think of experience in terms of associative learning, which certainly plays an important role.[68] For example, *Pieris rapae* butterflies have been shown to associate colored paper discs with the presence of oviposition stimulant.[69] However, the involvement of physiological effects has been suggested by diet-related changes in chemoreceptor sensitivities. Schoonhoven et al[70] concluded that chemoreceptor sensitivity of caterpillars may vary with diet, time of day or the state of hunger. Schoonhoven[71] showed that *Manduca sexta* larvae accepted a wider range of potential host plants after feeding on artificial diet, and van Loon[54] found that *Pieris* larvae reared on artificial diet had reduced chemosensory responses to a range of phenolic compounds. Even dietary water has been shown to affect the sensitivity of lepidopterous larvae to feeding deterrents.[72]

Recent experiments with *Pieris rapae* larvae have revealed an extreme case of diet-dependent sensitivity to deterrents in a plant.[46] When larvae were transferred from cabbage to nasturtium (*Troaeolum majus*), they refused to feed, although nasturtium-reared larvae readily accepted cabbage. Feeding deterrents were isolated from nasturtium and the major active constituent was identified as chlorogenic acid. However, larvae reared on wheat germ diet readily accepted

Table 2. Suppression of sensitivity to deterrents by exposure of *Pieris rapae* larvae to various plant compounds or diets

Dietary Exposure	Reduced Sensitivity to
Nasturtium Wheat germ diet Chlorogenic acid 2-O-β-D-glucosyl cucurbitacin E 2-O-β-D-glucosyl cucurbitacin I Strophanthidin Cymarin Erysimoside Digitoxin Digitoxigenin	Nasturtium
Nasturtium	Chlorogenic acid
Wheat germ diet	Chlorogenic acid Cymarin Erysimoside 2-O-β-D-glucosyl cucurbitacin E

Based on data from Ref. 73.

nasturtium and were insensitive to the deterrents under standard bioassay conditions. This phenomenon has been attributed to a type of habituation, or more correctly, "suppression of sensitivity development" as a result of exposure to deterrents in nasturtium or wheat germ diet.[46, 47] Larvae apparently develop sensitivity to deterrents as first instars while they feed on a normal host plant such as cabbage.

Subsequent studies have shown that exposure of larvae to a range of deterrents and related compounds can suppress the development of sensitivity to other deterrents, and this effect has been referred to as "cross habituation".[73] Nasturtium-reared larvae were less sensitive to strophanthidin-based cardenolides and the glucoside of cucurbitacin E, as well as to chlorogenic acid. When neonate larvae were fed on cabbage plants that were treated with various cardenolides or other deterrents, they were much more likely to accept nasturtium (Table 2). Even compounds such as 2-O-β-D-glucosyl cucurbitacin I and strophanthidin, which are not deterrents, caused larvae to accept nasturtium.[73]

Based on the results of van Loon,[54] it appears likely that these changes in behavior can be related to sensitivity of the chemoreceptors. However, considerable variation has been found among *P. rapae* larvae in their response to deterrents as a result of dietary exposure.[73] Electrophysiological experiments will be needed to determine whether variation in sensitivity exists among individuals at the peripheral sensory level. Genetic studies might then be designed to ask whether such differences are heritable.

CONCLUSIONS

As Feeny[6] suggested, many crucifers have developed the ability to combat glucosinolate-adapted insects with other classes of chemicals. Examination of the *Pieris*-crucifer model system has provided us with enough information to reach some preliminary conclusions about the compounds involved, the extent of their activity against different life stages of an insect, and differential effects on related butterfly species. Comparative analyses of *P. rapae* and *P. napi oleracea* adult and larval sensitivity to deterrents and deterrent extracts from a range of plants clearly indicate that we cannot generalize about the activity of any compound against adults and larvae of a species, or against related species (Table 3.).

Although larvae and adults of *P. rapae* are often deterred by the same compounds, some of these are much more effective against one developmental stage. The strophanthidin-based cardenolides, erysimoside and erychroside, are most effective as oviposition deterrents in *Erysimum cheiranthoides*, but the most active feeding deterrents are the digitoxigenin-based cardenolides. One of the oviposition deterrents from *Iberis amara*, 2-O-β-D-glucosyl cucurbitacin I, does not deter feeding, but on the other hand, chlorogenic acid in nasturtium is active only as a feeding deterrent. In general, it seems that *P. napi oleracea* is less responsive to the deterrents that have been identified so far. However, ovipositing females of this butterfly may be more strongly deterred by constituents of *Alliaria petiolata*. The effects of *A. petiolata* and *Barbarea vulgaris* allelochemicals on larvae of *P. napi oleracea* have yet to be determined. Larval mortality is probably due to toxic effects, but the involvement of feeding inhibition cannot be discounted.

Despite the fact that plants can produce deterrent chemicals that discourage oviposition or feeding, many factors obviously have profound effects on the success of this defense strategy. The balance between positive and negative cues perceived by the insect is critical, and this balance may be tipped in either direction by changes in environmental conditions or a variety of other external forces. Any increased production of chemicals could easily favor positive signals rather than negative ones, and the distribution of chemicals within a plant or at leaf surfaces can dramatically affect insect behavior. It is also conceivable that

Table 3. Deterrent effects of identified compounds and plant extracts on adults and larvae of *Pieris rapae* and *P. napi oleracea*

Test compound or extract	P. rapae Adults	P. rapae Larvae	P. napi oleracea Adults	P. napi oleracea Larvae
Erysimoside	D	d	d	–
Erychroside	D	d	d	–
Digitoxigenin 3-O-β-D-glucoside	d	D	–	–
Glucodigigulomethyloside	d	D	–	–
Glucodigifucoside	d	D	–	–
2-O-β-D-glucosyl cucurbitacin E	D	D	–	–
2-O-β-D-glucosyl cucurbitacin I	D	–	–	–
Chlorogenic acid	–	D	–	–
Alliaria petiolata BuOH extract	d	d	D	?
Barbarea vulgaris BuOH extract	d	?	d	?

(D) = strong deterrent (d) = weak deterrent (–) = no effect (?) = not tested

the allocation of a plant's resources to defense may favor production of chemicals that combat only one life stage or only one species of insect.

Although the dynamics of chemical production and distribution in a plant can explain much of the variability that we see in degree of protection from an insect species, the physiological state of the insect also plays a crucial role.[57] The effectiveness of a deterrent is dependent on the actual behavioral response of the insect, and this is a function of sensory perception as well as CNS coding of the information received. We now have evidence to suggest that the sensory system of an insect may be influenced by its diet. Thus the fate of a plant that is apparently well defended against an insect could conceivably depend on the previous dietary experience of that insect.

Plants are generally capable of producing an array of chemical weapons against insects that may come into play at any stage of the invasion process. The example of *Pieris rapae* and *P. napi oleracea* shows that dual defenses against larvae and adults of an insect can occur, but generally one class of compounds is active against both stages of the insect. Some economy in biosynthesis of deterrents is evident, and a single defense against either larvae or adults is quite common. This example may not be typical of insect-plant relationships, but it provides some insight into the muliplicity of plant and insect variables that must be considered.

ACKNOWLEDGMENTS

Thanks are due to John Austin and Nicole Tarnowsky for their invaluable help with graphics and references. This work was supported in part by NSF Grant DEB - 9419797.

REFERENCES

1. PAINTER, R.H. 1951. Insect resistance in crop plants. University Press of Kansas, Lawrence/London. p. 521
2. KOGAN, M. 1977. The role of chemical factors in insect/plant relationships. Proc. XV Int. Congress Entomol.,Washington, D.C. pp. 211–227.
3. RENWICK, J.A.A. 1983. Nonpreference mechanisms: plant characteristics influencing insect behavior. In: Plant Resistance to Insects, (P.A. Hedin, ed.), American Chemical Society. pp. 199–213.
4. BALDWIN, I. 1994. Chemical changes rapidly induced by folivory. In: Insect Plant Interactions, V: (E.A. Bernays, ed.), CRC Press: Boca Raton, Ann Arbor, London, Tokyo. pp. 1–23.
5. DICKE, M., SABELIS, M.W., TAKABAYASHI, J. 1990. Do Plants Cry For Help? Evidence Related to a Tritrophic System of Predatory Mites, Spider Mites and Their Host Plants. In: Symp. Biol. Hung. 39: (A. Szentesi, T. Jermy, eds.), Akademiai Kiado, Budapest, pp. 127–134.
6. FEENY, P. 1977. Defensive Ecology of the Cruciferae. Ann. Missouri Botanic Garden. 64: 221–234.
7. FENWICK, G.R., HEANEY, R.K., MULLIN, W.J. 1983. Glucosinolates and their breakdown products in food and food plants. CRC Critical Reviews in Food Science and Nutrition. 18(2): 123–201.
8. CHEW, F.S. 1988. Searching for defensive chemistry in the Cruciferae, or, do glucosinolates always control interactions of Cruciferae with their potential herbivores and symbionts? No! In: Chemical mediation of coevolution, (K.C. Spencer, ed.), Academic Press, Inc., San Diego, pp. 81–112.
9. STÄDLER, E. 1992. Behavioral responses of insects to plant secondary compounds. In: Herbivores: their interactions with secondary plant metabolites, II Ecological and evolutionary processes: (G.A. Rosenthal , M.R. Berenbaum, eds.), Academic Press, Inc., San Diego: pp. 45–88.
10. HEGNAUER, R. 1964. Chemotaxonomie der Pflanzen. (R. Hegnauer, series ed.) Birkhäuser Verlag, Basel.
11. NIELSEN, J.K. 1978. Host plant discrimination within the Cruciferae: feeding responses of four leaf beetles (Coleoptera: Chrysomelidae) to glucosinolates, cucurbitacins and cardenolides. Entomol. exp. appl. 24: 41–54.
12. NIELSEN, J.K. 1978. Host plant selection of monophagous and oligophagous flea beetles feeding on crucifers. Entomol. exp. appl. 24: 562–569.

13. USHER, B.F., FEENY, P. 1983. Atypical secondary compounds in the family cruciferae: Tests for toxicity to *Pieris rapae*, an adapted crucifer-feeding insect. Entomol. exp. appl. 34: 257–262.
14. VERSCHAFFELT, E. 1911. The cause determining the selection of food in some herbivorous insects. Proc. Acad. Sci. Amsterdam. 13: 536–542.
15. RENWICK, J.A.A., CHEW, F.S. 1994. Oviposition behavior in Lepidoptera. Ann. Rev. Entomol. 39: 377–400.
16. ILSE, D. 1956. Behavior of butterflies before oviposition. J. Bombay Natural History Soc. 53: 486–488.
17. KOLB, G., SCHERER, C. 1982. Experiments on wavelength specific behaviour of *Pieris brassicae* L. during drumming and egg-laying. J. Comp. Physiol. 149: 325–332.
18. RENWICK, J.A.A., RADKE, C.D. 1988. Sensory cues in host selection for oviposition by the cabbage butterfly, *Pieris rapae*. J. Insect Physiol. 34: 251–257.
19. MYERS, J. 1969. Distribution of foodplant chemoreceptors on the female Florida queen butterfly, *Danus gilippus berenice* (Nymphalidae). J. Lepid. Soc. 23: 196–198.
20. MA, W.C., SCHOONHOVEN, L.M. 1973. Tarsal contact chemosensory hairs of the large white butterfly, *Pieris brassicae* and their possible role in oviposition behaviour. Entomol. exp. appl. 16: 343–357.
21. STÄDLER, E., RENWICK, J.A.A., RADKE, C.D., SACHDEV-GUPTA, K. 1995. Tarsal contact chemoreceptor response to glucosinolates and cardenolides mediating oviposition in *Pieris rapae*. Physiol. Entomol. 20: 175–187.
22. ROTHSCHILD, M., ALBORN, H., STENHAGEN, G., SCHOONHOVEN, L.M. 1988. A strophanthidin glycoside in Siberian wallflower: a contact deterrent for the large white butterfly. Phytochemistry. 27: 101–108.
23. RENWICK, J.A.A., RADKE, C.D. 1987. Chemical stimulants and deterrents regulating acceptance or rejection of crucifers by cabbage butterflies. J. Chem. Ecol. 13: 1771–1776.
24. VAN LOON, J.J.A., BLAAKEMEER, A., GRIEPINK, F.C., VAN BEEK, T.A., SCHOONHOVEN, L.M., DE GROOT, A. 1992. Leaf surface compound from *Brassica oleracea* (Cruciferae) induces oviposition by *Pieris brassicae* (Lepidoptera: Pieridae). Chemoecology. 3: 39–44.
25. RENWICK, J.A.A., RADKE, C.D., SACHDEV-GUPTA, K. 1992. Leaf surface chemicals stimulating oviposition by *Pieris rapae* (Lepidoptera: Pieridae) on cabbage. Chemoecology. 3: 33–38.
26. TRAYNIER, R.M.M., TRUSCOTT, R.J.W. 1991. Potent natural egg-laying stimulant for cabbage butterfly *Pieris rapae*. J. Chem. Ecol. 17: 1371–1380.
27. HUANG, X., RENWICK, J A.A. 1994. Relative activities of glucosinolates as oviposition stimulants for *Pieris rapae* and *P. napi oleracea*. J. Chem. Ecol. 20: 1025–1037.
28. DU, Y.J., VAN LOON, J.J.A., RENWICK, J.A.A. 1995. Contact chemoreception of oviposition stimulating glucosinolates and an oviposition deterrent cardenolide in two subspecies of *Pieris napi*. Physiol. Entomol. 20: 164–174.
29. SACHDEV-GUPTA, K., RADKE, C.D., RENWICK, J.A.A. 1992. Chemical recognition of diverse hosts by *Pieris rapae* butterflies. In: Proceedings of the 8th International Symposium on Insect-Plant Relationships, (S.B.J. Menken, J.H Visser, P. Harrewijn, eds.), Kluwer, Dordrecht, pp. 136–138.
30. SCHOONHOVEN, L.M. 1972. Secondary plant substances and insects. Recent Advances in Phytochemistry. 5: 197–224.
31. DAVID, W.A.L., GARDINER, B.O.C. 1966. The effect of sinigrin on the feeding of *Pieris brassicae* L. larvae transferred from various diets. Entomol. exp. appl. 9: 95–98.
32. NIELSEN, J.K., LARSEN, L.M., SORENSEN, H. 1979. Host plant selection of the horseradish flea beetle *Phyllotreta armoraciae* (Coleoptera: Chrysomelidae): Identification of two flavonol glycosides stimulating feeding in combination with glucosinolates. Entomol. exp. appl. 26: 40–48.

33. LARSEN, L.M., NIELSEN, J.K., SORENSEN, H. 1982. Identification of 3-O-(2-O-(β-D-xylopyranosyl)-β-D-galactopyranosyl) flavonoids in horseradish leaves acting as feeding stimulants for a flea beetle. Phytochemistry. 21: 1029–1033.
34. CHEW, F.S. 1977. The effects of introduced mustards (Cruciferae) on some native North American cabbage butterflies (Lepidoptera: Pieridae). Atala. 5: 13–19.
35. CHEW, F.S. 1975. Coevolution of Pierid butterflies and their cruciferous foodplants. Oecologia (Berl.). 20: 117–127.
36. RODMAN, J.E., CHEW, F.S. 1980. Phytochemical correlates of herbivory in a community of native and naturalized Cruciferae. Biochem. Syst. & Ecol. 8: 43–50.
37. RENWICK, J.A.A., RADKE, C.D. 1985. Constituents of host- and non-host plants deterring oviposition by the cabbage butterfly, *Pieris rapae*. Entomol. exp. appl. 39: 21–26.
38. RENWICK, J.A.A., RADKE, C.D., SACHDEV-GUPTA, K. 1989. Chemical constituents of *Erysimum cheiranthoides* deterring oviposition by the cabbage butterfly, *Pieris rapae*. J. Chem. Ecol. 15: 2161–2169.
39. SACHDEV-GUPTA, K., RENWICK, J.A.A., RADKE, C.D. 1990. Isolation and identification of oviposition deterrents to cabbage butterfly, *Pieris rapae*, from *Erysimum cheiranthoides*. J. Chem. Ecol. 16: 1059–1067.
40. DIMOCK, M.B., RENWICK, J.A.A., RADKE, C.D., SACHDEV-GUPTA, K. 1991. Chemical constituents of an unacceptable crucifer, *Erysimum cheiranthoides*, deter feeding by *Pieris rapae*. J. Chem. Ecol. 17: 525–533.
41. SACHDEV-GUPTA, K., RADKE, C.D., RENWICK, J.A.A., DIMOCK, M.B. 1993. Cardenolides from *Erysimum cheiranthoides*: feeding deterrents to *Pieris rapae* larvae. J. Chem. Ecol. 19: 1355–1369.
42. FEENY, P., PAAUWE, K.L., DEMONG, N.J. 1970. Flea beetles and mustard oils: host plant specificity of *Phyllotreta cruciferae* and *P. striolata* adults (Coleoptera: Chrysomelidae). Ann. Entomol. Soc. Amer. 63: 832–841.
43. HUANG, X., RENWICK, J.A.A. 1993. Differential selection of host plants by two *Pieris* species: the role or oviposition stimulants and deterrents. Entomol. exp. appl. 68: 59–69.
44. SACHDEV-GUPTA, K., RADKE, C.D., RENWICK, J.A.A. 1993. Antifeedant activity of cucurbitacins from *Iberis amara* against larvae of *Pieris rapae*. Phytochemistry. 33: 1385–1388.
45. HUANG, X.P., RENWICK, J.A.A., SACHDEV-GUPTA, K. 1993. Oviposition stimulants and deterrents regulating differential acceptance of *Iberis amara* by *Pieris rapae* and *P. napi oleracea*. J. Chem. Ecol. 19: 1645–1663.
46. RENWICK, J.A.A., HUANG, X.P. 1995. Rejection of host plant by larvae of cabbage butterfly: diet dependent sensitivity to an antifeedant. J. Chem. Ecol. 21: 465–475.
47. HUANG, X.P., RENWICK, J.A.A. 1995. Chemical and experiential basis for rejection of *tropaeolum majus* by *Pieris rapae* larvae. J. Chem. Ecol. In press.
48. HUANG, X.P., RENWICK, J.A.A., SACHDEV-GUPTA, K. 1993. A chemical basis for differential acceptance of *Erysimum cheiranthoides* by two *Pieris* species. J. Chem. Ecol. 19: 195–210.
49. HUANG, X., RENWICK, J.A.A., SACHDEV-GUPTA, K. 1994. Oviposition stimulants in *Barbarea vulgaris* for *Pieris rapae* and *P. napi oleracea*: isolation, identification, and differential activity. J. Chem. Ecol. 20: 423–438.
50. HUANG, X., RENWICK, J.A.A., CHEW, F.S. 1995. Oviposition stimulants and deterrents control differential acceptance of *Alliaria petiolata* by *Pieris rapae* and *P. napi oleracea*. Chemoecology 5/6: 79–87.
51. CHEW, F.S., RENWICK, J.A.A. 1995. Chemical ecology of hostplant choice in *Pieris* butterflies. In: Chemical Ecology of Insects 2, (R.T.Cardé, W.J. Bell, eds.), Chapman and Hall, New York, pp. 214–248.
52. CHEW, F.S., ROBBINS, R.K. 1984. Egg Laying in Butterflies. In: The Biology of Butterflies,XIII: (R.I. Vane-Wright , P. Ackery, eds.), Academic Press, London, pp. 65–79.

53. VAN LOON, J.J.A., FRENTZ, W.H., VAN EEUWIJK, F.A. 1992. Electroantennogram responses to plant volatiles in two species of *Pieris* butterlies. Entomol. exp. appl. 62: 253–260.
54. VAN LOON, J.J.A. 1990. Chemoreception of phenolic acids and flavonoids in larvae of two species of *Pieris*. J. Comp. Physiol. A. 166: 889–899.
55. RICHARDS, O.W. 1940. The biology of the small white butterfly (*Pieris rapae*), with special reference to the factors controlling its abundance. J. Animal Ecology. 9: 243–288.
56. HUANG, X.P., RENWICK, J.A.A., SACHDEV-GUPTA, K. 1993. Oviposition stimulants in *Barbarea vulgaris* for *Pieris rapae* and *P. napi oleracea*: isolation, identification and differential sensitivity. J. Chem. Ecol. 20: 423–438.
57. DETHIER, V.G. 1982. Mechanisms of host-plant recognition. Entomol. exp. appl. 31: 49–56.
58. MILLER, J.R., STRICKLER, K.L. 1984. Finding and accepting host plants. In: Chemical Ecology of Insects, (W. Bell, R. Cardé, eds.), Chapman and Hall, London, pp. 127–155.
59. RENWICK, J.A.A., HUANG, X. 1994. Interacting chemical stimuli mediating oviposition by Lepidoptera. In: Functional Dynamics of Phytophagous Insects, (T.N. Ananthakrishnan, ed.), Oxford & IBH Publishing Co., New Delhi, pp. 79–94.
60. KORITSAS, V.M., LEWIS, J.A., FENWICK, G.R. 1989. Accumulation of indole glucosinolates in *Psylliodes chrysocephala* L. -infested, or -damaged tissues of oilseed rape (*Brassica napus* L.). Experientia. 45: 493–495.
61. BODNARYK, R.P. 1992. Effects of wounding on glucosinolates in the cotyledons of oilseed rape and mustard. Phytochemistry. 31: 2671–2677.
62. BODNARYK, R.P. 1994. Potent effect of jasmonates on indole glucosinolates in oilseed rape and mustard. Phytochemistry. 35: 301–305.
63. GRIFFITHS, D.W., BIRCH, A.N.E., MACFARLANE-SMITH, W.H. 1994. Induced changes in the indole glucosinolate content of oilseed and forage rape (*Brassica napus*) plants in response to either turnip root fly (*Delia floralis*) larval feeding or artificial root damage. J. Sci. Food & Agric. 65: 171–178.
64. WARING, G.L., COBB, N.S. 1992. The impact of plant stress on herbivore population dynamics. In: Insect-plant interactions, Vol. IV, (E.A. Bernays, ed.), CRC Press, Boca Raton, pp. 167–226.
65. WOLFSON, J.L. 1980. Oviposition response of *Pieris rapae* to environmentally induced variation in *Brassica nigra*. Entomol. exp. appl. 27: 223–232.
66. MYERS, J.H. 1985. Effects of physiological condition of the host plant on the ovipositional choice of the cabbage white butterfly, *Pieris rapae*. J. Anim. Ecol. 54: 193–204.
67. HUGENTOBLER, U., RENWICK, J.A.A. 1995. Effects of plant nutrition on the balance of insect relevant cardenolides and glucosinolates in *Erysimum cheiranthoides*. Oecologia. 102: 95–105.
68. SZENTESI, A., JERMY, T. 1989. The role of experience in host plant choice by phytophagous insects. In: Insect-Plant Interactions, II: (E.A. Bernays, ed.), CRC Press, Boca Raton, pp. 39–74.
69. TRAYNIER, R.M.M. 1986. Visual learning in assays of sinigrin solution as an oviposition releaser for the cabbage butterfly, *Pieris rapae*. Entomol. exp. appl. 40: 25–33.
70. SCHOONHOVEN, L.M., BLANEY, W.M., SIMMONDS, M.S.J. 1987. Inconsistancies of chemoreceptor sensitivites. In: Insects - Plants, (V. Labeyrie, G. Fabres, D. Lachaise, eds.), Dr. W. Junk, Dordrecht, Netherlands, pp. 141–145.
71. SCHOONHOVEN, L.M. 1969. Sensitivity changes in some insect chemoreceptors and their effect on food selection behaviour. Koninkl. Nederl. Akademie van Wetenschappen. 72: 491–98.
72. GLENDINNING, J., SLANSKY JR., F. 1994. Interactions of allelochemicals with dietary constituents: effects on deterrency. Physiol. Entomol. 19: 173–186.
73. HUANG, X.P., RENWICK, J.A.A. 1995. Cross habituation to feeding deterrents and acceptance of a marginal host plant by *Pieris rapae* larvae. Entomol. exp. appl. 76: 295–302.

Chapter Four

DEFENSIVE CHEMICALS IN GRASS-FUNGAL ENDOPHYTE ASSOCIATIONS

Malcolm R. Siegel[1] and Lowell P. Bush[2]

[1] Department of Plant Pathology
[2] Department of Agronomy
University of Kentucky
Lexington, Kentucky 40546

Introduction	82
Symbiosis and Defensive Chemicals	83
Biology	83
Symbiosis, Mutualism, Multitrophic Interactions and Defensive Chemicals	85
Chemistry of Toxic Metabolites	87
Pyrrolizidine Alkaloids	87
Ergot Alkaloids	91
Pyrrolopyrazine Alkaloid	92
Indole Diterpene Alkaloids	93
Location of Alkaloids within the Host	95
Biological Activity and Mode of Action	97
Biological Activity—Vertebrates	97
Biological Activity—Invertebrates	100
Biological Activity—Plants and Fungi	102
Mode of Action	102
Plant Environment and Host/Fungal Genome Interaction	104
Plant Environment	104
Host/Fungal Genome Interaction	105
Manipulations of Symbiota and Defensive Mutualism	106
Criteria for Manipulation	106
Use of Naturally Occurring Endophytes	106
Lolitrem Synthesis in the Perennial Ryegrass Symbiota	107
Ergovaline Synthesis in Tall Fescue Symbiota	107

Use of Endophytes Modified by Molecular Genetics 108
Conclusions 108

INTRODUCTION

The ability of organisms to form long term intimate and diverse relationships with each other (symbiosis) is now recognized as a common ecological phenomenon. Symbiosis, as a general term, does not imply detriment or benefit, but rather that the outcome (net effect) of species interaction exists within a symbiotic continuum or "species interaction grid" that includes agonism (predation and disease) and mutualism (benefits for both species).[1,2] The continuum also includes pleotropic symbiosis, where net effects of species interaction vary spatially or temporally in relative agonism or mutualism.[3] Grasses systematically infected with specific clavicipitaceous fungi are examples of species interactions that span the symbiotic continuum, profoundly affecting the ecological fitness of the hosts.[4-8]

The mycosymbionts are members of the tribe Balansieae (balansoid fungi), family Clavicipitaceae (Ascomycotina) that form biotrophic, perennial, endophytic (intercellular) or epiphytic associations with many invasive and agronomically important C3 and C4 grasses.[9-11] Of special interest, is the mutualistic symbiosis of endophytic fungi of the genus *Epichloë* and C3 cool season grass species (Poaceae), especially those used in forage, turf, and soil conservation.[12-14] Infected grasses are often asymptomatic and rarely show the external signs of the fungus. They may have enhanced ecological fitness.[6] Enhanced host fitness factors include increased plant growth[15-17] and fecundity,[18] which affects competitiveness,[16,19] and tolerance to biotic factors, primarily insect[20-22] and mammalian herbivores,[23-26] and, to lesser extents, nematodes,[27,28] fungi[29,30] and plants.[31] Tolerance to abiotic, drought stress may also be increased.[32] In many cases, tolerance to biotic stresses have been correlated with specific natural products produced by the grass/endophyte associations that act in a defensive mutualism.[33] While the defensive chemicals responsible for tolerance to grazing animals enhance plant ecological fitness, they are a detriment to livestock production and lead to significant economic losses.[34]

Classes of toxins produced by symbiota (grass/endophyte associations) affecting herbivory are well-defined and include the pyrrolizidine (lolines), ergot (clavines, lysergic acids and ergopeptide, and peptine derivatives), pyrrolopyrazine (peramine) and indole diterpene (lolitrems) alkaloids.[21,35] The classes of alkaloids, except the lolines, can be considered mycotoxins because they have been produced in culture by the fungi.[35-37] Host-endophyte compatibility and specific alkaloid production are influenced by the environment[15,38,39] and/or host and fungal genomes.[38,40-42] For example, symbiota produce different amounts and patterns of alkaloids with different herbivore responses. The production of toxins can be useful in improvement of grasses when used in forage, turf, and soil conservation.[43–45] This is illustrated by the *Festuca arundinacea* (tall fescue)/*Acremonium coenophialum* symbiotium which has a greater persistence and phytomass than non-infected grass.[16,19] The symbiotium produces ergot alkaloids[35] that are responsible for the debilitating symptoms suffered by animals grazing infected grass.[46,47] One useful manipulation of this symbiotum would be the reduction or elimination of the ergot alkaloids without significant loss of fitness components, such as seed dissemination of the endophyte, host tolerance to invertebrate pests, and drought tolerance.

The endophytic fungi infecting grasses will be discussed in this paper and the topics include: the biology and ecology of the symbioses and associated multitrophic interactions; chemistry, diversity of the defensive chemicals, their spectrum of biological activity and modes of action; the effects of environment and host and fungal genomes on toxin production; and how manipulations of symbiota affect levels of toxins and, hence, spectra of biological activity.

SYMBIOSIS AND DEFENSIVE CHEMICALS

Biology

Claviceps species, having a long history of causing ergot poisoning in mammals, are the best known members of the Clavicipitaceae.[48] They are parasites of grass localized in ovaries and having a short period between infection and disease expression. A sclerotium stage (containing the ergot alkaloids) replaces the ovary and separates from the host grass. The Balansieae fungi are a separate group within the Clavicipitaceae family and are characterized by their location (vegetative and systemic parasites), habit (perennial and endophytic or epiphytic), expression (symptomatic and asymptomatic) and host dependency (biotrophic).[4] Although the Balansieae are biotrophic (feed on living tissue) they are readily cultured[49] allowing for many of the studies that have clarified their biology, ecology, phylogeny, and diversity of toxic alkaloids.

The biology of the Balansieae fungi has been discussed in detail.[4,6,11,50] Of importance are the similar yet distinctive characteristics of the genus *Epichloë* to the other genera within the tribe (*Atkinsonella*, *Balansia*, *Balansiopsis*, *Myriogenospora*, see White[11] for alternative classification of species). Similar characteristics include the systemic perennial mode of colonization (endophytic) and location of the ectophytic fruiting stage on inflorescences nodes or leaf sheaths of culms, causing sterilization of the affected florets. The primary dissimilar characteristics are the formation of *Acremonium* microconidia,[11] identity of alkaloids they produce,[35,51] and the existence in some species of *Epichloë* of both symptomatic and/or asymptomatic life cycles.[52,53]

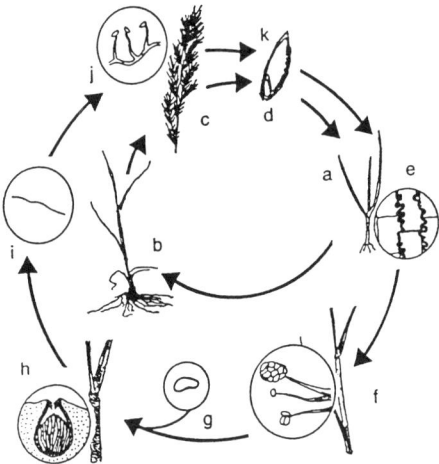

Figure 1. Life cycle of *Epichloë festucae* in *Festuca rubra*. Fungal structures are shown in circles. In the asexual cycle, highly convoluted hyphae grow intercellularly in leaf sheaths (a), floral meristems (b), and in the ovules of the florets (c) such that the fungus is transmitted in the next generation of seed (d). In the sexual cycle, the fungus also grows intercellularly in vegetative portions of the plant (meristems, crowns and leaf sheaths and blades) without causing symptoms (e), but then emerges from the leaf sheath surrounding the immature host inflorescence, as plectenchymatous stromata which produces spermatia, and arrests inflorescence maturation—choke disease (f). Fertilization occurs, by heterothallic outcrossing, and transfer of spermatia of opposite mating type (g) and is mediated by symbiotic flies (*Phorbia* spp.).[55] If the parents are conspecific (same mating population), perithecia asci develop (h) and filamentous ascospores are ejected (i). Germinating ascospores initiate cycles of asexual sporulation (conidiation) and are postulated to cause infection of host florets (j) and ultimately of seed (k). Endophytes of tall fescue and meadow fescue are only known to undergo the asexual cycle. Figure adapted from Tsai et al.[56]

The dual endophytic-ectophytic life cycle of an *Epichloë* sp. is illustrated in Fig. 1. Some *Epichloë* spp. produce stroma on every flowering panicle (choke disease) and exhibit only horizontal transmission. Other species are pleiotropic[53,54] exhibiting both life cycles. Expression of pleiotropic symbiosis is dependent on intrinsic (symbiont genomes) and/or extrinsic (environmental) factors. Finally, for certain endophytes there is no apparent ectophytic and spore producing stage. They are clonally seed disseminated entities exhibiting only vertical transmission. The various *Epichloë* species and important *Acremonium* anamorphs that have no apparent sexual state are listed in Table 1.

Endophytes may be detected qualitatively by histochemical[49] or serological[62] techniques in vegetative plant tissue and seeds. Morphology and mating population studies,[42,53,57,58] isozyme[63-65] and molecular analyses[56,66] are the basis for determining species and phylogenetic relationships of fungi. The asexual anamorphs are classified as *Acremonium* spp. section *Albo lanosa* in the Deuteromycetes.[59] This class is an artificial, utilitarian assemblage of asexual fungi and vegetative states of sexual fungi. While some controversy exists about the anamorph classification, they are clearly related to each other and to the genus. Endophytes, infecting tall fescue (*A. coenophialum*) and perennial ryegrass (*A. lolii*), have received the greatest attention because of the agronomic importance of the grasses, diversity of alkaloids identified and the array of host benefits characterized.

Table 1. Fungal species of the genus *Epichloë*

Fungal Species	Host Species	Symptoms[a]	Ref
Sexual			
Epichloë typhina	*Dactylis glomerata*	S	57
	Lolium perenne	S	
	Anthoxanthum odoratum	S	
E. festucae	*Festuca longifolia*	NS	53
	F. rubra rubra	PS	
	F. rubra commutata	PS, NS	
	F. valesiaca	NS	
E. baconii	*Agrostis capillaris*	S	57
	Ag. stolonifera	S	
	Ag. tenuis	S	
E. clarkii	*Holcus lanatus*	S	57
E. amerillans	*Ag. hiemalis*	S	58
	F. obtusata	S	
Epichloë sp.	*Elymus canadensis*	PS	53
Asexual[b]			
Acremonium coenophialum	*F. arundinacea*	NS	59
A. lolii	*L. perenne*	NS	60
A. uncinatum	*F. pratensis*	NS	61

[a] Expression of fungus on source host(s): S, completely stroma forming; PS, partially stroma forming (pleotropic); NS, nonstroma forming.
[b] Other *Acremonium* spp. have been described by white.[54]

Symbiosis, Mutualism, Multitrophic Interactions and Defensive Chemicals

The degree of reproductive expression of the endophyte (stroma) on flower panicles can be used as one indicator of the types of symbiosis and the effect on ecological fitness of the plant and mycosymbiont. However, valuation of the symbiosis by measuring detriments (costs) and benefits may not be easily accomplished. *Epichloë* spp. that completely choke the flowering panicles could be considered pathogenic because the loss of sexual reproduction affects the genetic diversity of the grass. Such infected grasses however may still have tolerances to biotic and abiotic stresses and, because of the loss of sexuality, can have a greater vegetative growth potential. These fitness factors may compensate, in community dynamics, for the disease and allow populations to survive for long periods within their ecological niches.[9] Conversely, endophytes that remain asymptomatic, asexual, seed-borne, disseminated entities are considered mutualistic.

Muller's ratchet hypothesis implies that in clonal organisms the accumulation of marginally deleterious mutations would, in time, reduce fitness.[67] However, because many endophytes are interspecific hybrids with *Epichloë* spp., the deleterious effects of mutations, impacting fitness may be overcome by periodic influx of entire genomes from sexual relatives.[56,66] Although hybridization events may be rare, they apparently provide such a selective advantage since hybrid endophytes are common. They are certainly a means for diversity which impinges on the many fitness components of symbiota, including alkaloid profiles. Pleiotropic symbiosis would appear to represent the optimum condition for genetic diversity as both symbionts reproduce sexually. However, fitness components should not be viewed individually with respect to the evaluation of costs and benefits associated with specific type of symbiosis. Identifying enhancement components, including plant growth factors, tolerances to biotic and abiotic stress, the effects of host and fungal genetic variation and environmental conditions are required to evaluate the overall spectrum of symbiotic interactions. Such evaluations indicate that certain pleiotropic and vegetative endophytes are ecologically highly beneficial mutualists, to such an extent that long term survival only occurs for infected grasses.

While the biochemical and physiological basis for most multitrophic interactions have not been identified, those associated with the toxic alkaloids and herbivory are generally better understood and offer a unique view of biosynthetic pathways and factors that control their expression. The spectrum and redundancy of alkaloids produced by symbiota are independent of the species of *Epichloë* and *Acremonium* (Table 2). Three alkaloid classes have been observed in three symbiota and many symbiota contain combinations of two classes of alkaloids, some only one class and a few none. Of the alkaloids evaluated in Table 2, ergot alkaloids, primarily ergovaline, were found in the greatest number of symbiota (55%), followed by peramine (52%), N-acetyl-, N-formyl-loline (32%) and the lolitrems (9%). Additionally, grass species can be hosts for more than one mycosymbiont (Table 1, 2), and that symbiota, naturally infected or derived from artificially introduced isolates, can produce a wide variety of alkaloid profiles. These data indicate that host and fungal genomes have a strong influence on alkaloid synthesis, and that manipulations of the genomes are means of changing alkaloid profiles. Additionally, variation in alkaloid profiles of symbiota, as well as persistence and compatibility of the symbionts, may be correlated with the phylogenetic relationships as indicated by cluster analysis of allozyme profiles[64] and molecular analysis of DNA β-tubulin and rRNA sequences.[56,66]

CHEMISTRY OF TOXIC METABOLITES

Since the discovery of fungal endophytes in tall fescue and perennial ryegrass and their significance in the observed toxicosis, reviews of toxins in the symbiota have focused on the lolines, ergots, peramine, lolitrems, and other miscellaneous metabolites with pharmacological activity. Greater detail for specific toxin groups or affected animals has been presented in earlier reviews.[21,24,25,35–37,46,68,70–73] Chemistry and isolation methodologies are presented here to provide a background for the biological activity of the unsaturated amino pyrrolizidine alkaloids, the ergot alkaloids, peramine and the lolitrems in symbiota.

Pyrrolizidine Alkaloids

Pyrrolizidine alkaloids of grass/fungal associations are commonly referred to as lolines. They are 1-amino derivatives of a saturated pyrrolizidine base with an oxygen bridge between carbons 2 and 7 (Fig. 2). Loline and several derivatives exist in symbiota, but N-formyl- and N-acetyl- loline exist in greatest concentrations. Lolines accumulate in symbiota to greater concentrations than any of the other alkaloids and have been measured in excess of 10,000 µg/g. Loline was first isolated from seed of *L. cuneatum* and confirmed by Yates and

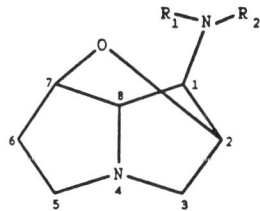

	R_1	R_2
Loline	H	CH_3
N-formylloline	CHO	CH_3
N-acetylloline	CH_3CO	CH_3
N-methylloline	CH_3	CH_3
Norloline	H	H
N-acetylnorloline	CH_3CO	H
N-formylnorloline	CHO	H

Figure 2. Saturated amino pyrrolizidine (loline) alkaloids.

Table 2. Toxic alkaloids produced by *Acremonium* or *Epichloë*-infected grasses

Grass	Endophyte	Alkaloid Class[a]				Ref.
		Pyr.	Er.	PyrPy	Idt	
Agrostis hiemalis	*Epichloë amarillans*	-	+	+	-	51
Bromus anomalus	*Acremonium starri*	-	+	+	-	51
Dactylis glomerata	*E. typhina*	-	-	-	-	[b]
Elymus canadensis	*Epichloë sp.*	-	-	+	-	51
Festuca arundinaceae	*A. coenophialum*	+	+	+	-	51,63,69
	A. coenophialum	+	+	+	-	63
	Acremonium sp. (FaTG-2)[c]	-	+	-	-	63
	Acremonium sp. (FaTG-2)	-	+	+	-	63
	Acremonium sp. (FaTG-2)	-	+	+	+	63
	Acremonium sp. (FaTG-3)	+	-	+	-	63
F. arizonica	*A. huerfanum*	-	-	-	-	51
F. gigantea	*Acremonium sp.*	+	+	-	-	51
F. glauca	*E. festucae*	-	+	+	-	51
F. longifolia	*E. festucae*	-	+	+	+	51
	E. festucae (AI[d]/*F. rubra commutata*)	-	+	-	-	51
F. obtusa	*A. starrii*	-	-	-	-	51
F. paradoxa	*Acremonium sp.*	-	-	+	-	51
F. pratensis	*A. uncinatum*	+	-	+	-	63
F. rubra commutata	*E. festucae*	-	+	+	-	51
	E. festucae	-	+	-	-	51
F. rubra litoralis	*E. festucae* (AI/*F. rubra commutata*)	-	+	-	-	51
F. rubra rubra	*E. festucae*	-	+	-	-	51
F. versuta	*Acremonium sp*	-	-	-	-	51
Lolium multiflorum	*Acremonium sp.*	+	-	ND	ND[e]	69

DEFENSIVE CHEMICALS IN GRASS-FUNGAL ENDOPHYTE ASSOCIATIONS

Grass	Endophyte	Pyr	Er	PyrPy	Idt	Ref
L. persicum	Acremonium sp.	+	-	ND	ND	69
L. perenne	A. lolii	-	+	+	+	51,63
	A. lolii	-	-	+	+	63
	A. lolii	-	+	-	-	63
	A. lolii	-	+	+	-	63
	Acremonium sp. (LpTG-2)[c]	-	+	+	-	63
	E. typhina	-	-	+	-	51
	A. coenophialum (AI/F. arundinaceae)	+	+	+	-	51
	E. festucae (AI/F. longifolia)	-	+	+	-	51
L. rigidum	Acremonium sp.	+	-	ND	ND	69
L. temulentum	Acremonium sp.	+	-	ND	ND	69
Hordeum bogdanii	Acremonium sp.	-/Tr[f]	-/Tr	ND	ND	69
H. brevisubulatum	Acremonium sp.	-	-	ND	ND	69
Poa alsodes	Acremonium sp.	Tr	+	ND	ND	69
	Acremonium sp.	+	+	ND	ND	69
	Acremonium sp.	-	+	ND	ND	69
P. ampla	Epichloë sp.	-	-	+	-	51
P. autumnalis	Acremonium sp.	+	-	+	-	51
Sitanion longifolium	Acremonium sp.	-	+	+	-	51
Stipa robusta	Acremonium sp.	+	+	ND	ND	69

[a] Pyr, pyrrolizidines, (lolines); Er, ergots (ergovaline); PyrPy, pyrrolopyrazine (peramine); Idt, indolediterpenoids (lolitrem B).
[b] Bush & Siegel, unpublished data.
[c] FaTG, F. arundinacea endophyte taxonomic grouping; LpTG, Lolium perenne endophyte taxonomic grouping.
[d] AI, artificially introduced isolate from a different grass species.
[e] ND, not determined.
[f] -/Tr, none to a trace; multiple symbiota analyzed.

Tookey[74] from tall fescue in 1965. Robins et al.[75] reported N-acetylloline and N-formylloline in tall fescue infected with *Balansia sp.* endophyte, and it is this report that has stimulated much of the fungal endophyte/grass toxin research and renewed interest in the biology of the symbiota.

Loline and derivatives found in symbiota have been synthesized.[76,77] Synthesis of loline is difficult and has not resulted in production of large amounts for biological studies.[76] Petroski et al.[77] isolated loline from plant tissue and prepared the derivatives by manipulation of the substitutions on the amino-N. N-formyl- and N-acetyl- loline are easily synthesized from loline and amounts for small bioassay studies have been obtained. We found that a crude alkaloid preparation of 60–80% N-acetyl- and N-formyl- loline can be obtained from a dried endophyte-infected grass powder sequentially extracted with hexane and

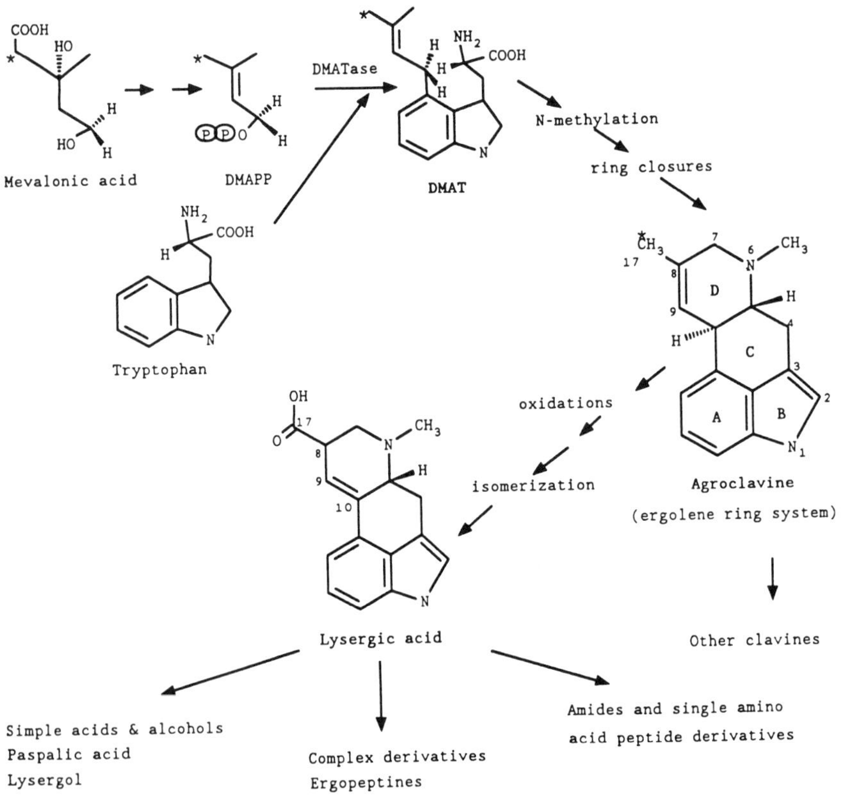

Figure 3. Biosynthesis of ergot alkaloids. DMAPP = dimethylallyl pyrophosphate, DMAT = 4-(γ,γ-dimethylallyl)tryptophan, DMATase = dimethylallyl diphosphate:L-tryptophan dimethylallyltransferase, * represents carbon 2 from mevalonic acid into the ergolene ring.

50% ethanol in water for 24 h each, followed by strong cation exchange chromatography and liquid/liquid extraction. This alkaloidal preparation may be treated with HCL to yield loline hydrochloride from which loline derivatives may be made.[36,77]

Ergot Alkaloids

This broad category of alkaloids in symbiota includes clavines with the ergoline ring system and lysergic acid and derivatives (Fig. 3). Lysergic acid derivatives includes simple acids and alcohols, such as paspalic acid and lysergol; simple amides such as lysergic acid amide; single amino acid peptide derivatives (ergopeptides) plus complex 'cyclol' (tricyclic peptide) derivatives from three amino acids to form ergopeptine alkaloids (Fig. 4). Initially the ergopeptine alkaloids (ergovaline and ergosine) received the most attention. Ergovaline is the predominant member of this group in most symbiota (Table 2). More recent reports have focused on the simple lysergic acid derivatives, especially in *Achnatherum inebrians* and *Stipa robusta*.[78,79] The ergopeptine alkaloids are divided into three groups depending upon the amino acid connecting the ergolene ring and the cyclol. Alanine, α-aminobutyric acid and valine function as the connecting unit in ergotamine, ergoxine and ergotoxine groups, respectively. Ergopeptine alkaloids are unstable and are sensitive to air, light,

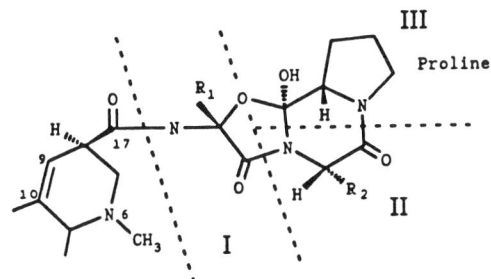

Alkaloid	Position I [R_1]	Position II [R_2]
Ergotamine[1]	Alanine [-CH_3]	Phenylalanine [-$CH_2C_6H_6$]
Ergovaline[1]	Alanine [-CH_3]	Valine [-$CH(CH_3)_2$]
Ergosine[1]	Alanine [-CH_3]	Leucine [-$CH_2CH(CH_3)_2$]
Ergonine[2]	Amino Butyric Acid	Valine [-$CH(CH_3)_2$]
Ergocrystine[3]	Valine [-$CH(CH_3)_2$]	Phenylalanine [-$CH_2C_6H_6$]
Ergocornine[3]	Valine [-$CH(CH_3)_2$]	Valine [-$CH(CH_3)_2$]

[1] Ergotamine group.
[2] Ergoxine group.
[3] Ergotoxine group.

Figure 4. Ergopeptine alkaloids and the amino acids of the tricyclic peptide.

acids and bases. Ergopeptines will undergo isomerization, hydrolysis and additions to form compounds that are difficult to resolve and that may be pharmacologically inactive.

Separation of ergot alkaloids has been by many different forms of chromatography and detection has been by colorimetric analysis, fluorescence detection, tandem mass spectrometry, immunoassay, and high performance capillary zone electrophoresis.[72,80,81] Routine separation and quantification is by HPLC with fluorescence detection,[82,83] but for detection of the presence of ergot alkaloids in symbiota the ELISA technique is the most preferred method.[80,84] Ergovaline is the predominant alkaloid in many symbiota, especially those with *L. perenne* and *F. arundinacea* as hosts. Therefore, many of the techniques have been designed to separate the epimeres, ergovaline and ergovalinine from other extracted materials or to detect these epimeres in the presence of other extracted materials. Shelby and Kelley[80] developed a competitive inhibition enzyme-linked immunosorbent assay (CI-ELISA) with a monoclonal antibody agonist ergonovine. The assay has different sensitivities to a broad array of the ergot alkaloids.

Because of the known biological activity of compounds in the ergopeptine biosynthetic pathway, a brief discussion of their biosynthesis is warranted. These alkaloids are indole derivatives of tryptophan and mevalonic acid (Fig. 4) and biosynthesis in the endophytes is expected to be similar to that determined in *Claviceps* species.[85] Prenylation of tryptophan, formation of the ergoline ring system and formation of the tricyclic peptide are important biosynthetic steps for ergopeptine formation. 4-(γ,γ-Dimethylallyl)tryptophan (DMAT) is the first intermediate in the pathway. Dimethylallyl diphosphate:L-tryptophan dimethylallyltransferase catalyzes the formation of DMAT. This enzyme has been purified and characterized.[85,86] This first committed step in the biosynthesis is a logical and appropriate target for gene manipulation to eliminate the entire class of ergot alkaloids.

Much effort has gone into finding naturally occurring symbiota that lack ergovaline or contain very low levels of the ergopeptine alkaloids. However, if symbiota contain low levels of ergopeptine alkaloids because of slow or inhibited assemblage of the cyclol tripeptide, significant amounts of clavines, lysergic acid or simple derivative of lysergic acid could accumulate. Accumulation of lysergic acid amide and/or the simple ergopeptide alkaloid, ergonovine, has been reported in some symbiota.[78,79] These symbiota produce toxins to some economically important herbivores and have a negative impact on agriculture.

Pyrrolopyrazine Alkaloid

Peramine is the only known pyrrolopyrazine or guanidinium alkaloid reported in symbiota (Fig. 5). It is found in a large number of grass/endophyte associations.[51] Peramine consists of the lipophilic pyrrolopyrazine ring and a hydrophilic guanidinyl side chain and is less lipophilic than the other alkaloids

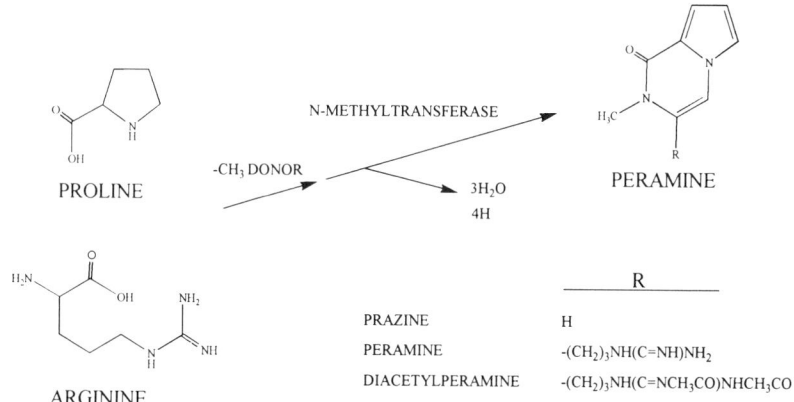

Figure 5. A proposed peramine biosynthesis.

found in symbiota. Routine peramine extraction is best done with 80% aqueous ethanol. Detection is by reversed- phase TLC[87] or HPLC.[37]

Biosynthesis of peramine probably occurs via cyclization and condensation reactions of the amino acids proline and arginine.[88] It has been hypothesized also that the N-methylation is catalyzed by a N-methyltransferase with S-adenosylmethionine as the methyl donor.[35] Appropriate investigations with labeled precursors and degradation of product need to be completed to fully elucidate the biosynthetic pathway.

Indole Diterpene Alkaloids

Indole diterpene alkaloids in fungal endophyte associations are known commonly as lolitrems. Lolitrems are found in few symbiota, *L. perenne/A. lolii*, *F. longifolia/E. festucae*, and *F. arundinacea/Acremonium* spp. (Table 2). Lolitrems are tremorgenic neurotoxins but not all of the compounds called lolitrems are as biologically active as lolitrem B, the predominant compound in *A. lolii*-infected perennial ryegrass. Paxilline is in the proposed pathway to lolitrem B biosynthesis and is also tremorgenic (Fig. 6). Lolitrems are lipophilic and readily extracted with chloroform:methanol and quantified by HPLC with fluorescence detection.[89]

Garthwaite et al.[90,91] described an immunological detection of the lolitrems. Antibodies to the precursors of lolitrem B had little cross reactivity to lolitrem B or to each other. These antibodies also have been used in thin-layer chromatography-ELISA grams and HPLC-EIA detection.[91] Identification of paxilline and lolitrem B in cultures of *A. coenophialum* and *A. uncinatum* has been made. Interestingly, lolitrem B has not been detected in symbiota of these two endophytes.[92] Sensitive

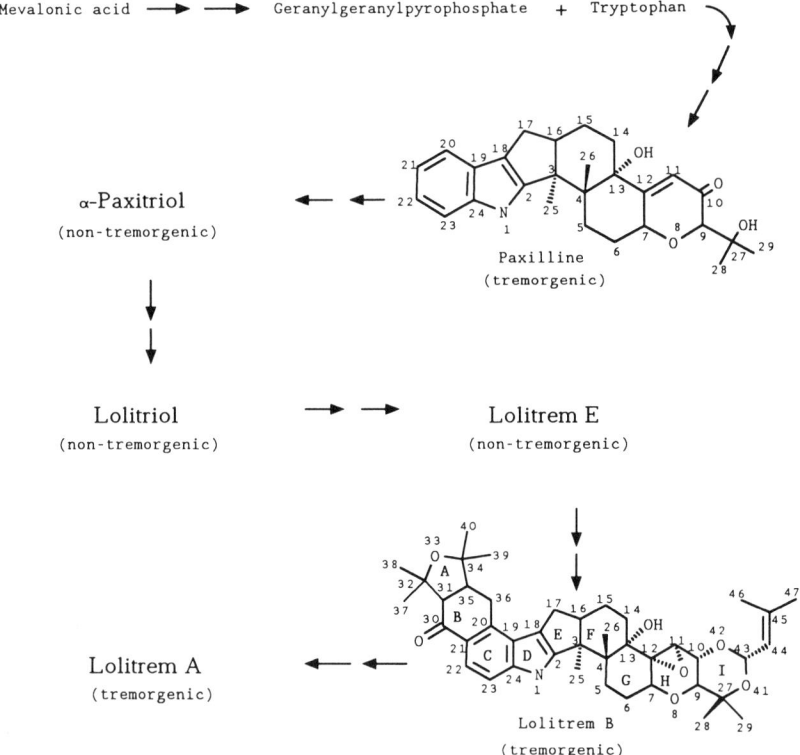

Figure 6. General scheme for lolitrem B biosynthesis.

highly specific assays such as this are extremely helpful for studies on mycosymbiont selection, genetic manipulation of toxin production in the mycosymbiont, and the accumulation of toxins in symbiota. Also, detailed studies on tremorgenic activity of the lolitrems with large animals was limited until Miles et al.[89] developed a large scale extraction procedure to yield sufficient lolitrem B for *in vivo* sheep studies. The development of an accurate precise analytical assay and isolation of quantities sufficient for pharmacological studies has enhanced the investigations into the lolitrems of these symbiota.

The biosynthetic pathway of lolitrem B has not been elucidated, but Miles et al.[93] proposed a very likely scheme (Fig. 6). Atoms in paxilline are found in corresponding positions in lolitrem B and consequently they proposed the starting precursor for lolitrem B to be mevalonic acid and tryptophan as found for paxilline.[94] The classical isoprenoid pathway from acetate would result in the formation of the diterpene geranylgeranylpyrophosphate. Tryptophan plus the diterpene would yield paxilline. Paxilline has been shown to be an important biosynthetic intermediate for several indole-diterpenoids.[95] It is interesting that paxilline is tremorgenic and some of the intermediates before lolitrem B and lolitrem A are not tremorgenic. However, it must be noted that paxilline is only 20% as active as lolitrem B in a mouse bioassay.[93]

Location of Alkaloids within the Host

Localization of alkaloids within the plant is dependent upon the stage of the life cycle of the plant. The endophyte is not uniformly present in all plant tissues and the alkaloids are translocated and sites of accumulation may or may not be sites of biosynthesis. The ergot alkaloids, lolitrems and peramine are assumed to be produced within the plant by the endophyte, as the fungus produces these alkaloids in culture. Sites of biosynthesis of lolines are less well understood because production of the lolines in fungal culture has not been reported.

The loline alkaloids are present in symbiota in concentrations 1000x greater than any of the other alkaloids. In vegetative shoot tissues, concentrations of 1000—4000 µg/g are frequently measured.[36] N-formylloline, N-acetylloline, and N-acetylnorloline are measured in greatest concentration in the tall fescue symbiotum with N-formylloline often present at levels 3-fold greater than the other loline alkaloids.[96] In reproductive plants, proportions of the lolines in different tissues are approximately 80, 5, 10, and 5 in the spikelet, flower stem, rachis, and leaf, respectively.[36] In vegetative tissues the relative accumulation is greatest in the pseudostem with lesser amounts in the leaf blade and senescent tissue (Table 3). Lolines accumulate in roots to a few hundred µg/g dry weight and are most likely responsible for the activity observed against root aphids.[97]

Table 3. Percent distribution of alkaloids in *Acremonium* infected perennial ryegrass or tall fescue

Alkaloid	Plant Parts		
	Leaves	Pseudostem	Senescent Leaf Sheath
	%	%	%
Lolines	36	59	5
Ergovaline	22	60	18
Lolitrem B	17	37	46
Paxilline	52	33	15
Peramine	56	37	7

Data adapted from.[20,98,99]

Ergovaline is also found to greatest concentration (1–2 µg/g dry weight) in the pseudostem of the perennial ryegrass and tall fescue symbiota[98] (Table 3). In sexually reproductive plants, ergovaline levels are greatest in the seed and lowest in the leaf, a pattern similar to that found for the loline alkaloids.[82,100,101] As with the loline alkaloids very low concentrations of ergovaline are found in roots.[100]

Lolitrem B has also been detected in roots of the perennial ryegrass symbiotum but at less than 1 µg/g.[102] In sexually reproductive plants the concentration was greatest in basal leaf sheath and in descending lesser amounts in upper leaf sheath, seed, stem, and leaf blade.[103] In the younger leaves of vegetative plants, portions of the leaf blade contained greater amounts of lolitrem B than the leaf sheath,[104] however, in the older leaves, the leaf sheath contained greater levels of lolitrem B than the leaf blade (Table 3). Within the leaf blade the concentration of lolitrem B increased from the tip to the base and was greatest in the upper portion of the leaf sheath. The range of lolitrem B concentration in perennial ryegrass seedlings infected with *A. lolii* is 2–4 µg/g[102] and in mature tillers the range is 1–3 µg/g dry weight in the leaf blade with the leaf sheath having 5–15 µg/g.[104,105]

Peramine concentrations in the perennial ryegrass symbiotum were generally higher in younger tissue[104,105] with concentration in regrowth tissue of mature tillers in the 20–30 µg/g dry weight range. Peramine was not recovered from roots of mature plants, but was found in roots of very young seedlings.[102] The authors speculated that this observation was the result of translocation of peramine from the seed into the developing seedling. In sexually reproductive plants of the tall fescue symbiotum the flower stem had greater peramine content than leaf blade or panicle.[101]

The observations of high *A. lolii* concentrations in the leaf sheaths of *L. perenne* and very low *A. lolii* in the tips of leaf blades suggest that peramine is translocated from site of biosynthesis, the endophyte, to other host tissues.[104] Older leaves have greater concentrations of *A. lolii* than younger leaves, but younger leaves have greatest concentration of peramine, also supporting the concept of long distance transport of peramine.[104,105] In contrast to peramine, lolitrem B accumulation is greatest in tissues with high endophyte concentration, suggesting little translocation from site of biosynthesis. On a whole plant basis, however, the yearly mean concentrations of lolitrem and peramine were closely related to the yearly mean concentration of *A. lolii*.[105] Thus, lolitrem B concentrations were positively and significantly correlated with peramine concentrations. In contrast, Roylance et al.[101] found no significant correlation between leaf peramine and ergovaline in the tall fescue symbiotum. Leaf blades contain little endophyte in either *A. lolii* or *A. coenophialum* symbiota, but the blades still accumulate significant amounts of ergovaline, indicating that translocation readily occurs into the leaf blade. Translocation of loline alkaloids into the leaf blade is also indicated by the great accumulation of alkaloids in regrowth tissue, which often did not contain detectable amounts of *A. coenophialum* endophyte. Biosynthesis of loline alkaloids is not known to occur in the endophyte, but these compounds are not detected in host plants without the endophyte. Consequently, it is assumed that a signal for synthesis or the alkaloid is translocated from the tissue region containing the endophyte into the tissue accumulating the alkaloid.

Translocation of alkaloids becomes important biologically when one considers the types of tissues consumed by herbivores. Large grazing mammals consume mainly leaf blades and the endophyte is present in least amounts in this portion of the shoot. Consequently, the toxins usually are present in lesser amounts in the highest quality tissue. If a significant proportion of above ground biomass is grazed, animals will consume more of the leaf sheath and pseudostem tissue which contains greater amounts of toxin. Peramine is found in relatively high levels in the pseudostem of plants of the perennial ryegrass symbiotum, and Argentine stem weevil is deterred from feeding and egg laying in these tissues. Many aphids and insect herbivores feed on the leaf blade of the tall fescue symbiotum, the vegetative tissue with greatest concentration of loline alkaloids. Aphids feed in the phloem, and phloem translocation is suggested for the toxin which subsequently leads to reduced aphid numbers found on infected tall fescue.

BIOLOGICAL ACTIVITY AND MODE OF ACTION

Tall fescue and perennial ryegrass are among the most important and widely grown pasture grasses for grazing livestock in the United States and New Zealand. The presence of endophytes in *Festuca* and *Lolium* spp. has been known for many decades,[54] but their association with antimammalian properties,[26,106] insecticidal activity,[28] and economic significance[34] are recent discoveries. Research has identified the anti-herbivory chemicals, their distribution within the plant, information on their biological activity, and mode of action. These discoveries have led to the introduction of endophytes into turf grasses for improved insect resistance and attempts at alleviating animal toxicoses by introduction of modified (low toxin producing) endophytes into pasture grasses.

Biological Activity—Vertebrates

Acremonium infected grasses are responsible for toxicoses symptoms of numerous ruminant and nonruminent herbivores (Table 4). The degree of toxicoses is dependent on the distribution patterns and levels of endophyte and alkaloids in the grasses, and the grazing habit of the herbivores. Toxin level and the response of animals to the toxins may also vary qualitatively and quantitatively with environmental variations. Approximately 58–65% of the plants grown on 12–14 million hectares of tall fescue in the USA are infected with *A.*

Table 4. Toxicosis symptoms of animals grazing endophyte-infected pasture grasses

Grass	Animal	Symptoms	Ref
Tall fescue			
	Cattle	Reduced feed intake and daily weight gains, elevated body temperatures and respiration, excessive salivation, rough hair coat, reduced serum prolactin and chloresterol levels, reduced lactation, agalactia, reduced conception	25,106,109
	Sheep	Reduced daily weight gains, reduced serum prolactin levels, reduced lactation, reduced conception	109,110
	Horses	Reduced serum prolactin, agalactia, reduced conception, laminitis	24,111
	Mice, Shews, Voles, Rats	Reduced population densities	23
Perennial ryegrass			
	Sheep Cattle Horse Deer	Reduced neuromuscular function, neck & limb tremors, reduced daily weight gains, reduced serum prolactin, agalactia, reduce conception	26,108,112

coenophialum.[107] All of the 10 million hectares of perennial ryegrass grown in New Zealand and much of the 5 million hectares in Australia are infected with *A. lolii*.[108] Hoveland[34] has estimated that fescue toxicosis costs the cattle industry in the USA 600 hundred million U. S. dollars annually. If horses and sheep are added to the list, the cost is considerably higher.

Fescue toxicosis and ryegrass staggers are less of a problem outside the USA and Australasia because infectivity levels of the grasses are lower in non-stressful cool season environments (e.g. United Kingdom and central and northern Europe).[113,114] There are at least two reasons for lower infectivity levels. In New Zealand, non-infected perennial ryegrass will not survive the Argentine stem weevil (*Listronotus bonariensis*) infestation,[115] which is not present in Europe. In areas of low abiotic stress survival of both infected and non-infected grasses occurs.[114] One further complication is that seed may be stored for up to 18 months prior to crop establishment and storage conditions can involve high seed moisture content and temperatures above 5°C. In these conditions, endophyte viability is reduced while seed germination is not.[116,117] This allows for the establishment of swards, in stress-free environments, that have and can maintain low levels of infected plants. In contrast, when swards containing mixed population of infected and non-infected grasses are under severe biotic and abiotic stresses, the numbers of surviving endophyte-free grass plants will be minimal.

Animal response to toxins is also an important consideration. Fescue toxicosis (summer toxicosis) of grazing cattle is exacerbated by ambient temperatures above 25ºC. However, different breeds of cattle also show variations in response to toxins when environmental conditions are similar.[106] Under high grazing pressures (drought or high stocking levels) herbivores are more likely to graze into the lower portions of fescue and ryegrass plants which contain the highest levels of toxic alkaloids. Sheep, horses, and deer are more susceptible than cattle to ryegrass staggers because of their close grazing habits. In environments that are cool and have adequate moisture for plant growth the symptomology of fescue and ryegrass toxicosis are less extreme, not only because toxin levels are often lower in the more succulent portions of the plant, but also because the incidence of infections may be lower and the animal is better able to dissipate heat.

Alkaloid profiles are the obvious consideration in determining causes of animal toxicoses. Only those profiles from symbiota that contain ergot and/or lolitrem alkaloids produce significant symptomology (Table 2). The tall fescue symbiota produces ergot alkaloids, lolines and peramine, while the perennial ryegrass symbiota produces ergot alkaloids, lolitrems and peramine. Consequently, when animals graze infected perennial ryegrass they are subjected to two major toxins, one involving neurotoxicity (tremorgenic lolitrems) and the other ergot toxicity, which probably results in mixture of symptoms (Table 4). Evidence is also available which suggests that loline alkaloids may be contributors to the fescue toxicosis syndrome (see mode of action section). However, the meadow fescue symbiota that produces only loline alkaloids has not yet been implicated in mammalian toxicosis.[118]

As previously mentioned, cost and benefits of symbiosis, as an evaluation within the symbiotic continuum, is an arbitrary one under certain circumstances. Antiherbivory is a fitness benefit for the host but may have a detrimental effect on ultimate utilization of grasses. In addition to livestock studies, investigations of small mammals have been conducted to determine the effects of alkaloids of infected plant material. Laboratory animals (rats, mice, rabbits) feeding on infected grass seed or plant extracts and animal chow regimes show many of the same symptoms (reduced feed intake, weight gain, reproduction and milk production, etc.) as the larger grazing animals (Table 4). Small insectivorous and herbivorous mammals, such as *Blarina brevicauda* (short tail shrews), *Microtus pinetorum* (pine voles), *Sigmodon hispidus* (cotton rats) and *Reithrodontomys humulis* (eastern harvest mice) had lower capture rates in endophyte-infected than non-infected tall fescue field plots.[23] Lower population densities can be viewed as a host and human benefit when symbiota are used in agriculture, conservation, recreation, home, commercial, and industrial situations. Conversely, perturbation of the ecosystem can adversely affect many species if the food chain is altered.

Biological Activity—Invertebrates

Deleterious effects (feeding deterrence and reduced populations, growth, development, and reproduction) have been reported for numerous insects feeding on endophyte-infected seed, vegetation, plant extracts, and with specific alkaloids incorporated into artificial diets. A number of reviews specifically address the various aspects of the insect-plant-endophyte interaction.[20,22,28,73,119,120] Clement et al.[120] listed 45 insect species from 6 different orders that were adversely affected by endophyte-infected grasses or their associated alkaloids. The insects had various modes of feeding and include stem borers, phloem, foliage and root feeders. Additionally, different insect stages respond differently to the various endophyte-infected plants and their alkaloids. Some of the economically important species affected are *Blissus leucopterus hirtus* (hairy chinch bug), Argentine stem weevil, *Herteromychus arator* (black beetle), *Sphenophorus parvulus* (bluegrass billbug), *Popilla japonica* (Japanese beetle), *Cyclocephala lurida* (Southern masked chafer), *Costelytra zealandica* (grass grub), *Crambus* spp. (sod web worm), *Parapediasia teterrella* (bluegrass webworm), *Spodoptera frugiperda* (fall armyworm) and leafhopper spp. of the family Cicadellidae. The same factors, previously discussed for vertebrate toxicosis, are also applicable to insects. Interestingly, individually all four classes of alkaloids have been shown to have invertebrate activity (deterrence and/or toxicity) to specific insects. Not all classes of alkaloids are toxic to the same insect or to different stages of development for the same insect. However, the array of alkaloids present have the potential for a wide spectrum of activity and multiple mechanisms of action, that could offer distinct advantages in defensive mutualisms. Laboratory studies with alkaloids on insect species that have no natural associations with grasses have demonstrated the effect of topical application of lolines against *Ctenocephalides felis* (cat flee), *Periplaneta americana* (American cockroach), *Musca autumnalis* (adult face fly), and eggs of *Heliothis virescens* (tobacco budworm).[20]

The complexity of insect-symbiota interactions is best illustrated by the activities of the Argentine stem weevil. This is an extreme example of the importance of defensive mutualism whereby the plant is incapable of survival without its mycosymbiont partner. The Argentine stem weevil adults feed on the foliage and oviposit eggs primarily in the stem via biting holes. On hatching the larvae mine the stems. New tillers can be infested by adults or larvae from old stems. Because of the close association of the weevil with its host, any effect on a part of the life cycle will greatly affect population densities. In fact, the life cycle stages are affected by one or more of the alkaloids present in the perennial ryegrass symbiotum. In feeding studies, adult weevils were deterred by peramine[121] and ergovaline[122] at diet concentrations of 0.1 µg/g. Oviposition by gravid females was much less sensitive. The reduced egg laying on infected grass then is probably due to the feeding deterrence effect of alkaloids on females which disrupts the insects life cycle. The effects of peramine on larvae feeding are also based on deterrence and not toxicity, as concentrations of 2 µg/g deterred the larvae, and concentrations as high as 25 µg/g were not toxic.[121] However, lolitrem B and its precursor paxilline do show toxicity in larval feeding studies at 5 and 10 µg/g, respectively.[123,124] There would appear to be potential for synergism among these three alkaloids, considering that they show seasonal and plant distributional fluctuations.[105] Thus, chronic exposures are likely to occur for adults and larvae.

Many of the insects affected by symbiota feed on above ground portions of the plant. Few root feeding insects, grubs and adults, and nematodes are affected by symbiota. Equivocal evidence has been reported for white grubs of *Popillia japonica* (Japanese beetle) and *Cyclocephala lurida* (southern masked chafer beetle). Results of greenhouse and field studies were variable and contradictory, depending upon the type of assay used.[125] Feeding studies of grubs on roots of infected tall fescue in pots or loline alkaloid supplemented diets indicated that first to third instar *P. japonica* larvae had reduced growth and survival. However, when studies were conducted in field plots over several years, no differences were noted in population densities or weights of grubs.[126]

Not all insects and nematodes assayed for antibiosis were affected by symbiota.[28] Interestingly, only the tall fescue, loline producing, symbiotum adversely affected six nematode species. Neither perennial ryegrass nor *F. rubra* (red fescue)/*E. festucae* symbiotum had any effect on nematode species.[27,127] In whole plant no choice greenhouse studies, the number of *Melodoidogyne marylandi* juveniles in the soil, an indicator of hatching, declined in the presence of *A. coenphialum* -infected tall fescue roots.[27] Invasion of infected roots was negligible when compared to non-infected roots. Three mechanisms were suggested for antibiosis activity. 1) Root exudates could either deter the hatching or act as a repellent, suppressing host-finding behavior. 2) Repellency could discourage root penetration or changes in root morphology that would effectively prevent penetration and/or feeding. 3) Endoderm cell wall thickness, especially on the inner side, is greater in endophyte-infected roots, providing a mechanism of resistance. Alkloid concentrations in roots are low or not detectable. Little information is available on whether levels are environment or genome sensitive. It is possible that fluctuating alkaloid levels could have permanent or transitory effects on root inhabiting insects and nematodes.

Tritrophic interactions have been reported. A beneficial interaction involves *P. japonica* grubs, the tall fescue symbiotum and an entomopathogenic (*Heterohabditis bacteriophora*) nematode. Third instar larvae feeding on endophyte-infected tall fescue or diets containing ergotamine tartrate had higher rates of mortality than controls.[128] It has been suggested that deterrence-induced starvation of larvae renders them more susceptible to entomopathogenic nematodes. Another tritrophic interaction involves the tall fescue symbiotum with *Spodoptera frugiperda* (fall armyworm) and the parasitoid wasps (*Euplectrus* spp.).[129] Fall armyworms fed infected tall fescue or synthetic diets containing loline alkaloids caused reduced pupal mass and survival on the parasitoid larvae. In this case, the benefits of protection appear to be reduced if natural enemies of insects are also negatively affected by symbiota.

Biological Activity—Plants and Fungi

A few reports indicate that symbiota have allelopathic properties affecting other plants. N-formylloline, but not N-acetylloline or N-methylloline, inhibited germination of *L. multiflorum* seed.[31] The amount of germination of endophyte-infected and endophyte-free seed of tall fescue was not affected by N-acetyl- or N-formyl-loline but the rate of germination was decreased.[36] Selectivity was demonstrated as alfalfa and canola seed germination was not inhibited by any of the loline alkaloids tested.

Antifungal activity of numerous *Acremonium* and *Epichloë* spp. has been demonstrated in agar plate assays against specific grass pathogens.[130–133] However, resistance of symbiota to pathogenic fungi has proven to be elusive. Only *Pucccinia coronata* (crown rust),[134] a seedling disease caused by *Rhizoctonia zeae*,[30] and the leaf pathogen *Cladosporium phlei*[135] have been reported to be inhibited by symbiota. Tall fescue symbiotum also affected colonization and reproduction of mycorrhizal fungi (*Glomus* spp.).[29] Peramine, lolines and several ergot alkaloids were not active in agar plate assays, indicating that those toxic alkaloids do not play a role in antifungal activity.[133] A group of fungitoxic compounds were isolated from the stoma of an *Epichloë* spp. on *Phleum pratensis* (timothy).[136] They include sesquiterpenoids, chokols A-G oxygenated fatty acids, and phenolic compounds.

Mode of Action

The compounds discussed above are known toxins in many different bioassays, however, their individual or combined mode of action and mechanism of action *in vivo* is not understood. Many observations of biological activity have been reported and together they have provided some information on the basic mode of action. Fescue toxicosis has been associated with an ambient temperature stress on the animal with gross symptoms including vasoconstriction, lameness, dry gangrene in extremities, and reduced weight gains, milk production and reproduction (Table 4).[137] Endocrine related effects include reduced prolactin and melatonin secretions and altered neurotransmitter metabolism in the hypothalamus, the pituitary and pineal glands.[70] These observations support a basic dopaminergic mechanism for toxicity and implicate the ergopeptine alkaloids as the significant toxin.[138–145] Moubarak et al.[146,147] provided a potential mechanism for the dopaminergic and serotoninergic action. Ergovaline inhibited the brain synaptosomal Na^+/K^+ ATPase enzyme system and this altered activity may be directly related to neural changes associated with consumption of the toxin.

The gross effects of vasoconstriction may also come from the ergot alkaloids.[148,149] Dryer[150] found ergovaline to be a powerful vasoconstrictor of isolated bovine uterine and umbilical arteries. Also, lysergic acid amide (ergine), the simple amide of lysergic acid and a precursor of ergopeptine alkaloid biosynthesis, has been shown to be a vasoconstrictor in cattle.[151] Ergot alkaloids may be the most significant toxins in the tall fescue symbiotum to mammals and they interact with other alkaloids present.[152] They also cause insect toxicity and feeding deterrence,[51,152,153] however, in most situations they are not as potent as other toxins found in the symbiotum. Virtually nothing is known about their mode or mechanism of action in invertebrates.[22,120]

Peramine (Fig 5) apparently functions only as an insect feeding deterrent in ryegrass symbiotum and has very low chronic toxicity.[22] Argentine stem weevil, the one insect extremely sensitive to peramine, may avoid more potent toxins because of antixenosis. Most insects are insensitive to peramine.[37] The mode of action of peramine is not known, but Rowan et al.[121] reported diacetylperamine has about 10% of the deterrent activity of peramine. In structure/activity relationship investigations it was found that if the guanidinyl moiety is intact and substitution on the prazine ring is altered, activity is about 300-fold less. Altering the side chain to a hydroxyl or nitrile makes the compounds biologically inactive to Argentine stem weevil. Adding a carbon in the guanidinyl side chain to form homoperamine did not alter deterrent activity. The authors concluded that a minimal heterocyclic ring system and a partial positive charge on the side chain was essential for insect feeding deterrence. It is noted that none of these analogs have been isolated from symbiota, and mammalian toxicity has not been measured.[118]

The mechanism of action of the lolitrems also is unknown, but enhanced release of excitatory amino acids to the central nervous system has been measured in sheep with symptoms of ryegrass staggers.[154] The pre- and post-synaptic step is generally associated with transient increased GABA levels in the synapse and antitremorgenic activity.[155] GABA is an inhibitory amino acid transmitter and blockage of GABA uptake would minimize the antitremorgenic activity and tremors could occur.

Loline alkaloids are found in the symbiota in greatest quantities, and they may be one of the caustive agents of fescue toxicosis. Loline dichlorohydrate decreased blood pressure, coronary blood flow,[156] and loline was lethal to mice when given intravenously.[74] N-acetylloline was vasoconstrictive in the lateral saphenous vein of cattle and equine arteriovenous tissues,[157,158] but at concentrations considerably less than the ergot alkaloids. Additionally, N-formylloline and N-acetylloline added to the diets of rats decreased feed intake and weight gain.[75] N-formylloline is degraded rapidly to loline *in vitro* ruminal assays, but N-acetylloline is generally recovered even after 48 h.[159] Tall fescue symbiotum diets caused reduced white cell counts, reduced serum titer to immunization and reduced response of spleen cells to mitogens, but increased T suppressor cells in spleens.[160] *In vitro* splenocytes of mice were more sensitive than cattle blood lymphocytes to the effect of loline alkaloids.[161] In this study, N-methylloline was most active and N-acetylloline had less activity than N-formylloline. Loline caused reduced prolactin release from rat pituitary cells and the reduction of prolactin release was reversed by domperidone, however, loline required a much higher concentration to obtain similar results with treatments of ergovaline.[138] Loline at high concentration inhibited prolactin release from rat hemipituitaries, and D_2 dopamine antagonists reversed the effect as well as for ergovaline, suggesting that both loline and ergovaline may act as D_2 dopamine receptor agonists. These results emphasize the potential for interaction and synergy among the toxins and the difficulty of extrapolating *in vitro* bioassays to *in vivo* observations.

The loline alkaloids have activity against many insects.[106,153,162] Microgram amounts of N-formylloline in drinking water or topical applications were toxic to large milkweed bug.[20,152,162] In structure/activity relationship tests with fall armyworm and European corn borer the longer acyl chain N'-loline derivatives significantly deterred feeding in "choice" tests and in "no choice" tests decreased larval weight.[163] Loline and N-propionylloline were not different from control treatments in the two studies, whereas ring substitution on the N' atom had high activity in these tests. N-acetylloline, N-formylloline, N-methylloline, N-propionylloline and N-myristoylloline were as toxic to greenbug aphids as the commercial pesticide nicotine sulfate.[163] The authors concluded that the loline derivatives act as metabolic toxins and not as feeding deterrents. This is supported by the observation that N-acetylloline has a tritrophic interaction with a parasitoid wasp of fall armyworm.[129] In a no choice diet situation, survival of the wasp was decreased.

PLANT ENVIRONMENT AND HOST/FUNGAL GENOME INTERACTION

Plant Environment

In *F. arundinacea/A. coenophialum* and *F. pratensis/A. uncinatum* symbiota a seasonal accumulation pattern for loline alkaloids occurs with increased accumulation during late summer.[36,164,165] Apparent accumulation of these alkaloids during the hot dry period of the growing season may simply be the result of reduced dry weight accumulation. Reduced leaf water potential increased loline alkaloids.[166,167] Water stress also increased accumulation of ergopeptine alkaloids.[39,167,168] No significant differences in peramine contents were observed between control and water-stressed plants although the xylem water was reduced and a difference in ergovaline content was measured.[168] In greenhouse grown endophyte-infected tall fescue associations, where the greatest seasonal environmental treatment would be photoperiod, there was no seasonal pattern to peramine accumulation.[101] Lolitrem B accumulation in perennial ryegrass symbiotum in New Zealand is low in winter and early spring, increases during summer, and reaches a maximum in autumn.[37,105] Important in the pattern of accumulation of lolitrem B is the observation that as lolitrem B levels decreased in autumn harvests, the levels of the tremorgenic, lolitrem B precursor, paxilline decreased only slightly or increased.[98] Ball et al.[105] found a positive and significant relationship between the accumulation of lolitrem B and peramine.

Host/Fungal Genome Interaction

Many naturally occurring, agronomically important symbiota have been screened for toxin accumulation. These investigations have the premise that different biotypes of fungal endophyte exist within endophyte species and that these biotypes differ in their capacity to accumulate toxin(s). This assumption is made knowing that the obligate *Acremonium* endophytes are only transmitted to subsequent generations through the maternal parent, and that any recent sexual mating has occurred only with the host plant genome and not with the fungal endophyte genome. Variability within the endophyte has to come from sexual mating or hybridization early in the evolutionary process toward becoming obligate endophytes. Symbiota with significantly different accumulation of toxins have been found, and much of the research and optimism of success for manipulation of toxin level through natural selection or molecular biology is based on these differences.

High levels of loline alkaloids are found primarily in associations with *A. coenophialum*,[51] however, highest levels have been found in *F. pratensis/A. uncinatum* associations.[36,165] Low concentrations have been found in several other symbiota, but the associated fungal endophyte often has not been identified. Loline alkaloid accumulation showed little environmental influence at two diverse locations.[165] Levels of ergovaline are less than that for the loline alkaloids. Differences have been reported for different host and endophyte combinations[51] and even within the tall fescue symbiotum selected from one grass cultivar large differences in ergovaline accumulation occurred.[169] High accumulations of ergovaline occur in grasses infected with endophyte strains 187BB and 196 from France and Spain compared to a New Zealand wild type.[63,98] Low and high ergovaline accumulating symbiota have been found within *F. arundinacea/A. coenophialum* populations and it appears that ergovaline biosynthesis is influenced by both the endophyte and host genomes.[38,41] A range from 2 to 132 μg/g was reported in symbiota that contained peramine,[51] indicating a high degree of natural variation. Results from other studies have suggested that the host genome is more important than the endophyte genome in determining peramine accumulation.[98,101] Peramine accumulation is relatively equal within shoot tissue and shows uniform seasonal accumulation. This suggests that manipulation of other alkaloids will not alter the accumulation and insect feeding deterrence of peramine. Low concentrations of lolitrems have been detected in symbiota other than *L. perenne*/*A. lolii*.[51,170] Investigations with *A. lolii* genotypes in *L. perenne* genotypes suggest that both genomes are important for accumulation of lolitrem B.[63,98] As described earlier, some of the precursors of lolitrem B are tremorgenic and, of these, paxilline and lolitrem B have been found in *A. coenophialum* cultures[92] indicating the potential for accumulation of tremorgens in tall fescue symbiota. This would be significant to individual herbivores. Fungal endophyte and host grass genomes appear to significantly alter accumulation of lolitrem B. The above data suggest that changes in alkaloid profiles due to manipulation of one symbiont cannot be accurately predicted in the absence of the other.

MANIPULATIONS OF SYMBIOTA AND DEFENSIVE MUTUALISM

Because C3 grasses have agronomic and recreational value any manipulation of symbiota that selectively affects their diversity and ecological significance is of economic importance. Manipulation of symbiota may eliminate or reduce the amount of an alkaloid, alleviating vertebrate toxicoses.[43–45] Conversely, endophytes introduced into endophyte-free hosts to improve tolerance for invertebrate pests would be especially useful for turf grasses.[12] Endophyte manipulation is only advantageous so long as it does not come at the cost of other host fitness enhancements.

Criteria for Manipulation

Symbiota are world-wide in distribution and collected infected seed should be considered a valuable source of microbial germplasm.[120] The use of this germplasm, as sources for improvement of symbiota, is dependent on a number of interrelated factors. They include types of alkaloid profiles desired, method of manipulation, and host-fungus compatibility. Endophytes from naturally-occurring symbiota or those modified by molecular genetics may be used. Once a suitable endophyte candidate is available the techniques required for introduction of endophytes into hosts, testing for compatibility, and production of new cultivars, are undertaken.[12,43–45]

Use of Naturally Occurring Endophytes

Naturally occurring endophytes have been introduced into plants by artificial mycelial inoculation of seedling meristems,[171] somatic embryo cultures,[172] plantlets derived from meristem cultures[173] and by plant maternal line breeding.[38,174] Indigenous endophytes, those that naturally infect their own host and related congeneric host species, are the primary candidates for alkaloid manipulations because high, but not absolute, interspecific host and fungus compatibility has been demonstrated.[12,13,38,40,41,51,171,175,176]

Lolitrem Synthesis in the Perennial Ryegrass Symbiota

A lolitrem minus isolate (187BB) from a perennial ryegrass symbiota[63] was chosen as a candidate for introduction into non-infected perennial ryegrass cultivars to alleviate ryegrass staggers.[168] Peramine and ergovaline were detected in all endophyte improved cultivars in varying amounts and their synthesis appeared to involve host-endophyte genome interactions.[98] Field testing of lolitrem minus cultivars showed the grasses were still protected from Argentine stem weevil predation, and animals grazing them did not develop ryegrass staggers.[14,177] However, some of the cultivars were shown to produce ergovaline concentrations that were much higher than in other endophyte-infected ryegrass, resulting in animals displaying some symptoms similar to fescue toxicosis.[177] Because lolitrem and ergovaline are responsible for ryegrass toxicoses, both must be eliminated or greatly reduced. The resulting peramine containing symbiotum may reflect a loss of activity spectrum for invertebrate pests, as well as, the potential for development of insect resistance to peramine.

Ergovaline Synthesis in Tall Fescue Symbiota

Christensen et al.[63] determined that some of the naturally-infected tall fescue symbiota produced no apparent ergovaline, and they would appear to be ideal candidates for introduction into current tall fescue cultivars and breeding lines. However, not only must ergovaline remain unexpressed, but all other toxic precursors and derivatives should be eliminated or greatly reduced. Because ergot and lolitrem synthesis have a common intermediate, tryptophan, it will be important to screen the modified symbiota for lolitrem and precursors.

The incorporation of one genotype of an alkaloid-deficient endophyte isolate into a new cultivar with multiple host genomes could result in changes in gene frequencies within the host population and a decrease in genotypic and phenotypic variability of the new symbiotum.[43] Multiple isolates may be necessary to preserve genotypic and phenotypic variability for host fitness across the broad region of adaptation of pasture grasses. Another approach to plant modification of endophyte-regulated alkaloid synthesis is to screen and breed plants for low alkaloid content.[38,43,101] This approach may partially overcome the problem of single endophyte genome introductions.

Use of Endophytes Modified by Molecular Genetics

The potential to use molecular genetics to generate modified endophytes and control the biosynthesis of important alkaloid toxins, especially ergot alkaloids, has been discussed by Schardl.[44] Before these studies could be undertaken knowledge of endophyte transformation systems were elucidated (via hygromycin resistance) using isolates from perennial ryegrass[178] and tall fescue.[179] Knowledge of the biosynthetic pathway of the alkaloids is a prerequisite to develop molecularly modified endophytes. This includes identification of critical gene(s) that encode the enzymes responsible for synthesis of the important alkaloid toxins. At present, only ergot alkaloid biosynthesis fulfills this requirement. While several steps in ergoline synthesis may be appropriate as targets to block the formation of ergot alkaloids, the first rate-limiting step is the logical choice because if this step is eliminated even potentially toxic intermediates will not accumulate. As was discussed earlier, this step involves the prenylation of tryptophan to the first intermediate in the committed pathway, 4-(γ,γ dimethylallyl) tryptophan (DMAT) and involves DMAT synthase. Because ergot producing endophytes are related to *Claviceps* spp., it is presumed that they have this enzyme as well. A cDNA clone from *C. purpurea* mRNA, expressed in yeast, directs production of the DMAT synthase gene (*dmaW*).[180] This clone is being used as a probe to identify the putative DMAT synthase gene(s) of *A. coenophialum*.

Two techniques, gene disruption and antisense RNA expression, are being used to prevent *dmaW* activity. The techniques and associated problems with their use are described by Schardl.[44] Because of the limitations of techniques used in molecular genetic manipulation, more than just the intended change may occur in the endophyte's genome. Such changes would not only affect alkaloid synthesis, but a number of other expressed characteristics of the mycosymbiont alone, as well as the modified symbiotum. Consequently, careful screening and characterization of new symbiotum is paramount.

CONCLUSIONS

Chemicals used in host defense against parasites and predators are not novel in themselves. Co-evolution between plant and pest has resulted, in many cases, in the development of natural or induced host defense mechanisms. The grass-fungal endophyte associations described in this paper and the resulting antiherbivore defensive chemicals produced represent only one aspect of a highly unique ecological adaptation by the symbionts. Coupled to improved plant growth and tolerance to abiotic stresses, certain symbiota are more ecologically fit than their non-infected counterparts. For the important forage grasses (tall fescue and perennial ryegrass) where the mycosymbionts are seed disseminated entities, neither hosts nor fungi can exist apart if biotic and abiotic stresses are severe. Consequently, the symbiosis is viewed as mutualistic. It is not as clear whether all symbiota should be judged similarly, as a number of *Epichloë*-infected grasses display symptoms which are antagonistic to host reproduction even though the symbiota may exhibit other beneficial fitness enhancements. The biochemical and physiological diversity of symbiota has been demonstrated by the array of potent toxins produced, their numbers, levels and position in the plants, and the responses of vertebrates and invertebrates to the alkaloids. Ergot and lolitrem alkaloids cause serious mammalian toxicoses and have widespread anti-insect activity. The potential for manipulating the symbiota by eliminating the appropriate alkaloid(s) without causing major changes in other fitness enhancements is a distinct future possibility. Certainly the manipulation involving the introduction of endophytes and associated alkaloids into grasses used for turf and soil conservation to improve insect resistance is easily accomplished.

ACKNOWLEDGMENT

This is publication number 95–12–130 of the Kentucky Agricultural Experiment Station, published with the approval of the director.

REFERENCES

1. BRONSTEIN, J.L. 1994. Our current understanding of mutualism. Quart. Rev. Biol. 69: 31–51.
2. LEWIS, D.H. 1985. Symbiosis and mutualism: crisp concepts and soggy semantics. In: The Biology of Mutualism. (D.H. Boucher, ed.), Oxford University Press, New York, pp. 29–39.
3. MICHALAKIS, Y., OLIVIERI, I., RENAUD, F., RAYMOND, M. 1992. Pleiotropic action of parasites: how to be good for hosts. Trends Ecolog. Evol. 7: 59–62.
4. BACON, C.W. 1994. Fungal endophytes, other fungi, and their metabolites as extrinsic factors of grass quality. In: Forage Quality, Evaluation, and Utilization. (C.C. Fahery, ed.), American Society of Agronomy, Madison, pp. 318–336.
5. CLAY, K. 1988. Clavicepitaceous fungal endophytes of grasses: Coevolution and the change from parasitism to mutualism. In: Coevolution of Fungi with Plants and Animals. (K.A. Pirozynski, D. Hawksworth, eds.), Academic Press, London, pp. 79–105.
6. CLAY, K. 1994. The potential role of endophytes in ecosystems. In: Biotechnology of Endophytic Fungi of Grasses. (C.W. Bacon, J.F. White, Jr. eds.), CRC Press, Boca Raton, pp. 73–86.
7. SCHARDL, C.L. 1996. Interaction of grasses with endophytic *Epichloë* species and hybrids. In: Plant-Microbe Interaction, Vol. 1. (G. Stacey, N.T.Keen, eds), Chapman & Hall, New York, pp. 107–140.
8. SIEGEL, M.R., SCHARDL, C.L. 1991. Fungal endophytes of grasses: detrimental and beneficial associations. In: Microbial Ecology of Leaves. (J.H. Andrew, S.S. Hirano, eds.), Springer Verlag, Berlin, pp. 198–221.
9. BACON, C.W., DE BATTISTA, J. 1990. Endophytic fungi of grasses. In: Soil and Plants. (D.K. Avora, B. Rai, K.G. Mukerji, G.R. Knudsen, eds.), Marcel Dekker, Inc., New York, pp. 231–256.
10. DIEHL, W.W. 1950. Balansia and the Balansiae in America. U.S. Department of Agricultural Monograph 4. U.S. Government Printing Office, Washington, D.C. p. 82.

11. WHITE, J.F., JR. 1994. Taxonomic relationships among the members of the Balansieae (Clavicipitales). In: Biotechnology of Endophytic Fungi of Grasses. (C.W. Bacon, J.F. White, Jr., eds.), CRC Press, Boca Rotan, pp. 3–20.
12. FUNK, C.R., BELANGER, F.C., MURPHY, J.A., 1994. Role of endophytes in grasses used for turf and soil conservation. In: Biotechnology of Endophytic Fungi of Grasses. (C.W. Bacon, J.F. Ehite, Jr., eds.), CRC Press, Boca Raton, pp. 201–210.
13. JOOST, R.E. 1995. *Acremonium* in fescue and ryegrass: Boon or bane? A review. J. Anim. Sci. 73: 881–888.
14. LATCH, G.C.M. 1994. Influence of *Acremonium* endophytes on perennial grass improvement. N.Z. J. Agric. Res. 37: 311–318.
15. BACON, C.W. 1993. Abiotic stress tolerances (moisture, nutrients) and photosynthesis in endophyte-infected tall fescue. Agric. Ecosyst. Environ. 44: 123–141.
16. BOUTON, J.H., GATES, R.N., BELESKY, D.P., OWSLEY, M. 1993. Yield and persistence of tall fescue in the southeastern coastal plain after removal of its endophyte. Agron. J. 85: 52–55.
17. LATCH, G.C.M., HUNT, W.F., MUSGRAVE, D.R. 1985. Endophytic fungi affect growth of perennial ryegrass. N. Z. Agric. Res. 28: 165–168.
18. RICE, J.S., PINKERTON, B.W., STRINGER, W.C., UNDERSANDER, D.J. 1990. Seed production in tall fescue as affected by fungal endophyte. Crop Sci. 30: 1303–1305.
19. HILL, N.S., BELESKY, D.P., STRINGER, W.C. 1991. Competitiveness of tall fescue as influenced by *Acremonium coenophialum*. Crop Sci. 31: 185–195.
20. DAHLMAN, D.L., EICHENSEER, H., SIEGEL, M.R. 1991. Chemical perspectives on endophyte-grass interactions and their implications to insect herbivory. In: Microbial Mediation of Plant-Herbivore Interactions. (P. Barbosa, V.A. Krischik, C.L. Jones, eds.), Wiley-Interscience, New York, pp. 227–252.
21. POPAY, A., ROWAN, D.D. 1994. Endophytic fungi as mediators of plant-insect interactions. In: Insect-Plant Interactions. (E.A. Bernays, ed.), CRC Press, Boca Raton, pp. 83–103.
22. ROWAN, D.D., LATCH, G.C.M. 1994. Utilization of endophyte-infected perennial ryegrass for increased insect resistance. In: Biotechnology of Endophytic Fungi of Grasses. (C.W. Bacon, J.F. White, Jr., eds.), CRC Press, Boca Raton, pp. 169–183.
23. COLEY, A.B., FRIBOURG, H.A., PELTON, M.R., GWINN, K.D. 1995. Effects of tall fescue endophyte infestation on relative abundance of small mammals. J. Envir. Qual. 24: 472–475.
24. CROSS, D.L., REDMOND, L.M., STRICKLAND, J.R. 1995. Equine fescue toxicosis: Signs and solutions. J. Anim. Sci. 73: 899–908.
25. PATERSON, J., FORCHERIO, C., LARSON, B., SAMFORD, M., KERLEY, M. 1995. The effects of fescue toxicosis on beef cattle productivity. J. Anim. Sci. 73: 889–898.
26. PRESTIDGE, R.A. 1993. Causes and control of perennial ryegrass staggers in New-Zealand. Agric. Ecosyst. Environ. 44: 283–300.
27. GWINN, K.D., BERNARD, E.C. 1993. Interaction of endophyte-infected grasses with the nematodes *Meloidogyne marylandi* and *Protylenchus scribneri*. In: Proceedings of the Second International Symposium on *Acremonium*/grass Interactions. (D.E. Hume, G.C.M. Latch, H.S. Easton, eds.), AgResearch, Grasslands Research Centre, Palmerston North, NZ, pp. 156–160.
28. LATCH, G.C.M. 1993. Physiological interactions of endophytic fungi and their hosts—biotic stress tolerance imparted to grasses by endophytes. Agric. Ecosyst. Environ. 44: 143–156.
29. GUO, B.Z., HENDRIX, J.W., AN, Z.Q., FERRISS, R.S. 1992. Role of *Acremonium* endophyte of fescue on inhibition of colonization and reproduction of mycorrhizal fungi. Mycologia 84: 882–885.
30. GWINN, K.D., GAVIN, A.M. 1992. Relationship between endophyte infestation level of tall fescue seed lots and *Rhizoctonia zeae* seedling disease. Plant Disease 76: 911–914.
31. PETROSKI, R.J., DORNBOS, D.L., JR., POWELL, R.G. 1990. Germination and growth inhibition of annual ryegrass (*Lolium multiflorum* L.) and alfalfa (*Medicago sativa* L.) by loline alkaloids and synthetic N- acylloline derivatives. J. Agric. Food Chem. 38: 1716–1718.

32. WEST, C.P. 1994. Physiology and drought tolerance of endophyte-infected grasses. In: Biotechnology of Endophytic Fungi of Grasses. (C.W. Bacon, J.F. White, Jr., eds.), CRC Press, Boca Raton, pp. 87–102.
33. CLAY, K. 1988b. Fungal Endophytes of grasses: a defensive mutualism between plants and fungi. Ecol. 69: 10–16.
34. HOVELAND, C.S. 1993. Importance and economic significance of the *Acremonium* endophytes to performance of animals and grass plant. Agric. Ecosyst. Environ. 44: 3–12.
35. PORTER, J.K. 1994. Chemical constituents of grass endophytes. In: Biotechnology of Endophytic Fungi of Grasses. CRC Press, Boca Raton, pp. 103–124.
36. BUSH, L.P., FANNIN, F.F., SIEGEL, M.R., DAHLMAN, D.L., BURTON, H.R. 1993. Chemistry, occurrence and biological effects of saturated pyrrolizidine alkaloids associated with endophyte grass interactions. Agric. Ecosyst. Environ. 44: 81–102.
37. ROWAN, D.D. 1993. Lolitrems, peramine and paxilline—mycotoxins of the ryegrass endophyte interaction. Agric. Ecosyst. Environ. 44: 103–122.
38. AGEE, C.S., HILL, N.S. 1994. Ergovaline variability in *Acremonium*-infected tall fescue due to environment and plant genotype. Crop. Sci. 34: 221–226.
39. ARECHAVALETA, M., BACON, C.W., PLATTNER, R.D., HOVELAND, C.S., RADCLIFFE, D.E. 1992. Accumulation of ergopeptide alkaloids in symbiotic tall fescue grown under deficits of soil water and nitrogen fertilizer. Appl. and Environ. Microbiol. 58: 857–861.
40. CHRISTENSEN, M.J. 1995. Variation in the ability of *Acremonium* endophytes of perennial rye-grass (*Lolium perenne*), tall fescue (*Festuca arundinacea*) and meadow fescue (*Festuca pratensis*) to form compatible associations in the three grasses. Mycol. Res. 99: 466–470.
41. HILL, N.S., PARROTT, W.A., POPE, D.D. 1991. Ergopeptine alkaloid production by endophytes in a common tall fescue genotype. Crop Sci. 31: 1545–1547.
42. LEUCHTMANN, A., CLAY, K. 1993. Nonreciprocal compatibility between *Epichloë typhina* and 4 host grasses. Mycologia 85: 157–163.
43. HILL, N.S. 1993. Physiology of plant-endophyte interaction: Implications and use of endophytes in plant breeding. In: Proceedings of the Second International Symposium on *Acremonium*/grass Interactions. (D.E. Hume, G.C.M. Latch, H.S. Easton, eds.), AgResearch, Grasslands Centre, Palmerston North, NZ, pp. 161–170.
44. SCHARDL, C.L. 1994. Molecular and genetic methodologies and transformation of grass endophytes. In: Biotechnology of Endophytic Fungi of Grasses. (C.W. Bacon, J.F. White, eds.), CRC Press, Boca Raton, pp. 151–168.
45. SIEGEL, M.R., BUSH, L.P. 1994. Importance of endophytes in forage grasses, a statement of problems and selection of endophytes. In: Biotechnology of Endophytic Fungi of Grasses. (C.W. Bacon, J.F. White, Jr., eds.) CRC Press, Boca Raton, pp. 135–150.
46. PORTER, J.K. 1995. Analysis of endophyte toxins: Fescue and other grasses toxic to livestock. J. Anim. Sci. 73: 871–880.
47. STRICKLAND, J.R., OLIVER, J.W., SR., CROSS, D.L. 1993. Fescue toxicosis and its impact on animal agriculture. Vet Human Toxicol 35: 454–464.
48. BERDE, B., STRUMER, E. 1978. Introduction to the pharmacology of ergot alkaloids and related compounds as a basis of their therapeutic application. In: Ergot Alkaloids and Related Compounds. (B. Berde, H.O. Schilid, eds.), Springer-Verlag, New York, pp. 1–28.
49. BACON, C.W., WHITE, J.F., JR. 1994. Stains, media, and procedures for analyzing endophytes. In: Biotechnology of Endophytic Fungi of Grasses. (C.W. Bacon, J.F. White, Jr., eds.), CRC Press, Boca Raton, pp. 47–58.
50. HILL, N.S. 1994. Ecological relationships of Balansiae-infected Graminoids. In: Biotechnology of Endophytic Fungi of Grasses. (C.W. Bacon, F.F. White, Jr., eds.), CRC Press, Boca Raton, pp. 59–72.
51. SIEGEL, M.R., LATCH, G.C.M., BUSH, L.P., FANNIN, F.F., ROWAN, D.D., TAPPER, B.A., BACON, C.W., JOHNSON, M.C. 1990. Fungal endophyte-infected grasses: Alkaloid accumulation and aphid response. J. Chem. Ecol. 16: 3301–3315.

52. WHITE, J.F., JR. 1988. Endophyte-host associations in forage grasses. XI. A proposal concerning origin and evolution. Mycologia 80: 442–446.
53. LEUCHTMANN, A., SCHARDL, C.L., SIEGEL, M.R. 1994. Sexual compatibility and taxonomy of a new species of *Epichloë* symbiotic with fine fescue grasses. Mycologia 86: 802–812.
54. WHITE, J.F., JR. 1993. Taxonomy, life cycle, reproduction and detection of *Acremonium* endophytes. Agric. Ecosyst. Environ. 44: 13–38.
55. BULTMAN, T.L., WHITE, J.F., JR., BOWDISH, T.I., WELCH, A.M., JOHNSTON, J. 1995. Mutualistic transfer of *Epichloë* spermatia by *Phorbia* flies. Mycologia 87: 182–189.
56. TSAI, H.-F., LIU, J.-S., CHRISTENSEN, M.J., LATCH, G.C.M., SIEGEL, M.R, SCHARDL, C.L. 1994. Evolutionary diversification of fungal endophytes of tall fescue grass by hybridization with *Epichloë* species. Proc. Natl. Acad. Sci. 91: 2542–2546.
57. WHITE, J.F., JR. 1993. Endophyte-host associations in grasses .19. a systematic study of some sympatric species of *Epichloë* in England. Mycologia 85: 444–455.
58. WHITE, J.F., JR. 1994. Endophyte-host associations in grasses .20. Structural and reproductive studies of *Epichloë* amarillans sp nov and comparisons to *Epichloë* typhina. Mycologia 86: 571–580.
59. MORGAN-JONES, G., GAMS, W. 1982. Notes on Hyphomycetes, XLI. An endophyte of *Festuca arundinacea* and the anamorph of *Epichloë* typhina, new taxa in one of two new sections of *Acremonium*. Mycotaxon 15: 311–318.
60. LATCH, G.C.M., CHRISTENSEN, M.J., SAMUELS, G.J. 1984. Five endophytes of *Lolium* and *Festuca* in New Zealand. Mycotaxon 20: 535–550.
61. GAMS, W., PETRINI, O., SCHMIDT, D. 1990. *Acremonium uncinatum*, a new endophyte in *Festuca pratensis*. Mycotaxon 37: 67–71.
62. GWINN, K.D., SHEPARD-COLLINS, M.H., REDDICK, B.B. 1991. Tissue print-immunoblot: an accurate method for the detection of *Acremonium coenophialum* in tall fescue. Phytopathology 81: 747–748.
63. CHRISTENSEN, M.J., LEUCHTMANN, D.D., ROWAN, D.D., TAPPER, B.A. 1993. Taxonomy of *Acremonium* endophytes of tall fescue (*Festuca arundinacea*), meadow fescue (*F. pratensis*), and perennial ryegrass (*Lolium perenne*). Mycol. Res. 97: 1083–1092.
64. LEUCHTMANN, A. 1994. Isozyme characaterization, persistance, and compatibility of fungal and grass mutualists. In: Biotechnology of Endophytic Fungi of Grasses. (C.W. Bacon, J.F. White, Jr., eds.) CRC Press, Boca Raton. pp. 21–33.
65. LEUCHTMANN, A., CLAY, K. 1990. Isozyme variation in the *Acremonium/Epichloë* fungal endophyte complex. Phytopathology 80: 1133–1139.
66. SCHARDL, C.L., LEUCHTMANN, A., TSAI, H.-F., COLLETT, M.A., WATT, D.M., SCOTT, D.B. 1994. Origin of a fungal symbiont of perennial ryegrass by interspecific hybridization of a mutualist with the ryegrass choke pathogen, *Epichloë* typhina. Genetics 136: 1307–1317.
67. MULLER, H.J. 1964. The relation of recombination to mutational advance. Mutation Res. 1: 2–9.
68. POWELL, R.G., PETROSKI, R.J. 1992. Alkaloid toxins in endophyte-infected grasses. Natural toxins 1: 163–170.
69. TEPASKE, M.R., POWELL, R.G. 1993. Analyes of selected endophyte-infected grasses for the presence of loline-type and ergot-type alkaloids. J. Agric. Food Chem. 41: 2299–2303.
70. PORTER, J.K., THOMPSON, JR., F.N. 1992. Effects of fescue toxicosis on reproduction in livestock. J. Anim. Sci. 70: 1594–1603.
71. POWELL, R.G., PETROSKI, R.J. 1993. The loline group of pyrrolizidine alkaloids. In: The alkaloids: Chemical and Biological Perspectives. (S.W. Pelletier, ed.), Springer-Verlag, 8: 320–338.
72. GARNER, G.B., ROTTINGHAUS, G.E., CORNELL, C.N., TESTERECI, H. 1993. Chemistry of compounds associated with endophyte grass interaction—ergovaline-related and ergopeptine-related alkaloids. Agric. Ecosyst. Environ. 44: 65–80.

73. BREEN, J.P. 1994. *Acremonium* endophyte interactions with enhanced plant resistance. Ann. Rev. Entomol. 39: 401–423.
74. YATES, S.G., TOOKEY, H.L. 1965. Festucine, an alkaloid from tall fescue (*Festuca arundinacea* Schreb.): chemistry of the functional groups. Aust. J. Chem. 18: 53–60.
75. ROBBINS, J.D., SWEENY. J.G., WILKINSON, S.R., BURDICK, D. 1972. Volatile alkaloids of Kentucky 31 tall fescue seed (*Festuca arundinacea* Schreb.). J. Agric. Food Chem. 29: 653–657.
76. TUFARIELLO, J.J., MECKLER, H., WINZENBERG, K. 1986. Synthesis of the *Lolium* alkaloids. J. Org. Chem. 51: 3356–3357.
77. PETROSKI, R.J., YATES, S.G., WEISLEDER, D., POWELL, R.G. 1989. Isolation, semi-synthesis, and NMR spectral studies of loline alkaloids. J. Nat. Prod. 52: 810–817.
78. PETROSKI, R.J., POWELL, R.G., CLAY, K. 1992. Alkaloids of *Stipa robusta* (sleepy grass) infected with an *Acremonium* endophyte. Nat. Toxins 1: 84–88.
79. MILES, C.O., LANE, G.A., DI MENNA, M.E., GARTHWAITE, I., PIPER, E.L., BALL. O. J.-P., LATCH, G.C.M., BUSH, L.P., MIN, F.K., FLETCHER, I., HARRIS, P.S. High levels of ergonovine and lysergic acid amide in toxic *Achnatherum inebrians* accompany infection by an *Acremonium*-like endophytic fungus. J. Agric. Food Chem. in press.
80. SHELBY, R.A., KELLEY, V.C. 1992. Detection of ergot alkaloids from *Claviceps* species in agricultural products by competitive ELISA using a monoclonal antibody. J. Agric. Food Chem. 40: 1090–1092.
81. MA, Y., MEYER, K.G., AFZAL, D., AGENA, E.A. 1993. Isolation and quantification of ergovaline from *Festuca arundinacea* (tall fescue) infected with the fungus *Acremonium coenophialum* by high-performance capillary electrophoresis. J. Chromatog. 652: 535–538.
82. ROTTINGHAUS, G.E., GARNER, G.B., CORNELL, C.N., ELLIS, J.L. 1991. HPLC method for quantitating ergovaline in endophyte-infected tall fescue: variation of ergovaline levels in stems with leaf sheaths, leaf blades, and seed heads. J. Agric. Food Chem. 39: 112–115.
83. CRAIG, M.M., BILICH, D., HOVERMALE, J.T., WELTY, R.E. 1994. Improved extraction and HPLC methods for ergovaline from plant material and rumen fluid. J. Vet. Diagn. Invest. 6: 348–352.
84. HILL, N.S., AGEE, C.S. 1994. Detection of ergoline alkaloids in endophyte-infected tall fescue by immunoassay. Crop Sci. 34: 530–534.
85. SHIBUYA, M., CHOU, H.-M., FOUNTOULAKIS, M., HASSAM, M., KIM, S.-U., KOBAYASHI, K., OTSUKA, H., ROGALSKA, E., CASSADY, J.M., FLOSS, H.G. 1990. Sterochemistry of the isoprenylation of typtophan catalyzed by 4-(_,_-dimethylally) typtophan synthase from *Claviceps*, 1st pathway-specific enzyme in ergot alkaloid biosynthesis. J. Amer. Chem. Soc. 112: 297–304.
86. GEBLER, J.C., POULTER, C.D. 1992. Purification and characterization of dimethylallyltryptophan synthase from *Clavicepes purpurea*. Arch. Biochem. Biophy. 296: 308–313.
87. FANNIN, F.F., BUSH, L.P., SIEGEL, M.R. ROWAN D.D. 1990. Analysis of peramine in fungal endophyte-infected grasses by reversed-phase thin-layer chromatography. J. Chromatog. 503: 288–292.
88. ROWAN, D.D., HUNT, M.B., GAYNOR. D.L. 1986. Peramine, a novel insect feeding deterrent from ryegrass infected with the endophyte *Acremonium loliae*. J. Chem. Soc. Chem. Commun. pp. 935–936.
89. MILES, C.O., MUNDAY, S.C., WILKINS, A.L., EDE, P.M., TOWERS, N.R. 1994. Large scale isolation of lolitrem B and structure determination of lolitrem E. J. Agric. Food Chem. 42: 1488–1492.
90. GARTHWAITE I, MILES C.O., TOWERS, N.R. (1993) Immunological detection of the indole diterpenoid tremorgenic mycotoxins. In: Proceedings of the Second International Symposium on *Acremonium*/grass Interactions. (D.E. Hume, G.C.M. Latch, H.S. Easton, eds.), AgResearch, Grasslands Research Centre, Palmerston North, NZ, pp. 77–80.

91. GARTHWAITE, I., MILES, C.O., TOWERS, N.R. 1994. An immunological approach to the study of the ryegrass staggers syndrome. In: International Conference on Harmful and Beneficial Microorganisms in Grassland, Pasture and Turf. (K. Krohn, V.H. Paul, J. Thomas, eds) 17 International Organization Biological and Integrated Control of Noxious Animals and Plants, Gent, Belgium, pp. 175–180.
92. PENN, J., GARTHWAITE, I., CHRISTENSEN, M.J., JOHNSON, C.M., TOWERS, N.R. 1993. The importance of paxilline in screening for potentially tremorgenic *Acremonium* isolates. In: Proceedings of the Second International Symposium on *Acremonium*/grass Interactions. (D.E. Hume, G.C.M. Latch, H.S. Easton, eds.), AgResearch, Grasslands Research Centre, Palmerston North, NZ, pp. 88–92.
93. MILES, C.O., WILKINS, A.L., GALLAGHER, R.T., HAWKES, A.D., MUNDAY, S.C., TOWERS, N.R. 1992. Synthesis and tremorgenicity of paxitriols and lolitriol: possible biosynthetic precursors of lolitrem B. J. Agric. Food Chem. 40: 234–238.
94. LAWS, I., MANTLE, P.G. 1989. Experimental constraints in the study of the biosynthesis of indole alkaloids in fungi. J. Gener. Micro. 135: 2679–2692.
95. PENN, J., MANTLE, P.G. 1994. Biosynthetic intermediates of indole-diterpenoid mycotoxins from selected transformations at C-10 of paxilline. Phytochem. 35: 921–926.
96. YATES, S.G., PETROSKI, R.J., POWELL, R.G. 1990. Analysis of loline alkaloids in endophyte-infected tall fescue by capillary gas chromatography. J. Agric. Food Chem. 38: 182–185.
97. SCHMIDT, D. 1993. Effects of *Acremonium uncinatum* and a *Phialophora*-like endophyte on vigour, insect and disease resistance of meadow fescue. In: Proceedings of the Second International Symposium on *Acremonium*/grass Interactions. (D.E. Hume, G.C.M. Latch, H.S. Easton, eds.), AgResearch, Grasslands Research Centre, Palmerston North, NZ, pp. 185–188.
98. DAVIES, E., LANE, G.A., LATCH, G.C.M., TAPPER, B.A., GARTHWAITE, I., TOWERS, N.R., FLETCHER, L.R., POWNALL, D.B. 1993. Alkaloid concentrations in field-grown synthetic perennial ryegrass endophyte associations. In: Second International Symposium on *Acremonium*/grass Interactions. (D.E. Hume, G.C.M. Latch, H.S. Easton, eds.), AgResearch, Grasslands Research Centre, Palmerston North, NZ, pp. 72–76.
99. BURHAN, W. 1984. Development of *Acremonium coenophialum* and accumulation of N-acetyl and N-formyl loline in tall fescue (*Festuca arundinacea* Schreb.). Master's Thesis, University of Kentucky, Lexington, KY. p. 64.
100. AZEVEDO, M.D., WELTY, R.E., CRAIG, A.M., BARTLETT, J. 1993. Ergovaline disribution, total nitrogen and phosphorus content of two endophyte-infected clones. In: Proceedings of the Second International Symposium on *Acremonium*/grass Interactions. (D.E. Hume, G.C.M. Latch, H.S. Easton, eds.), AgResearch, Grasslands Research Centre, Palmerston North, NZ, pp. 59–62.
101. ROYLANCE, J.T., HILL, N.S., AGEE, C.S. 1994. Ergovaline and peramine production in endophyte-infected tall fescue: Independent regulation and effects of plant and endophyte genotype. J. Chem. Ecol. 20: 2171–2183.
102. BALL, O.J-P., PRESTIDGE, R.A., SPROSEN, J.M. 1993. Effect of plant age and endophyte viability on peramine and lolitrem B concentration in perennial ryegrass seedlings. In: Proceedings of the Second International Symposium on *Acremonium*/grass Interactions. (D.E. Hume, G.C.M. Latch, H.S. Easton, eds.), AgResearch, Grasslands Research Centre, Palmerston North, NZ, pp. 63–66.
103. DI MENNA, M.E., MORTIMER, P.H., PRESTIDGE, R.A., HAWKES, A.D., SPROSEN, J.M. 1992. Lolitrem B concentrations, counts of *Acremonium lolii* hyphae, and the incidence of ryegrass staggers in lambs on plots of *A. lolii*-infected perennial ryegrass. N.Z. J. Agric. Res. 35: 211–217.
104. KEOGH, R.G., TAPPER, B.A. 1993. *Acremonium lolii*, lolitrem B, and peramine concentrations within vegetative tillers of perennial ryegrass. In: Proceedings of the Second International Symposium on *Acremonium*/grass Interactions. (D.E. Hume, G.C.M. Latch, H.S. Easton, eds.), AgResearch, Grasslands Research Centre, Palmerston North, NZ, pp. 81–84.

105. BALL, O.J-P., PRESTIDGE, R.A., SPROSEN, J.M. 1995. Interrelationships between *Acremonium lolii*, peramine, and lolitrem B in perennial ryegrass. Appl. Environ. Microbiol. 61: 1527–1533.
106. SCHMIDT, S.P., OSBORN, T.G. 1993. Effects of endophyte-infected tall fescue on animal performance. Agric. Ecosyst. Environ. 44: 233–262.
107. SHELBY, R.A., DALRYMPLE, L.W. 1987. Incidence and distribution of the tall fescue endophyte in the United States. Plant Disease 71: 783–786.
108. HEESWIJCK, R.V., MCDONALD, G. 1992. *Acremonium* endophytes in perennial ryegrass and other pasture grasses in Australia and New Zealand. Austra. J. Agric. Res. 43: 1683–1709.
109. THOMPSON, F.N., STUEDEMANN, J.A. 1993. Pathophysiology of fescue toxicosis. Agric. Ecosyst. Environ. 44: 263–281.
110. CHESTNUT, A.B., BERNARD, J.K., HARSTIN, J.B., REDDICK, B.B. 1992. Performance of growing lambs fed *Acremonium coenophialum* infested tall fescue (*Festuca arundinacea* Schreb) Hay. Small Rum. Res. 7: 9–19.
111. ROHRBACH, B.W., GREEN, E.M., OLIVER, J.W., SCHNEIDER, J.F. 1995. Aggregate risk study of exposure to endophyte-infected (*Acremonium coenophialum*) tall fescue as a risk factor for laminitis in horses. Amer. J. Vet. Res. 56: 22–26.
112. FLETCHER, L.R., HOGLUND, J.H., SUTHERLAND, B.L. 1990. The Impact of *Acremonium* endophytes in New Zealand. Proc. N.Z. Grassland Assoc. 52: 227–235.
113. LEWIS, G.C., WHITE, J.F., JR., BONNEFONT, J. 1993. Evaluation of grasses infected with fungal endophytes against locusts. Ann. Appl. Biol. 122: 142–143.
114. RAVEL, C., CHARMET, G., BALFOURIER, F. 1995. Influence of the fungal endophyte *Acremonium lolii* on agronomic traits of perennial ryegrass in France. Grass Forage Sci. 50: 75–80.
115. BARKER, G.M., ADDISON, P.J. 1993. Argentine stem weevil and damage in ryegrass swards of contrasting *Acremonium* infection. In: Proc. 6th Australasian Conf. Grassland Invertebrate Ecol.. (R.A. Prestidge, ed.), pp. 161–168.
116. HARE, M.D. 1990. Viabilty of *Lolium* endophyte fungus in seed and germination of *Lolium perenne* seed during five years of storage. In: Proceedings of the Second International Symposium on *Acremonium*/grass Interactions. (S.S. Quisenberry, J. Joost, eds.), Louisiana Agric. Exper. Stn., Baton Rouge, pp.147–149.
117. WELTY, R.E., AZEVEDO, M.D., COOPER, T.M. 1987. Influence of moisture content, temperature, and length of storage on seed germination and survival of endophytic fungi in seeds of tall fescue and perennial ryerass. Phytopathology 77: 893–900.
118. DACCORD, R., ARRIGO, Y., GUTZWILLER, A., SCHMIDT, D. 1995. Les endophytes: un facteur limitant les performances du ruminant? Revue Suisse Agric. 27: 197–199.
119. CLAY, K., MARKS, S., CHEPLICK, G.P. 1993. Effects of insect herbivory and fungal endophyte infection on competitive interactions among grasses. Ecol. 74: 1767–1777
120. CLEMENT, S.L., KAISER, W.J., EICHENSEER, H. 1994. *Acremonium* endophytes in germplasms of major grasses and their utilization for insect resistance. In: Biotechnology of Endophytic Fungi of Grasses. (C.W. Bacon, J.F. White Jr., eds.), CRC Press, Boca Raton, pp. 185–200.
121. ROWAN, D.D., DYMOCK, J.J., BRIMBLE, M.A. 1990. Effect of fungal metabolite peramine and analogues on feeding and development of Argentine stem weevil (*Listronotus bonariensis*). J. Chem. Ecol. 16: 1683–1695.
122. POPAY, A.J., PRESTIDGE, R.A., ROWAN, D.D., DYMOCK, J.J. 1990. The role of *Acremonium lolii* mycotoxins in insect resistance of perennial ryegrass (*Lolium perenne*). In: Proceedings of the Second International Symposium on *Acremonium*/grass Interactions. (S.S. Quisenberry, R.E. Joost, eds.), Louisiana Agric. Exper. Sta., Baton Rouge, pp. 44–47.
123. DYMOCK, J.J., PRESTIDGE, R.A., ROWAN, D.D. 1989. The effect of lolitrem B on Argintine stem weevil. Proc. N.Z. Weed Pest Control. Conf. 42: 73–75.

124. PRESTIDGE, R.A., GALLAGHER, R.T. 1988. Endophyte fungus confers resistance to ryegrass: Argentine stem weevil studies. Ecolog. Ent. 13: 429–435.
125. POTTER, D.A., PATTERSON, C.G., REDMOND, C.T. 1992. Influence of turfgrass species and tall fescue endophyte on feeding ecology of japanese beetle and southern masked chafer grubs (Coleoptera, Scarabaeidae). J. Econ. Ent. 85: 900–909.
126. DAVIDSON, A.W., POTTER, D.A. 1995. Response of plant-feeding, predatory, and soil-inhabiting invertebrates to *Acremonium* endophyte and nitrogen fertilization in tall fescue turf. J. Econ. Entomol. 88: 367–379.
127. COOK, R., LEWIS, G.C., MIZEN, K.A. 1991. Effects on plant-parasitic nematodes on infection of perennial ryegrass, *Lolium perenne*, by endophytic fungus, *Acremonium lolii*. Crop Protec.10: 403–407.
128. GREWAL, S.K., GREWAL, P.S., GAUGLER, R. 1995. Endophytes of fescue grasses enhance susceptibility of *Popillia japonica* larvae to an entomopathogenic nematode. Entomol. Exp. Appl. 74: 219–224.
129. BULTMAN, T.L., BOROWICZ, K.L., SCHNEBLE, R.M., COUDRON, T.A., BUSH, L.P. Effect of a fungal endophyte and loline alkaloids on the growth and survival of two *Euplectrus* parasitiods. Entomologia, in press.
130. CHRISTENSEN, M.J., LATCH, G.C.M. 1991. Variation among isolates of *Acremonium* endophytes (*A. coenophialum* and possibly *A. typhinum*) from tall fescue (*Festuca arundinacea*). Mycol. Res. 95: 1123–1126.
131. CHRISTENSEN, M.J., LATCH, G.C.M., TAPPER, B.A. 1991. Variation within isolates of *Acremonium* endophytes from perennial ryegrass. Mycol. Res. 95: 918–923.
132. SCHMIDT, D. 1994. Du nouveau sur les endophytes se la fetuque des pres. Revue Suisse Agric. 26: 57–63.
133. SIEGEL, M.R., LATCH, G.C.M. 1991. Expression of antifungal activity in agar culture by isolates of grass endophytes. Mycologia 83: 525–537.
134. FORD, V.L., KIRKPATRICK, T.L. 1989. Effects of *Acremonium coenophialum* in tall fescue on host disease and insect resistance and allelopathy to *Pinus taeda* seedlings. In: Proc. Arkansas Fescue Toxicosis Confer. (C.P. West, ed.), University of Arkansas, Arkansas Agric. Exper. Sta., pp. 29–34.
135. SHIMANUKI, T. 1987. Studies on the mechanisms of the infection of timothy with purple spot disease caused by *Cladosporium phlsi* (Gregory) de Vries. Res. Bull. Hokkaido Nat. Agric. Exper. Sta. 148: 1–56.
136. KOSHINO, H., YOSHIHARA, T., ICHIHARA, A., TAJIMI, A., SHIMANUKI, T. 1992. Two sphingoid derivatives from stromata of *Epichloë* typhina on *Phleum pratense*. Phytochem. 31: 3757–3759.
137. MAHMOOD, T., OTT, R.S., FOLEY, G.L., ZINN, G.M., SCHAEFFER, D.J., KESLER, D.J. 1994. Growth and ovarian function of weanling and yearling beef heifers grazing endophyte-infected tall fescue pastures. Therioigenology 42: 1149–1158.
138. STRICKLAND, J.R., CROSS, D.L., BIRRENKOTT, G.P., GRIMES, L.W. 1994. Effect of ergovaline, loline, and dopamine antagonists on rat pituitary cell prolactin release in Vitro. Amer. J. Vet. Res. 55: 716–721.
139. STRICKLAND, J.R., CROSS, D.L., JENKINS, T.C., PETROSKI, R.J., POWELL, R.G. 1992. The effect of alkaloids and seed extracts of endophyte-infected tall fescue on prolactin secretion in an in vitro rat pituitary perfusion system. J. Anim. Sci. 70: 2779–2786.
140. LIPHAM, L.B., THOMPSON, F.N., STUEDEMANN, J.A., SARTIN, J.L. 1989. Effects of metoclopramide on steers grazing endophyte-infected fescue. J. Anim. Sci. 67: 1090–1097.
141. PORTER, J.K., STUEDEMANN, J.A., THOMPSON, JR., F.N., BUCHANAN, B.A., TUCKER, H.A. 1993. Melatonin and pineal neurochemicals in steers grazed on endophyte-infected tall fescue: Effects of metoclopramide. J. Anim. Sci. 71: 1526–1531.

142. LARSEN, B.T., SULLIVAN, D.M., SAMFORD, M.D., KERLEY, M.S., PATERSON, J.A., TURNER, J.T. 1994. D_2 Dopamine receptor response to endophyte-infected tall fescue and an antagonist in the rat. J. Anim. Sci. 72: 2905–2910.
143. HEMKEN, R.W., BOLING, J.A., BULL, L.S., HATTON, R.H., BUCKNER, R.C., BUSH, L.P. 1981. Interaction of environmental temperature and anti-quality factors on the severity of summer fescue toxicosis. J. Anim. Sci. 52: 710–714.
144. REDMOND, L.M., CROSS, D.L., STRICKLAND, J.R., KENNEDY, S.W. 1994. Efficacy of domperidone and sulpiride as treatments for fescue toxicosis in horses. Am. J. Vet. Res. 55: 722–729.
145. CHOUDHARY, M.S., SACHS, M., ULUER, A., GLENNON, R.A., WESTKAEMPER, R.B., ROTH, B.L. 1995. Differential ergoline and ergopeptine binding to 5-hydroxytryptamine$_{2A}$ receptors: Ergolines require an aromatic residue at position 340 for high affinity binding. Mol. Pharmacol. 47: 450–457.
146. MOUBARAK, A.S., PIPER, E.L., WEST, C.P., JOHNSON, Z.B. 1993. Interaction of purified ergovaline from endophyte-infected tall fescue with synaptosomal ATPase enzyme system. J. Agric. Food Chem. 41: 407–409.
147. MOURBARAK, A.S., PIPER, E.L., WEST, C.P. 1993. Effects of ergotamine and ergonovine on renal ATPase system in rat. Med. Sci. Res. 21: 289–290.
148. GARNER, G.B., SPIERS, D.E., CORNELL, C.N., ROTTINGHAUS, G.E. 1993. Bovine thermal response to cold stress with and without *Acremonium coenophialum* infested tall fescue seed in the ration measured in a climate controlled chamber. In: Proceedings of the Second International Symposium on *Acremonium*/grass Interactions. (D.E. Hume, G.C.M. Latch, H.S. Easton, eds.), AgResearch, Grasslands Research Centre, Palmerston North, NZ, p. 128.
149. RHODES, M.T., PATERSON, J.A., KERLEY, M.S., GARNER, H.E., LAUGHLIN, M.H. 1991. Reduced blood flow to peripheral and core body tissues in sheep and cattle induced by endophyte-infected tall fescue. J. Anim. Sci. 69: 2033–2043.
150. DYER, D.C. 1993. Evidence that ergovaline acts on serotonin receptors. Life Sciences 53: 223–228.
151. OLIVER, J.W., ABNEY, L.K., STRICKLAND, J.R., LINNABARY, R.D. 1993. Vasoconstriction in bovine vasculature induced by the tall fescue alkaloid lysergamide. J. Anim. Sci. 71: 2708–2713.
152. YATES, S.G., FENSTER, J.C., BARTELT, R.J. 1989. Assay of tall fescue seed extracts, fractions, and alkaloids using the large milkweed bug. J. Agric. Food Chem. 37: 354–357.
153. PATTERSON, C.G., POTTER, D.A., FANNIN, F.F. 1991. Feeding deterrency of alkaloids from endophyte-infected grasses to Japanese beetle grub. Entomol. Exp. Appl. 61: 285–289.
154. MANTLE, P.G. 1983. Amino acid neurotransmitter release from cerebrocortical synaptosomes of sheep with severe ryegrass staggers in New Zealand. Res. Vet. Sci. 34: 373–375.
155. ELDEFRAWI, M.E., GANT, D.B., ELDEFRWI, A.T. 1990. The GABA receptor and the action of tremorgenic mycotoxins. In: Microbial Toxins in Foods and Feeds (A.E. Pohland, V.R. Dowell, J.L. Richard, eds.), Plenum Press, NY, pp. 291–295.
156. KARIMOV, V.A., KAMILOV, I.K. 1961. Pharmacology of the new loline alkaloid and of its derivataive. Dok. Akad. Nauk. Uzb. SSR. 12: 43–47.
157. OLIVER, J.W., POWELL, R.G., ABNEY, L.K., LINNABARY, R.D., PETROSKI, R.J. 1990. N-acetyl loline-induced vasoconstriction of the lateral saphenous vein (cranial branch) of cattle. In: Proceedings of the International Symposium on *Acremonium*/grass Interactions. (S.S. Quisenberry, R.E. Joost, eds), Louisiana Agric. Exp. Stn., Baton Rouge, LA. pp. 239–243.
158. ABNEY, L.K., OLIVER, J.W., REINEMEYER, C.R. 1993. Vasoconstrictive effects of tall fescue alkaloids on equine vasculature. J. Equine Vet. Sci. 13: 334–340.
159. WESTENDORF, M.L., MITCHELL, JR., G.E., TUCKEAR, R.E., BUSH, L.P., PETROSKI, R.J., POWELL, R.G. 1993. In vitro and in vivo runinal and physiological responses to endophyte-infected tall fescue. J. Dairy Sci. 76: 555–563.

160. DEW, R.K., BOISSONNEAULT, G.A., GAY, N., BOLINE, J.A., CROSS, R.J., COHEN, D.A. 1990. The effect of the endophyte (*Acremonium coenophialum*) and associated toxin(s) of tall fescue on serum titer response to immunization and spleen cell flow cytometry analysis and response to mitogens. Vet. Immunol. Immunopath. 26: 285–295.
161. HAYEK, M.G., BOISSONNEAULT, G.A., MITCHELL, JR. G.E., BUSH, L.P., POWELL, R.G. 1991. Effect of pyrrolizidine alkaloids (loline, N-methyl-loline, N-acetyl-loline, N-formyl-loline) on the mitogen response of bovine and murine lymphocytes. FASEB J. 5: A567, Abst. #1142.
162. SIEGEL, M.R., DAHLMAN, D.L., BUSH, L.P. 1989. The role of endophytic fungi in grasses: new approaches to biological control of pests. In: Integrated Pest Management for Turfgrass and Ornamentals. (A.R. Leslie, R.L. Metcalf, eds.),U. S. Environmental Protection Agency, Washington, D. C., pp. 169–186.
163. RIEDELL, W.E., KIECKHEFER, R.E., PETROSKI, R.J., POWELL, R.G. 1991. Naturally-occurring and synthetic loline alkaloid derivatives: insect feeding behavior modification and toxicity. J. Entomol. Sci. 26: 122–129.
164. PUTNAM, M.R., BRANSBY, D.I., SCHUMACHER, J., BOOSINGER, T.R., BUSH, L.P., SHELBY, R.A., VAUGHAN, J.T., BALL, D. 1991. The effects of the fungal endophyte *Acremonium coenophialum* in fescue on pregnant mares and foal viability. Am. J. Vet. Res. 52: 2071–2074.
165. BUSH, L.P., SCHMIDT, D. 1994. Alkaloid content of meadow fescue and tall fescue grown with natural endophytes. In: International Conference on Harmful and Beneficial Microorganisms in Grassland, Pasture and Turf. (K. Krohn, V.H. Paul, J. Thomas, eds). vol 17 International Organization Biological and Integrated Control of Noxious Animals and Plants, Gent, Belgium pp. 259–265.
166. KENNEDY, C.W., BUSH, L.P. 1983. Effect of environmental and management factors on the accumulation of N-acetyl and N-formyl loline alkaloids in tall fescue. Crop. Sci. 23: 547–552.
167. BELESKY, D.P., STRINGER, W.C., PLATTNER, R.D. 1989. Influence of endophyte and water regime upon tall fescue accessions. II. Pyrrolizidine and ergopeptine alkaloids. Ann. Bot. 64: 343–349.
168. BARKER, D.J., DAVIES, E., LANE, G.A., LATCH, G.C.M., NOTT, H.M., TAPPER, B.A. 1993. Effect of water deficit on alkaloid concentrations in perennial ryegrass endophyte associations. In: Proceedings of the Second International Symposium on *Acremonium*/grass Interactions. (D.E. Hume, G.C.M. Latch, H.S. Easton, eds.), AgResearch, Grasslands Research Centre, Palmerston North, NZ, pp. 67–71.
169. WELTY, R.E., CRAIG, A.M., AZEVEDO, M.D. 1994. Variability of ergovaline in seeds and straw and endophyte infection in seeds among endophyte-infested genotypes of tall fescue. Plant Dis. 78: 845–849.
170. PRESTIDGE, R.A. 1991. Susceptibility of Italian ryegrass (*Lolium multiflorum* Lam) to Argentine stem weevil (*Listronotus bonariensis* Kuschel) feeding and oviposition. N.Z. J. Agric. Res. 34: 119–125.
171. LATCH, G.C.M., CHRISTENSEN, M.J. 1985. Artifical infections of grasses with endophytes. Annals Appl. Biol. 107: 17–24.
172. KEARNEY, J.F., PARROTT, W.A., HILL, N.S. 1991. Infection of somatic embryos of tall fescue with *Acremonium coenophialum*. Crop Sci. 31: 979–984.
173. O'SULLIVAN, B.D., LATCH, G.C.M. 1993. Infection of plantlets, derived from ryegrass and fescue meristems, with *Acremonium* endophytes. In: Proceedings of the Second International Symposium on *Acremonium*/grass Interactions. (D.E. Hume, G.C.M. Latch, H.S. Easton, eds.), AgResearch, Grasslands Research Centre, Palmerston North, NZ, pp. 16–17.
174. FUNK, C.R., WHITE, R.H., BREEN, J.P. 1993. Importance of *Acremonium* endophytes in turf-grass breeding and management. Agric. Ecosyst. Environ. 44: 215–232.

175. KOGA, H., CHRISTENSEN, M.J., BENNETT, R.J. 1993. Incompatibility of some grass-*Acremonium* endophyte associations. Mycol. Res. 97: 1237–1244.
176. LEUCHTMANN, A. 1992. Systematics, distribution, and host specificity of grass endophytes. Natural toxins 1: 150–162.
177. FLETCHER, L.R. 1993. Grazing ryegrass/endophyte associations and their effect on animal health and performance. In: Proceedings of the Second International Symposium on *Acremonium* /grass Interactions: Plenary Papers. (D.E. Hume, G.C.M. Latch, H.S. Easton, eds.), AgResearch, Grasslands Research Centre, Palmerston North, NZ, pp. 115–120.
178. MURRAY, F.R., LATCH, G.C.M., SCOTT, D.B. 1992. Surrogate transformation of perennial ryegrass, *Lolium perenne*, using modified *Acremonium* endophyte. Mol. Gen. Genetics 233: 1–9.
179. TSAI, H.-F., SIEGEL, M.R., SCHARDL, C.L. 1992. Transformation of *Acremonium coenophialum*, a protective fungal symbiont of the grass *Festuca arundinacea*. Curr. Genet. 22: 399–406.
180. TSAI, H-F., WANG, H., GEBLER, J.C., POULTER, C.D., SCHARDL, C.L. 1995. The *Claviceps purpurea* gene encoding dimethylalltryptophan synthase, the committed step for ergot alkaloid biosynthesis. Biochem. Biophys. Res. Comm. 216: 119–125.

Chapter Five

MULTIPLE DEFENSES AND SIGNALS IN PLANT DEFENSE AGAINST PATHOGENS AND HERBIVORES

Ray Hammerschmidt[1] and Jack C. Schultz[2]

[1] Department of Botany and Plant Pathology
Michigan State University
East Lansing, Michigan 48824–1312
[2] Pesticide Research Lab
Penn State University
University Park, Pennsylvania 16802

Introduction	122
Constitutive Defenses against Pathogens	122
Constitutive Defenses against Herbivores	125
Biochemical Traits	125
Physical Traits	127
Induced Defenses against Pathogens	129
Localized Responses	129
Systemic Responses	134
Induced Defenses against Herbivores	136
Localized Responses	136
Systemic Responses	136
The Nature of Putative Systemic Signals	139
Salicylic Acid	140
Jasmonates	141
Peptides	142
Oligogalacturonides	143
Interactions between Resistance to Pathogens and to Herbivores	144
New Approaches to Understanding Complex Plant-Pest Interactions	145

Phytochemical Diversity and Redundancy in Ecological Interactions,
edited by Romeo et al., Plenum Press, New York, 1996

INTRODUCTION

Because they are attacked by many fungal, bacterial and viral pathogens as well as by insects and other herbivores, it is not surprising that plants have evolved constitutive and induced defense mechanisms that deal effectively with a range of enemies.[1-5] *Constitutive defense* comprises preformed compounds that are antimicrobial and/or toxic.[2,4] These are usually natural products, but may include proteins or structural barriers. In *induced defense*, toxic or physical barriers are produced only upon infection or attack. Induced resistance responses can be either *localized* or *systemic*. *Localized responses* may be confined to one or a few cells around the point of attack. In *systemic responses*, attack by a pathogen, feeding by an herbivore, or wounding can produce changes throughout the plant resulting in elevated constitutive defenses and/or increased responsiveness upon further infection or attack.[1,3,5]

Because many of the natural products implicated in herbivore defense have also been reported to play a role in pathogen resistance,[2] it seems possible that a plant may utilize the same or part of the same set of chemicals to protect themselves against a wide range of pests and pathogens. This is indirectly supported by observations that locally and systemically induced defenses are often non-specific and multicomponent.[1,5] Thus it is possible that plant defenses may exhibit redundancy at two levels: Use of multiple types of defenses to protect a plant against a specific pathogen; and use of all or a subset of induced and preformed defenses to protect the plant against both pathogens and herbivores. In this review, we provide an overview of some of the chemical processes used by plants to defend themselves in an attempt to see if there are common defenses and, thus, support the possibility that defenses are redundant.

CONSTITUTIVE DEFENSES AGAINST PATHOGENS

Definitive evidence for an important role of preformed natural products in disease resistance is often lacking, and is much less than for plant-herbivore interactions. In addition to toxic compounds, plants have structural barriers such as the cuticle or a periderm that must be breached to allow infection to occur. Thus, preformed defenses may be either chemical or structural.

Many plant species have been examined for the presence of preformed antifungal or antibacterial activity.[2,4] One of the first reports concerned the resistance of onion bulbs to the pathogen *Colletotrichum circinans*.[4] It was known that onions that had colored outer scales were resistant to this pathogen, while those varieties with white outer scales were susceptible. The dead outer scales of the colored onions contained sufficient quantities of catechol (**1**) and protocatechuic acid (**2**) to inhibit germination, and thus infection, but the living tissues of the resistant varieties were as susceptible as the white colored susceptible varieties. Most cases of resistance based on preformed compounds are, however, not so well defined.

1. Catechol   **2. Protocatechuic Acid**

Many of the steroidal glycoalkaloids (SGAs) of solanaceous species that have been implicated in disease resistance are O-glycosides of solanidine (**3**) and tomatidine (**4**). The evidence has been based primarily on the presence these compounds in concentrations that effectively inhibit pathogens *in vitro*. SGAs also are toxic to some insects. Thus, it is possible that they play a role in host plant defense against both insects and pathogens.[6,7] However, their negative impact on insect enemies, including parasites and diseases, may reduce or eliminate any antiherbivore benefit to the plant.[8]

3. Solanidine   **4. Tomatidine**

Because SGAs are antimicrobial, attempts have been made to correlate their levels in potato foliage and tubers with resistance to several pathogens.[6,9] However, most studies have failed to find such a correlation. Zeng recently found that concentrations of steroid glycoalkaloids in the upper 0.5 mm of wounded potato tuber tissue accumulated to levels that were fungistatic to *Fusarium sambucinum*, the potato dry rot pathogen, within 3 days of wounding.[9] This increase in SGA content correlated with the development of resistance to infection. Interestingly, infection of freshly wounded potato tuber tissue with *F. sambucinum* inhibits the accumulation of SGAs.[9]

A role for the SGA tomatine, a glycoside of tomatidine, in green tomato resistance has been more firmly established in studies using mutants of the fungus *Fusarium solani* that were insensitive to tomatine.[10] These mutants were more virulent on green tomato fruit containing tomatine than were the wild type sensitive fungus. Both the wild type and the tomatine-insensitive mutants exhibited similar pathogenicity on red tomato fruits lacking tomatine. Crosses between the tomatine insensitive/virulent isolates and the tomatine insensitive/avirulent wild type isolates further confirmed the relationship between pathogenicity and insensitivity to tomatine. It was concluded that tomatine was a preformed resistance factor of green tomato to *F. solani*.

The presence of a group of saponins called avenacins (e.g., **5**) is associated with resistance to infection of oats roots by *Gaumanomyces graminis* var. *tritici*.[2] When a large number of *Avena* species were screened, only the avenacin-free *A. longiglumis* was susceptible to *G. graminis* var. *tritici*.[11] The oat pathogen, *G. graminis* var. *avenae*, is less sensitive to avenacin than is *G. graminis* var. *tritici* and produces an enzyme (avenacinase) that detoxifies avenacin.[12] The role of avenacin as a resistance factor was further demonstrated by disrupting expression of the avenacinase gene in *G. graminis avenae*. These mutants lost their ability to infect oats, but were still able to infect the non-saponin containing roots of wheat. Similar additional studies are needed to demonstrate the relative roles that preformed toxic compounds play in disease resistance.

5. Avenacin

Other preformed compounds implicated in disease resistance require modification prior to expressing activity.[4] Activation frequently involves a glycosidase that releases a toxic aglycone (or rearranges the aglycone into a more toxic compound). Examples include the cyanogenic glycosides found in many plant families and the glucosinolates found in crucifers. In both cases, damage mixes the glucoside from one cellular compartment (or type of cell) and the glucosidase from another cellular compartment (or cell type), producing the toxic material.[4] Although *in vitro* toxicity to pathogens can be demonstrated readily, the evidence that these compounds are determinants of resistance is, mostly correlative.

Phenolic compounds also have been implicated as preformed defensive substances.[4,13] Some are toxic to pathogens, as described above for catechol and protocatechuic acid in onion. However, presence of preformed phenolic compounds is probably not sufficient to account for resistance to most pathogens. Based on observations of cellular browning during the active resistance response of host cells to pathogens, it seems more likely that preformed phenols are most active in defense as reactive quinones or free radicals produced as a result of decompartmentalization of host cells and subsequent action of phenoloxidases and peroxidases on the phenol.[3,13] Responses like these may be mistaken for *de novo* synthesis of the final products or precursors if analytical methods are not carefully designed.

In summary, most plant species contain constitutive natural products that are antimicrobial *in vitro*. However, there is little direct evidence that these compounds actually protect plants *in vivo*. This may be the case because the majority of research has focused on *induced* defenses against disease.

CONSTITUTIVE DEFENSES AGAINST HERBIVORES

Biochemical Traits

In apparent contrast to plant pathology's history, the great bulk of research done on plant defenses against herbivores has until recently focused on the role(s) of constitutive traits, mainly natural products.[14] There is ample *in vivo* evidence to support a defensive role for chemical constitutive plant traits, usually including correlative dose-response relationships. Recent reviews list at least 25 studies establishing such relationships.[15,16] Fewer than half this number include the kind of genetic analysis seen in plant-pathogen studies, although there are many more studies illustrating heritable resistance without identifying the constitutive mechanism responsible.

Constitutive defenses generally have one of several effects on herbivores. They may be antifeedant, which provides the plant with the greatest potential protection, since damage may be prevented almost before it begins. They may be acutely toxic, which has the potential to stop damage quickly, or they may have multiple chronic effects, including reducing digestion, chronic toxicity, etc. The last offers the least benefit to the plant, since considerable consumption can occur. In addition, there are 'indirect' effects such as altered susceptibility to the herbivore's enemies (predators, parasites, disease). These effects are less well understood than others, but may be the most important of all.

One of the best-studied examples of largely constitutive defense against herbivores is the production of furanocoumarins by *Pastinaca sativa*, wild parsnip. Berenbaum and coworkers have shown that resistance to the parsnip webworm (*Depressaria pastinacella*) is determined quantitatively by the concentration relationships among three furanocoumarins (bergapten (**6**) and its isomer xanthotoxin (**7**), and sphondin (**8**)) and nitrogen in flower head tissues.[15] They isolated the effects of these components using artificial insect diets, demonstrating this mechanism of resistance. The furanocoumarin levels are highly heritable and comprise 'absolute' barriers to exploitation of this plant by other nonadapted herbivores,[17] so that more- and less-resistant genotypes occur in nature.[18,19] However, levels of furanocoumarins and nitrogen are also profoundly influenced by light and nutrient availability, so that resistance of an individual parsnip plant may only partially reflect its genotype.[20] Moreover, Zangerl et al. subsequently discovered that increases in some *P. sativa* furanocoumarins are induced by insect damage.[21] Now we are left to estimate how much resistance is constitutive, how much induced.

6. Bergapten   **7. Xanthotoxin**   **8. Sphondin**

Alkaloid production is generally strongly heritable and often responsible for resistance to various herbivores.[22] Lupine populations differing in alkaloid composition are differentially susceptible to particular herbivores.[23] Much of this variation is thought to represent genotypic variation.[20] Constitutive nicotine concentration in tobacco is controlled in large part by a single gene regulating degradation, and is responsible for the plant's resistance to all but specific nicotine-adapted herbivores,[24] as well as relative resistance to adapted herbivores like the tobacco hornworm (*Manduca sexta*), whose detoxification metabolism is saturable.[25] Tomatine content in wild and cultivated tomato, in contrast, is controlled by two codominant alleles,[26] while glycoalkaloid content in potato appears to be under polygenic control.[7] The latter, however, is a mixture of diverse molecules, the production of which may be simply regulated. In all cases, alkaloid content is profoundly altered by environmental factors.

Production of individual monoterpenes is generally controlled by few genes in several genera across several families, from mints (*Mentha*) to trees (*Pinus*). Genotypic variation in monoterpene composition has been associated with differential plant susceptibility to various herbivores.[27] Environmental and induced sources of variation are also significant in most of these systems.

Presence/absence of the biologically-active tetracyclic terpenoids cucurbitacin B (**9**) and the structurally related cucurbitacin E is under control of one or few genes in cucumber (*Cucumis sativus*), squash (*Cucurbita pepo*), and watermelon (*Citrullus vulgares*), (reviewed in Kennedy and Barbour).[22] These chemicals are repellant to all but specifically adapted herbivore species, several of which perceive them as attractants and feeding stimulants. Hence, their presence confers plant resistance to a wide range of potential herbivores, but their levels determine relative attractiveness to adapted species. Since there remains some evident resistance variation among high-cucurbitacin plants to even the "best-adapted" insect species, other resistance mechanisms must also be present. Cucurbitacin content is also influenced by herbivory and environment. The system illustrates the complexity of assigning "resistance" levels to even a well-known plant defense system. Although constitutive cucurbitacin concentrations are clearly a factor in cucurbit resistance to herbivores, they are not the only factor involved, and their levels are not fixed, but vary with environmental factors, including damage.

9. Cucurbitacin B

Most classes of plant natural products have at least some compounds that have either significant antimicrobial or antiherbivore activity, or both, at least *in vitro*.[1-5] Because plants are simultaneously dealing with pathogens and herbivores, and because herbivores are often infected by their own pathogens, the realized impact of these constitutive "defenses" is difficult to predict.[8] To the extent that their modes of action are understood, some of these molecules can act against microbial and herbivore cells in common ways. Hence, it is difficult to determine which chemical may be primarily or exclusively a defense against pathogens, and which against herbivores. The evidence suggests that many are both.

Physical Traits

Although physical defenses in plants are evident to anyone tangling with a rose bush, there have been few studies of structural resistance mechanisms at the cellular/subcellular levels. Because many animals feed in compensatory fashion, often consuming more of tissues lacking nutritional value, increased lignification/suberization and resulting decreased nutritional content are likely to exacerbate a plant's problems with herbivores, not solve them. Only smaller, invasive herbivores, such as internal-feeding stem-borers and leaf-miners, or phloem-feeding aphids, seem susceptible to physical exclusion.

Nonetheless, general tissue toughness, deposition of silica, calcium carbonate, or lignin around vascular bundles or throughout tissues, and the woody habit have all been explained at least in part as antiherbivore devices.[28] Caswell and Reed argued that indigestible lignified vascular bundles make C4 plants less valuable to many insects and thus such plants are exploited less.[29] The effectiveness of these traits in protecting plants is controversial, in part because they are often co-correlated with biochemical traits that also have defensive properties and few have been subjected to genetic analyses.

Stem toughness is a heritable trait conferring protection on wheat plants against the wheat stem sawfly.[30] Similarly, lignification, number of vascular bundles, and related traits provide resistance in sugarcane to the sugarcane borer, and tough lignified vascular bundles block invasion of *Cucurbita* stems by the squash vine borer.[28] All of these traits are modified by environmental conditions, and demonstrations of their singular effectiveness are rare.

Leaf hairs and trichomes are plant structures that have a heritable basis and can be shown to confer protection on their bearers. Very small insects may be unable to reach valuable plant tissues without consuming leaf hairs; young cereal leaf beetle larvae are killed by doing so.[31] Several pubescent agricultural cultivars are more resistant to various insect pests than are glabrous genotypes, because of either feeding or oviposition inhibition, or entrapment by hooked trichomes.[24] On the other hand, these effects can be worse for predators of parasites than for their herbivore hosts, increasing the susceptibility of plants to their pests.[32]

Perhaps the best-studied "physical" defenses are actually "physico-chemical" in nature. Density and type of glandular trichomes are heritable traits, as is the nature of their chemical contents. Various genotypes of tomato, tobacco, potato, and cotton produce large numbers of glandular trichomes that rupture when contacted by insects, producing a rapidly-oxidized phenolic mixture that darkens and hardens upon exposure to air, immobilizing even-moderate-sized insects.[18] Some of the glandular contents may also have direct toxic activity. In geranium (*Pelargonium* spp.), glandular trichomes contain a metabolic elaboration of anacardic acid having fatty acid side chains of variable length (e.g. **10**).[33] Various forms of the molecule are differentially sticky upon exposure to air, depending upon ambient temperature and side-chain length.[33] Biosynthesis of these anacardic acid derivatives and their sidechain length are under genetic control, and the genes responsible for alternative forms have been isolated and cloned.[33] Because small insects and mites are trapped and killed by these sticky exudates, alternative geranium genotypes are differentially resistant, each within a different temperature range. This provides an excellent example of mechanisms underlying gene-by-environment interactions.

Glandular trichomes also exemplify a problem with analysis of "physical" defenses. Glandular contents often may have direct biochemical antibiotic activity and it is extremely difficult to separate chemical effects from those of purely physical barriers. For this reason alone, the role of strictly physical barriers to herbivory remains in doubt.

10. Anacardic Acid

INDUCED DEFENSES AGAINST PATHOGENS

Although preformed defense compounds play a role in resistance in a few plant-pathogen interactions, post-infection induced responses appear to be most common.[1,3] A brief description of the various types of localized defenses will be described followed by a limited discussion of the induction and expression of systemic resistance.

Localized Responses

Resistance to specific pathogens is often controlled by one or a few "major" genes in the host plant and a gene for avirulence in the pathogen.[3] The interaction of the products of the resistance and the avirulence genes allows the host plant to recognize the presence and identity of an attacking pathogen. Recognition then results in the expression of the various defense responses.

Hypersensitivity. Upon infection by an "avirulent" form of a pathogen, the infected plant cell often undergoes rapid death that is accompanied by the induction of various local defense responses. This type of cell death is the *hypersensitive response* (HR) and the biology of this phenomenon has recently been reviewed extensively by Goodman and Novacki.[34] In general, it is believed that HR is triggered after recognition of the pathogen by the host cell. Associated with the HR is the localized induction of an array of defenses that include phytoalexin production, accumulation of active oxygen species, *pathogenesis-related* (PR) protein synthesis and cell wall modifications. The death of the plant cell itself may be an effective defense against obligate pathogens such as mildews and rusts that need a living plant cell for growth and reproduction.

Phytoalexins. Phytoalexins are low molecular weight, antimicrobial compounds that accumulate after infection.[35] They have been isolated from numerous plant families and represent a variety of chemical structures.[36] In general, each plant family produces phytoalexins of one chemical class. For example, members of the Solanaceae produce sesquiterpenoids, legumes produce isoflavonoid or pterocarpan phytoalexins, and crucifers produce indoles substituted at the 3-position. This uniformity of phytoalexin type within families, however, is not universal. For example, the graminaceous monocots produce phytoalexins of rather diverse chemistry (terpenoids, phenylpropanoid derivatives, deoxyanthocyanidins).[36]

The precise timing, location and concentration of phytoalexin content is important in establishing a role for the compounds in defense.[13,36] This is because greater total amounts of phytoalexins may accumulate in compatible as compared to incompatible interactions. A recent example of the localization of phytoalexins in relation to the cessation of fungal growth comes from the work of Nicholson and co-workers on the sorghum-*Colletotrichum graminicola* interaction. In the resistance response the host produces orange- to red-colored phytoalexin luteolindin (**11**).[37] Combined microspectrophotometry and HPLC analysis, demonstrated that more than sufficient quantities of luteolindin accumulated in the infected host cell to account for cessation of fungal growth.[37] The fluorescence of some of the phytoalexins produced in the interaction of cotton leaf cells with *Xanthomonas campestris*[38] and oat cells with *Puccinia coronata*[39] have allowed similar types of quantification.

11. Luteolindin

One line of evidence that supports a role for phytoalexins in resistance is based on the correlation of tolerance or insensitivity of virulent strains of pathogen to phytoalexins produced by their host plant. There are two excellent examples of this. The virulence of *Nectria haematococca* (*Fusarium solani* f.sp. *pisi*) on pea (*Pisum sativum*) has been linked to its tolerance of and ability to detoxify the major pea phytoalexin, pisatin.[40] Similarly, virulence of the potato tuber dry rot pathogen *Gibberella pulicaris* (*Fusarium sambucinum*) is dependent upon the ability of this fungus to detoxify the phytoalexins, rishitin and lubimin.[41] In both cases, the relationship between virulence and the ability to detoxify the phytoalexins has been confirmed by genetic analysis. These studies demonstrate that unless an isolate of the pathogen can detoxify the phytoalexin, the fungus is avirulent.

Another approach is to test the role of phytoalexins using mutants of plants unable to produce phytoalexins. Tsuji et al. recently reported that *Arabidopsis* produces the indole derivative, camalexin (**12**), as a putative phytoalexin after challenge with an incompatible bacterium or treatment with silver nitrate.[42] The kinetics of camalexin accumulation in relation to growth of the bacterium in the plant suggested a defensive role. Glazebrook and Ausubel[43] have attempted to test the role of camalexin in the resistance of *Arabidopsis* to bacterial pathogens by producing mutants that are no longer capable of producing camalexin or which produce lower amounts after inoculation. They identified three distinct phytoalexin deficient (*pad*) mutants (*pad1, pad2* and *pad3*). Of these, *pad3* is a non-producer of camalexin while the other two produce less of this phytoalexin than wild type. Inoculation of the mutants with an avirulent race of *Pseudomonas* did not result in compatibility. However, it is possible that other microbial defenses were enhanced to compensate for the lack of camalexin, that antimicrobial intermediates in the camalexin pathway accumulated, or that camalexin is not important in the resistance to *Pseudomonas*.

12. Camalexin

Cell Wall Changes. Pathogens also must contend with the cell wall of the host. The cell wall would appear to be a good barrier to infection, but most pathogens either have an arsenal of enzymes to degrade the wall polysaccharides and/or use mechanical pressure to penetrate the wall or pass through the middle lamella.[44] Plant cells often respond to infection by modifying cell walls near the pathogen with materials that are thought to make the walls more resistant to penetration or enzymatic degradation.[45] Ride has proposed several possible means by which cell wall modifications may play a role in defense, and each of these should be testable.[46] These include: increasing the mechanical strength of the wall; increasing the resistance of the wall to enzymatic degradation; producing of antimicrobial precursors of wall modifying agents; providing diffusion barriers against nutrient flow to the pathogen and/or diffusion of toxins or enzymes from the pathogen to the host cell protoplast. In addition to static resistance mechanisms provided by a modified wall, the biochemical processes involved in cell wall modification may be important in defense.

One of the first cell wall modifications studied was the deposition of lignin or a lignin-like phenolic material in the wall after infection.[47] Although no direct measurements of increases in the mechanical properties of pathogen-induced lignin have been made, cell walls lignified in response to pathogen invasion are more resistant to degradation by cell wall-degrading enzymes.[46] Thus lignification may play a role in resistance by inhibiting cell wall and middle lamellar degradation needed for pathogen growth through the tissue.

Lignification also may comprise an active defense. The lignin precursors coniferyl alcohol and coniferyl aldehyde have antifungal properties and appear to function as phytoalexins in flax.[48] Hydrogen peroxide and phenolic free radicals produced during lignification also may directly inhibit pathogen growth.[46] Peroxidases were capable of generating toxic levels of hydrogen peroxide *in vitro*.[49] Fungal hyphae can be lignified *in vitro*,[50] and cytological studies showed that fungal hyphae can be encased in a lignin like material after infection into plant tissues.[50,51]

Oxidases and Activated Oxygen. Active oxygen species such as hydrogen peroxide, hydroxyl radical and superoxide, appear to be important factors in resistance to pathogens.[52–54] One of the earliest biochemical events in the hypersensitive response is an increase in oxidative potential and production of active oxygen species,[48] the so-called "oxidative burst".[52] Active oxygen species are believed to function in resistance by: increasing host cell wall resistance to hydrolytic enzymes by the crosslinking cell wall polymers;[55] acting as antimicrobial factors;[49,52] and acting as local signals involved in the induction of defense genes.[56,57] This will be discussed later in this review.

Induced defense reactions in plants usually include increases in oxidative enzymes such as peroxidase (POD)[58] and polyphenoloxidase (PPO).[59] Polymerization of lignin precursors into lignin and crosslinking of hydroxyproline-rich glycoproteins into cell wall are two possible functions for POD.[58] In addition, cell wall-associated PODs are also involved in the production of the hydrogen peroxide needed for lignin formation and wall protein crosslinking. Peroxidases, however, often increase in activity after many pathogenic and non-pathogenic stresses and may occur as numerous isozymes.[58] Thus, it is important to establish which and how many different PODs are being elicited by pathogen exposure, whether or not the substrate needed for a particular POD is also produced, and what alternative enzyme products may be in evidence. Without this information, evidence for *in vivo* POD activity may say little about the function of these enzymes in defense. This type of analysis is rare.

Phenoloxidase (PO) activity also increases after infection, but its role in defense is unclear.[59] POs, which are often sequestered in the cell away from the phenolic compounds that they can oxidize, may be important in some of the early defense responses that are associated with cellular browning. It is likely that the brown material that accumulates in cells undergoing HR is the result of previously compartmentalized phenols and POs mixing together as the internal membranes of the cell lose integrity early in the response. In support of this, Lazorovitz and Ward demonstrated an increase in PO activity in soybean cells undergoing HR even though there was no increase in extractable phenoloxidase.[60] How this brown material functions in defense is not known, but it would seem that this end product of phenol oxidation may be less important than the reactive quinones that can react with proteins and other quinones as well as being more toxic to microbes than the unoxidized parent phenolic compounds. This is an area that needs more study.

Pathogenesis Related and Other Proteins. Active resistance responses also have been associated with the accumulation of a number of proteins called pathogenesis-related or PR proteins.[61,62] These proteins initially were thought to be involved in resistance because their appearance correlated with the expression of hypersensitive resistance. More recently, various defensive functions have been ascribed to some PR proteins.[62]

Two of the PR protein classes (PR2 and PR3) have been identified as β1,3-glucanases and chitinases.[61,62] Because the true fungi contain chitin and glucans in their cell walls and the oomycetes contain predominantly β-1,3-glucans, it has been hypothesized that these enzymes may be active against fungi. *In vitro* assays have demonstrated that the enzymes exhibit antifungal activity.[62] Transgenic plants expressing higher levels of chitinases are more resistant to certain fungal infections.[63]

PR1 is frequently associated with induced defenses and has served as a marker for the induction of resistance in tobacco.[64,65] Alexander et al. reported that transgenic tobacco plants expressing PR-1 are more resistant to oomycete pathogens than the non-transformed controls.[66] The basis for this enhanced resistance may be related to the observation that a basic PR-1 from tobacco was inhibitory to the oomycete *Phytophthora infestans* in *vitro* assays.[67]

Systemic Responses

Challenging plants with pathogens that cause necrotic lesions often results in systemic acquired resistance (SAR) to a broad range of subsequent pathogen types, irrespective of the challenger's identity.[1,64,65,68,69] These observations suggests that multiple defense mechanisms or a single mechanism with overlapping effects have been induced. There systemic mechanisms appear to be the same as those used in localized resistance responses. Some may be activated as a result of the inducing inoculation, while others are rapidly induced only after a subsequent inoculation with a virulent pathogen. For example, in tobacco, inducing SAR results in the systemic expression of PR proteins and a plant-wide increase in POD and PO activity.[62,65] This increase is viewed as an *elevation of "constitutive" defenses* because the PR proteins are expressed as a result of induction and before any subsequent infection. In bean, inducing SAR results in the ability of the invaded host cells to undergo a hypersensitive response to subsequent fungal infection and more rapidly synthesize phytoalexins.[70,71] In cucumber, the inducing SAR causes expression of certain extracellular enzymes,[72–76] as well as the ability to modify cell walls more rapidly upon subsequent infection.[50,51] In these cases, SAR also includes *enhanced induceability*. Systemic acquired resistance employs the same types of defenses that are used in localized resistance responses,[1,68,69] and the acquired resistance phenomenon demonstrates that functional defenses exist even in plants without "genes for resistance" recognition.[1] Systemic acquired resistance has been studied in detail in only a few plants, which include green bean,[70] cucumber,[72] tobacco,[77] and *Arabidopsis*.[78–81] We will concentrate on cucumber as an example because much is known about the biology of SAR in this plant.

SAR was first reported in cucumber by Kuć et al.[82] The inoculation of one leaf of a plant susceptible to the anthracnose fungus *Colletotrichum lagenarium* resulted in the systemic development of resistance to subsequent infection by the same pathogen. This study was followed by others that showed resistance could be induced by and against a number of cucumber pathogens (reviewed in Hammerschmidt and Yang-Cashman).[72] The only common feature in this spectrum of diseases is that the effective resistance-inducing pathogens caused necrotic lesions.

The defense mechanisms utilized in cucumber SAR have been studied for only a few of the pathogens. A histological study by Richmond et al.[83] revealed that at least part of the resistance expressed against *C. lagenarium* was based on a failure of the pathogen to infect the host tissue successfully. Although fungal conidia germinated and appressoria were formed, few were successful in penetrating the epidermal layers of plants with acquired resistance. No obvious host response was observed. Thus, the lack of disease development could be attributed to a lack of infection, but the study did not explain how the induced plant acted to prevent penetration. Histochemical techniques were used to show that a-lignin-like polymer was deposited under many of the appressoria that did not penetrate, and concluded that this prevented the fungus from penetrating the host.[50] Rapid epidermal lignification was also found associated with induced resistance to other fungal pathogens.[72,84,85] The structures that appear to block penetration of induced tissues contain callose,[86] a common defense-related polymeric cell wall glucan[45] and silicon.[51] Thus, at least part of acquired resistance appears to involve multiple cell wall modifications and could be considered to be a type of defensive redundancy.

How lignin and the other wall modifications function in this resistance response is not known, but it is likely that changes in the mechanical strength or ability to be enzymatically degraded may be involved.[46] Both possibilities need to be tested. The inhibitor of cinnamyl alcohol dehydrogenase and lignification inhibitor, OH-PAS,[87] was used to test the role of lignification. Treatment of tissue with OH-PAS resulted in a failure of the host cells to lignify rapidly upon challenge inoculation, and resulted in more extensive colonization of the tissues by penetrating hyphae (Yang-Cashman and Hammerschmidt, unpublished results).

Lignification may also slow the development of hyphae that have successfully penetrated the outer epidermal wall. Lignin-positive materials can be seen in both the cell wall and the cytoplasm of the cell (Yang-Cashman and Hammerschmidt, unpublished results). This process may act to slow the pathogen by trapping it in the invaded cell. We have also observed that the hyphal tips in the host cells often become lignified.[50,51] Lignification of the hyphal tip would effectively stop growth of the fungus, and thus inhibit further disease development.

As a result of induced systemic resistance in cucumber, there is also an increase in an apoplastic chitinase (a PR protein)[75] and a group of peroxidases.[73,74,76] These enzymes are useful as markers to indicate when systemic resistance is developing. The actual purpose of the enhanced activity of these enzymes has yet to be determined conclusively[72]. Lipoxygenase (LOX) activity also increases in cucumbers expressing SAR.[88] This enzyme may be important in the generation of antifungal lipid peroxides or lipid oxidation products.[89] Additionally, increased lipoxygenase activity may result in the synthesis of other signal molecules such as jasmonic acid.[90]

Systemic induction of POD and LOX activity suggests that activated oxygen species may also be part of SAR expression. This is supported by induction of local and systemic resistance in potato foliage to *P. infestans*, which is accompanied by an increase in superoxide-generating activity and in superoxide dismutase, which may be involved in the conversion of superoxide into hydrogen peroxide.[91] Klessig and co-workers have concluded that the putative SAR signalling molecule, salicylate, inhibits catalase, permitting an accumulation of hydrogen peroxide and other active-oxygen species which then act as second messengers.[56,57] This idea, however, recently has been challenged.[92-94] Although not addressed by Chai and Doke,[91] in tobacco, the systemic increase in activated oxygen also may function in the strengthening of cell walls by crosslinking hydroxyproline-rich glycoproteins.[95]

INDUCED DEFENSES AGAINST HERBIVORES

Localized Responses

There are many examples of changes in plant biochemistry elicited by herbivore damage (or wounding) *via* the transformation of inactive metabolites into more active forms. As described above, many natural products exist in plant cells in glycosylated form, and are converted to aglycones upon contact with appropriate enzymes when cell organization is disrupted. One of the best-studied examples is cyanogenesis, in which cyanogenic glycosides are converted by glycosidases of plant or herbivore origin to highly toxic hydrogen cyanide when cells are crushed.[96] Similar examples include hydrolysis of glucosinolates in the Cruciferae to form thiocyanates, isothiocyanates and related compounds,[97] hydrolysis of cucurbitacin glycosides to cucurbitacins in cucumber,[98] and conversion of the phenolic glycosides salicortin and tremulacin into the feeding deterrents and toxins salicin, tremuloidin and related compounds in *Populus* species.[99] There is no evidence that these "responses" spread beyond the site of wounding.

The phenolic contents of many leaves are exposed to oxidative conditions when cells are disrupted.[100] Contact with air, POD, PO and other oxidative enzymes certainly must generate significant, localized oxidative transformation of these molecules.[100] The consequences could range from lignification, cell wall toughening, and formation of polyphenolic polymers, to the creation of toxic or distasteful oxygen radical and quinone forms. These reactions have not been studied on a localized basis, although Felton et al. have evidence that such transformations take place in soybean leaves upon herbivore attack, including an "oxidative burst" resembling that seen in response to pathogens.[101]

Bioassay evidence is equivocal on the significance of localized responses. Some investigators find that some herbivore species avoid recently-damaged leaf material, while others disagree.[102,103] This is not surprising, given the number of reactions that are possible, even within one wounded leaf,[104] and the diversity of adaptations among herbivore species. Some responses may make wounded leaves more attractive to herbivores, and less attractive to others. We are not aware of studies showing that highly localized lignification, as seen in pathogen defense, creates a physical barrier to further feeding. Mauricio et al.[105] and Edwards et al.[106] showed that plants tolerate widely dispersed leaf damage, as would result from locally-induced antifeedants, better than they do concentrated damage. Overall, localized reactions to wounding are probably common, but their biological significance has received little attention, and they may frequently be mistaken for *de novo* synthesis of the final products.

Systemic Responses

Although systemic responses to herbivory have been hypothesized for decades, detailed study has accelerated in recent years. Most species studied have been shown to exhibit either some widespread biochemical change or decreased value to herbivores. Hence, the types of responses are extremely diverse, and the details and redundancy of responses are known for few systems. No herbivore-response system is as well studied as the excellent SAR models.

A specific-elicitor model of plant-herbivore interactions has not become as well established as its equivalent in plant-microbe interactions. An alternative model based on damage-caused shifts in source-sink relationships and in allocation of biochemical substrates, however, has been widely adapted.[107] According to this view, the loss of tissues to herbivores results in a reduction in resources available for growth or maintenance. Compensation for these losses is achieved through increased resource acquisition by remaining tissues and by mobilizing stored reserves. If damage limits growth, the newly-gained or remobilized resources are invested in putatively-defensive secondary metabolites.[107] According to this "resource allocation theory", the plant's response to defoliation (for example) is determined largely by biochemical or metabolic constraints. Increases in resistance arise because of inevitable allocation shifts, not as specific responses targeting a particular threat.

The alternative view, sometimes called "optimal defense theory", is that plants respond to herbivores and wounding in an active, presumably adaptive way.[108] Such a response would require the plant to recognize that it has been wounded and to organize and mobilize systemic changes that provide resistance. This model closely resembles the SAR resulting from pathogen attack.

It is unclear which of these views is most accurate, but each probably contains a kernel of truth. It is possible that the "truth" may depend on the plant species and type of metabolic changes observed. For example, increases in leaf phenolics seen following defoliation are consistent with "resource allocation theory", since portions of phenolic metabolism in plants are thought to be regulated by the availability of substrates unused in primary metabolism.[109] Hence, loss of leaves might: 1) remove carbon and nitrogen necessary for photosynthesis and growth; 2) reduce growth; and 3) provide additional "overflow" carbon substrates for secondary metabolism from enhanced photosynthesis or from stored reserves. This could explain responses by trees to heavy defoliation or browsing (e.g., Schultz and Baldwin[110] and Tuomi et al.[107]).

In contrast, carefully plucking leaves at the petiole usually fails to elicit wound responses, and many wound responses, including increased phenolic metabolism, can be elicited by minor damage. Baldwin and Schultz observed altered phenolic metabolism in poplars receiving less than 5% leaf area removal by tearing but no changes in 'plucked' trees.[110] Even less damage is needed to induce alkaloid synthesis in tobacco.[111] "Wound-induced genes" (*win* genes) are expressed in response to squeezing poplar leaves with pliers.[112] The induction observed in these and similar studies cannot be explained solely by altered allocation patterns and source-sink relationships.

Some plant responses are elicited specifically by insects and many plant tissues release volatiles into the air when damaged. However, Turlings et al.[113] have shown that the suite of volatiles emitted by insect-damaged plants is distinct from the normal 'green leaf odor' release. When various arthropods are the agents, or when insect regurgitant is added to an artificial wound, volatiles attractive to parasites or predators of the damaging herbivores are enhanced in the emitted mix.[113,114,115] The widespread enzyme, β-glucosidase, has been identified as a potential elicitor in insects.[116] Although this work suggests that plants recognize and respond specifically to herbivores, none of the studies has completely eliminated the possibility of contamination—and hence elicitation—by microbial products. Perhaps some of the responses are actually components of classical HR or SAR. Hartley and Lawton reported greater-up-regulation of PAL in birch leaves adjacent to leaves damaged by insects or scissors than in leaves adjacent to leaves wounded with sterilized scissors.[117] In any case, much responses are not readily explained by "resource allocation theory".

Only one wound response has been described in as detailed fashion as SAR against pathogens. Many plants synthesize protease inhibitors (PIs) in response to wounding. In tomato, transcriptional regulation of PI genes is accomplished *via* a complex web of elicitors and signals involving pectic fragments, upregulation of lipoxygenase (LOX), accumulation of jasmonic acid (JA, **13**), and enhanced oxidative activity throughout the plant.[90] Accumulation of JA or its methyl ester (MeJA, **14**) is related (as cause or effect) to expression of a gene coding for a small protein, "prosystemin", which is enzymatically degraded to a small peptide, "systemin".[69,90] Systemin is mobile in the vascular system and appears responsible for the eventual increased expression of PI genes and increased PI concentrations. Systemin is the only protein identified as a systemic signal in plants. Its role in PI induction seems clear but early events are less clear. Neither oligosaccharides nor JA appear to be very mobile in vascular tissues, and regulation of the LOX pathway leading to JA production is complex, with several other products of unknown activity. All the steps leading to production of prosystemin are not clearly defined. Overall, this response resembles SAR in many ways, and may have several biochemical steps in common with it. The PI response clearly is not a resource reallocation phenomenon.

Several biochemical phenomena appear common to wound/herbivore responses. Not only is systemic signalling a common theme, but a limited set of signals (e.g. jasmonates, ethylene, abscisic acid) are common to many systems.[69] As might be expected in any gene regulation phenomenon, cell membrane depolarization and rapid cation (especially Ca^{++}) fluxes are observed frequently, together with "oxidative bursts" and upregulation of LOX, POD, PO, PPO, and various phenylpropanoid enzymes. Many of the same mechanisms also are central to SAR. The functional similarities lead us to anticipate potential linked consequences of plant responses to either microbes or herbivores.

Many known natural products pathways can be influenced by wounding, and produce various proteins, such as mentioned above.[118] The biological activities resulting from these changes can be diverse, and may include positive, neutral, and negative consequences for a plant. For example, the terpenoid cucurbitacins (Cucurbitaceae), among the first to be shown to increase in response to insect damage,[119] are highly antifeedant to most insects, but are attractive to those species adapted to them.[98,120] Increased cucurbitacin concentrations can be elicited by wounding, bacterial infection by *Erwinia tracheiphila*, and drought.[121] As a consequence, the plant becomes more attractive to cucurbitacin-adapted insects.[122] It seems reasonable to anticipate an antipathogen role for cucurbitacins. Besides being elicited by *E. tracheiphila*, they are active against fungi such as *Botrytis cinerea*.[123]

The cucumber system illustrates our uncertainty about the causes, mechanisms, and consequences of multiple responses. Cucumber's interactions with pathogens comprise one of the best-studied SAR systems,[72] and its cucurbitacin induction is one of the oldest and most intensely-studied responses to wounding.[118,120] Much also is known about several phytopathogen-vector relationships.[121] However, other common anti-phytopathogen or anti-herbivore responses such as proteinase inhibitors, lipoxygenase, saponin production, and "oxidative bursts" are also present, but relatively unstudied. It is not clear that wound-induced increases in cucurbitacins are systemic.[98]

One can conclude that systemic plant responses to herbivores probably share many steps with SAR involving pathogens, but too few systems have been studied in enough detail to understand the relationships among them. Systemic wound responses are organized by signals, as is SAR, but the protein signal involved in the tomato PI response is unique. Because similar pathways can yield overlapping results, because so many secondary metabolites induced by herbivory are also antimicrobial, and because elicitors for wound responses are generally unknown, it is not clear how many "wound- or herbivore-induced" responses may actually be elicited by microbial products. It also is not clear that all of these responses benefit the plant. Since they may influence microbial and other agents controlling herbivores, wound responses may actually benefit herbivores more than host plants.[8,124]

THE NATURE OF PUTATIVE SYSTEMIC SIGNALS

The fact that localized injury or pathogen attack can result in systemic changes in resistance to herbivores or pathogens indicates that a signal must be generated at the site of the initial injury or infection which is then translocated through the plant. Induction of resistance both above and below the inoculation or wounding site suggests that the signal is translocated in the phloem. This is supported by the observation that killing the phloem of the leaf used for induction by heat girdling the petiole blocks the development of systemic resistance.[125] Steam girdling tobacco stems prevents systemic increases in nicotine content following wounding.[111] Similarly, killing phloem of petioles of other leaves blocks development of resistance in those leaves.[126]

Support for the pathogen-inoculated leaf as the source of the systemically translocated signal came from the work of Dean and Kuć.[127,128] Excising the inoculated leaf before necrotic lesions appeared resulted in no expression of SAR. Furthermore, if the inoculated leaf was removed after lesions appeared, systemic resistance was still expressed. If non-induced scions were grafted onto induced plants with an inoculated leaf, the scions developed acquired resistance. However, if the inoculated leaf was removed prior to the grafting, the scion did not develop resistance even though the rootstock remained resistant. Collectively, these studies indicated that the inoculated leaf was the source of the translocated signal for acquired resistance. Baldwin used similar means to show that a signal exits tobacco leaves several hours after wounding, which results in enhanced nicotine synthesis in the roots.[111] Davis et al. have demonstrated evident signal translocation from wounded to unwounded poplar leaves, and expression of "wound-induced genes" is stimulated only in target leaves on the same phloem trace.[129]

An abundance of research on systemic responses to both pathogens and wounding indicates participation of systemic and local signals. Several candidate signals have been identified.

Salicylic Acid

Treatment of tobacco plants with acetylsalicylic acid (aspirin) was first demonstrated by White to induce resistance to TMV in tobacco.[130] Later, it was demonstrated that salicylic acid (SA, (15)) could induce resistance in cucumber to *C. lagenarium*.[131] Exogenous application of SA is also a well-known inducer of PR proteins.[132,133] This suggests that SA may be an endogenous signal for resistance, as was reported for both cucumber[134,135] and tobacco.[136,137] In both examples, the amount of SA in non-infected tissues rose prior to the onset of resistance.

Treatment of cucumber plants with exogenous SA was also shown to induce two markers for systemic acquired resistance: an apoplastic acidic chitinase and three apoplastic acidic peroxidase isoforms. Chitinase transcripts are strongly induced by SA at concentrations that induce resistance.[138] Similarly, Rasmussen et al. have demonstrated that SA will induce both peroxidase activity and peroxidase transcript accumulation.[74] In addition, the increase in SA in the phloem of cucumber precedes the increase in acidic peroxidase activity.[140]

Some of the most convincing evidence for a role for SA in the induction of resistance was reported in a series of papers by Ryals and co-workers. Gaffney et al. transformed tobacco with a bacterial gene that converts SA to catechol (SA hydroxylase encoded by the *nah*G gene).[139] They found that the plants could no longer be systemically induced and did not accumulate SA after infection with TMV. Interestingly, TMV infection resulted in larger necrotic lesions. It was subsequently found that *nah*G-transformed tobacco and *Arabidopsis* plants were more susceptible to a number of pathogens.[140]

The fact that the signal for systemic resistance appears to be phloem-mobile[125,126] and the observation that enhanced amounts of SA are found in phloem exudates during the induction of resistance[134–136] suggested that SA might be the translocated signal. It was proposed instead that SA accumulates in the cucumber phloem in response to another signal translocated throughout the plant.[125] It has been previously found that SAR developed in cucumber within twenty-four hours of inoculation with the non-pathogen of *P. syringae* pv. *syringae* if the inoculated first leaf was left on the plant for as little as six hours.[141] Using *P. syringae* pv. *syringae* to induce systemic resistance, Rasmussen et al. reported that detaching the *P. syringae* pv. *syringae*-inoculated leaf at four hours after inoculation resulted in the systemic accumulation of SA in the phloem, even though no increases in SA were detected in the phloem exudate of inoculated leaf at the time it was detached.[135] Allowing the *P. syringae* pv. *syringae*-inoculated leaf to remain on the plant for as little as six hours produced increases of over 200 μM salicylate at twenty-four hours post inoculation. Salicylate levels in phloem exudates of the detached *P. syringae* pv. *syringae*-inoculated leaves did not increase in inoculated leaves until 8 hours after inoculation.[135] These results indicated that SA is induced throughout the plant by another signal generated at the infection site.

Evidence that SA is not the translocated signal was further supported by a recent study by VerNooij et al. who grafted wild type tobacco scions onto *nah*G transformed tobacco rootstock.[142] They hypothesized that if SA was the translocated signal, then inoculation of the *nah*G rootstocks would not systemically protect the wild type scions. TMV infection of the *nah*G root stocks did, however, result in the development of resistance. These results indicated that SA was not the translocated signal because the rootstocks could not accumulate SA. Shulaev et al., however, recently reported that much of the salicylic acid that accumulated systemically in tobacco inoculated with TMV could be produced in the infected leaf.[143] These studies were based on labelling of salicylic acid with $^{18}O_2$. Thus, it is possible, at least in the case of tobacco, that salicylic acid may function as a mobile signal as well as being induced by another translocated signal.

Jasmonates

Linolenic acid in plant membranes can be converted *via* a LOX-mediated pathway to several biologically active molecules, the octadecanoids.[90] These compounds include jasmonic acid (**13**), methyl jasmonate (**14**), and traumatin (**16**). Jasmonic acid is probably ubiquitous in higher plants and can induce many biochemical and physiological changes.[90,144] Exogenous jasmonates induce proteinase inhibitors and phenoloxidases in tomato,[90] nicotine synthesis in tobacco,[145] terpene synthesis in many plants[90] and phenylpropanoid metabolism in parsley,[146] soybean[90] and oak (Schultz, unpublished). The fact that JA synthesis is wound-induced and can mimic wound-induced changes in plant tissue supports a role in wound-induced signalling.[90] These results suggests that lipid-derived molecules may act as systemic signals in plants. Moreover, JA up-regulates LOX, which in turn increases JA's own production as well as that of several other products in the same pathway.[144] Exogenous JA often induces plant responses only or mainly in tandem with other stimuli, such as wounding or fungal elicitors (W.R. Curtis, pers. comm.; Schultz, unpub. data).[90,144-146] Thus it may act more as an amplifier or conditioner than as an elicitor. Because JA also inhibits growth and synthesis of growth- and maintenance-related proteins (often interacting with abscisic acid),[146] it could influence plant responses directly or *via* resource allocation. These results along with the fact that SA and JA interact in antagonistic fashion,[90] make it clear that we are far from understanding the roles of these two putative signals in plant responses.

13. R=H   Jasmonic Acid
14. R=OMe Methyl Jasmonate

15. Salicylic Acid

16. Traumatin

Peptides

A small (18 amino acids) peptide given the trivial name "systemin" has been reported to be involved in wound-induced systemic signalling in tomato.[147] Systemin is produced by enzymatic cleavage of a larger protein (prosystemin),[148] and application of systemin to tomato leaf tissue results in the rapid translocation of the peptide out of the treated leaf and in the systemic induction of proteinase inhibitors.[148,149] This suggests that systemin may be the systemically translocated wound signal involved in the systemic expression of proteinase inhibitors. Construction of transgenic tomatoes expressing systemin antisense mRNA were not able to synthesize proteinase inhibitors upon wounding, further supporting a role for this peptide in systemic wound signalling.

Low levels of prosystemin mRNA are constitutively expressed in plant tissues except for roots.[148] This suggests that prosystemin is always present in the plant and wounding results in the release of systemin through protease action. Grafting of wild-type tomato plants onto transgenic tomato plants overexpressing systemin resulted in the expression of high levels of proteinase inhibitors in the rootstock as well as the scion. The induction of proteinase inhibitors in the scion provided evidence that systemin is the translocated wound signal.

Oligogalacturonides

Fragments of cell wall galacturonic acid polymers that could originate from the primary plant cell wall or middle lamella (oligogalacturonides) have been shown to elicit both local and systemic defense responses.[149,150] Local effects include the synthesis of phytoalexins. Because the application of oligoglacturonides can also mimic the wound-induced synthesis and accumulation of PIs in tomato, it has been suggested that these materials may be systemically translocated signals.[151] However, oligogalacturonides containing more than six galacturonide residues are not systemically transported.[152] This observation, along with the lack of pectolytic enzymes in leaf tissue that could release small mobile fragments, suggests that pectic fragments are not systemically mobile signals in response to microbes.

However, the role that oligogalacturonides play in systemic responses to wounding has been re-evaluated recently. MacDougall et al. showed that hexameric oligogalacturonides that were end-reduced readily moved through tomato seedlings.[153] Thin layer chromatographic analysis demonstrated that the applied oligogalacturonides were modified in the plant, but these results also indicated that radiolabelled oligogalacturonides were mobile (presumably *via* the transpiration stream). More recently, it was reported that end-reduced oligogalacturonides were translocated out of treated tissues, were phloem-mobile, and that the oligogalacturonides isolated from phloem exudate were similar to the material applied to the plant.[154] Thus, it is possible that certain oligogalacturonides may function as systemic wound signals.

It is obvious that there are many potentially important signals implicated in plant responses to pathogens and herbivores. Several, if not all, may play a role in both kinds of response. Some appear to be required for a particular response, others appear multifunctional. Some may work cooperatively, others are clearly antagonistic. The target receptors and endpoint activities are mostly unknown. However, because few investigators have studied two or more signalling systems at once, and no study has attempted to integrate all the known or putative signals in even one response, the roles of signals, and their interactions, remain a mystery.

INTERACTIONS BETWEEN RESISTANCE TO PATHOGENS AND TO HERBIVORES

Because systemic acquired resistance to pathogens and herbivores is often non-specific and is characterized by systemic induction of multiple defenses, and because so many induced secondary metabolites have both anti-microbial and antiherbivore activity, it is possible that acquired resistance to pathogens may also be effective against insects. The evidence for this, however, is not extensive and is often contradictory.

Infecting cotton with *Verticillium albo-atrum* increased resistance to infestation by a spider mite,[155] and infestation with the mite was shown to increase resistance to *Verticillium*, and subsequent mite attack.[156] In cotton, therefore, induced pathogen resistance can result in increased resistance to a mite and *vice versa*. Since cotton uses compounds like gossypols and condensed tannins as both insect and pathogen defenses,[157] this type of cross-resistance is not surprising. Similarly, isoflavonoid phytoalexins (induced by plant pathogens) frequently have been shown to reduce insect feeding and/or growth in several legumes.[158,159]

The TMV-induced hypersensitive response in tobacco induces some resistance to aphids.[160] However, Ajlan and Potter showed that tobacco previously induced for diseaes resistance with TMV did not express increased resistance to either aphids (*Myzus nicotianae*) or tobacco hornworm larvae.[161]

Significant induced resistance in cucumber has been achieved with almost a dozen different pathogens.[72] However, induction of systemic disease resistance with *C. lagenarium* or tobacco necrosis virus does not increase levels of resistance to a spider mite (*Tetranychus urticae*), an aphid (*Myzus melonis*), or the larvae of the fall armyworm (*Spodoptera frugiperda*).[122,161] Potter and colleagues concluded that these data indicate a lack of "cross resistance" to phytopathogens and insects. However, some of their results are not completely consistent with this conclusion. For example, they found that the cucurbitacin-specialist beetle, *Acalymma vittatum*, fed more on plants rubbed with tobacco necrosis virus (TNV), despite failing to detect increased cucurbitacin contents (the major beetle feeding stimulant) in these plants.[122] In addition, preferences shown by fall armyworm were actually highly variable on TNV-infected plants, and mites showed population growth decreases on the inoculated leaves of such plants.[122,162] Moreover, Apriyanto and Potter acknowledged that the induced pathogen treatments are insufficient to induce accumulation of specific antimicrobial agents to an extent that could affect insects.[122] Since no study has included substantial analyses of typical biochemical responses to insects after pathogen infection, or typical antipathogen responses after insect damage, and none has involved more than pairwise interactions, any mechanistic basis for interaction between responses to insects and responses to pathogens of cucumber or other plants remain poorly understood.

Considering that many insects are known to vector plant pathogens while others clearly interact with microbes that condition host plant quality by manipulating plant responses,[163–166] it would be surprising if plant responses to microbes failed to influence insects and *vice versa*. The literature reveals all possible outcomes for plants pests, but few studies have included both microbes and herbivores, or detailed characterization of the plant responses. Similarities between plant responses to the two groups of enemies provide a logical, mechanistic basis for such multitrophic interactions.

NEW APPROACHES TO UNDERSTANDING COMPLEX PLANT-PEST INTERACTIONS

It is clear that both constitutive and inducible plant traits that are protective against pathogens and herbivores have much in common. Many endpoint 'defenses' and the mechanisms for producing them are shared by both interactions. There is evidence that this produces overlapping consequences, which include (positively and negatively) altered plant resistance, as well as altered interactions among herbivores and microbes. Our understanding is too incomplete to determine whether or not these mechanisms are redundant, or to what degree the multiplicity of mechanisms can be linked to specific uses. Considerable defense specificity is achieved with pathogens but much less is evident in the case of against herbivores. Defensive responses to herbivores may actually reflect selection by microbes. Specificity of HR and SAR to microbes appears to lie at the recognition (elicitation) level. Many, if not most, subsequent biochemical steps leading to resistance are the same irrespective of the microbe's identity. Mechanisms that allow plants to recognize herbivores, have not been elucidated. If such defense systems contain redundancy, it is likely to be among signals, the targets of which remain to be discovered.

Despite the functional and structural similarities of plant defenses, the study of each individual plant-enemy interaction has not led to obvious generalizations. Plant pathologists have not capitalized on the large body of information on constitutive secondary metabolites developed by plant-herbivore workers. Neither have plant-herbivore workers capitalized on the wealth of genetic, biochemical, and molecular information unearthed by plant pathologists. As a result, we do not understand how pathogenesis is actually halted or prevented by phytochemicals. Neither do we understand very well how plant responses to herbivory are organized, nor do we know the heritable bases of such defenses. Debate about the adaptive value of these plant traits continues, in part because of these shortcomings.

Increased collaboration across classical disciplinary lines still is needed to integrate and advance sister subdisciplines. Genetic analysis and modern techniques are needed to elucidate the heritability and utility of traits thought to be defensive against herbivores and pathogens. One direct way to determine "utility" is to employ transgenic plants and/or specific mutants lacking or overexpressing various defensive traits to assess their impact on fitness. We now have the ability to estimate the realized relative success and cost of expressing a particular trait using these techniques.

Ecologists need to consider plant and herbivore physiology, biochemistry, and genetics just as central to their work as are natural history or taxonomy. Without a clear understanding of mechanisms underlying resistance, it is unlikely that useful generalizations or explanations of interaction patterns will be forthcoming. Moreover, the ability to develop resistant cultivars quickly depends on a solid understanding of underlying biochemical and molecular mechanisms. Much can be learned from plant pathology in this regard.

For their part, plant pathologists should focus their attention on the endpoint gene products studied by the plant-herbivore scientists. Many signals and PR proteins have uncertain, overlapping, or apparently redundant activities. The fundamental mechanistic basis of resistance, however, has not been clarified. It is important to remember that plant-pathogen interactions do not happen *in vacuo*; plants usually must deal with many stresses simultaneously. These other stressors must certainly condition plant responses to pathogens. Predictive models of resistance and susceptibility require an understanding of these interactions.

Advances in understanding will come through greater collaboration. Pathologists tend to treat plant traits as if they are under selection primarily or exclusively from microbes. Those working in plant-herbivore interactions tend to treat the same traits as shaped entirely by herbivory. Neither view can be remotely correct. Understanding the evolution of both constitutive and plastic plant traits requires the broadest, most comprehensive appreciation of selective influences. Neither plant pathology nor plant-herbivore studies has produced the tightly-integrated view of genetic variation, gene expression, metabolic structure, and ecological/evolutionary consequences needed for this understanding. Investigators need to adopt a broader, more integrative viewpoint, and to collaborate actively outside their classical disciplines to achieve this understanding.

REFERENCES

1. HAMMERSCHMIDT, R., KUĆ, J. (eds.) 1995. Induced Resistance to Disease in Plants. Kluwer Academic Publishers. Dordrecht. p. 186.
2. HARBORNE, J.B. 1993. Introduction to Biochemical Ecology, 4 ed. Academic Press. p. 318
3. DEVERALL, B.J. 1976. Defense Machanims of Plants. Cambridge University Press, London. p. 110.
4. SCHONBECK, F., SCHLOSSER E. 1976. Preformed substances as potential protectants. In: Physiological Plant Pathology. (R. Heitfuss and P.H. Williams, eds.) Springer-Verlag, Berlin. pp. 653–678.

5. KARBAN, R., MYERS, J.H. 1989. Induced plant responses to herbivory. Ann. Rev. Ecol. Syst. 20: 331–348.
6. KUĆ, J. 1984. Steroid glycoalkaloids and related compounds as potato quality factors. Am. Potato J. 61: 123–140.
7. SINDEN, S.L., SANFORD, L.L., WEBB, R.E. 1984. Genetic and environmental control of potato glycoalkaloids. Am. Potato J. 61: 141–156.
8. SCHULTZ, J.C., KEATING, S.J. 1991. Host plant-mediated inteactions between the gypsy moth and a baculovirus. In: Microbial. Mediation of Plant-Herbivore Interactions. (P. Barbosa, V.A. Krischik and C.G. Jones, eds.), John Wiley, New York. pp. 489–506.
9. ZENG, Y. 1993. Biochemical and physiological aspects of the plant host response in the Fusarium dry rot disease of potato. Ph.D. Dissertation, Michigan State University, East Lansing. p. 107.
10. DEFAGO, G., KERN, H. 1983. Induction of *Fusarium solani* mutants insensitive to tomatine and aggressiveness to tomato fruits and pea plants. Physiol. Plant Pathol. 22: 29–37.
11. OSBOURN, A.E., CLARKE, B.R., LUNNES, P., SCOTT, P.R., DANIELS, M.J. 1994. An oat species lacking avenacin is susceptible to infection by *Gaumanomyces graminis* var. *tritici*. Physiol. Mol. Plant Pathol. 45: 457–467.
12. BOWYER, P., CLARKE, B.R., LUNNESS, P., DANIEL, M.J., OSBOURN, A.E. 1995. Host range of a plant pathogenic fungus determined by a saponin detoxifying enzyme. Science 267: 371–374.
13. NICHOLSON, R.L. HAMMERSCHMIDT, R. 1992. Phenolic compounds and their role in disease resistance. Ann. Rev. Phytopath. 30: 369–389.
14. ROSENTHALL, G.A., BERENBAUM, M.R. (eds.). 1992. Herbivores:Their interactions with secondary metabolites. Academic Press, NY p. 961.
15. BERENBAUM, M.R., ZANGERL, A.R. 1992. Quantification of Chemical Coevolution. In: Plant Resistance to Herbivores and Pathogens, (R.S. Fritz and E.L. Simms, eds.), Ecology. Evolution, and Genetics, Univ. of Chicago Press, Chicago. pp. 69–90.
16. MARQUIES, R.J. 1991. Selective impact of herbivores. In: Plant Resistance to Herbivores and Pathogens: Ecology, Evolution and Genetics, (R.S. Fritz and E.L. Simms, eds.), University of Chicago Press, Chicago pp. 301–325.
17. BERENBAUM, M.R. 1995. The chemistry of defense: theory and practice. Proc. Natl. Acad. Sci. USA 92: 2–8.
18. BERENBAUM, M.R., ZANGERL, A.R., NITAO, J.K. 1986. Constraints on chemical coevolution: wild parsnips and the parsnip webworm. Evolution 40: 1215–1228.
19. ZANGERL, A.R., BERENBAUM, M.R., LEVINE, E. 1989. Genetic control of seed chemistry and morphology in wild parsnip (*Pastinica sativa*). J. Hered. 80: 404–407.
20. ZANGERL, A.R., BERENBAUM, M.R. 1986. Furanocoumarins in wild parsnip: evidence of photosynthetically active radiation, ultraviolet light, and nutrients. Ecology 68: 516–520.
21. ZANGERL, A.R. 1990. Furanocoumarin induction in wild parsnip: evidence for an induced defense against herbivores. Ecology 71: 1926–1932.
22. KENNEDY, G.G., BARBOUR, J.D. 1992. Resistance variation in natural and managed systems. In: Plant Resistance to Herbivores and Pathogens; Ecology, Evolution and Genetics, (R.S. Fritz and E.L. Simms, eds.), Univ. of Chicago Press, Chicago, pp. 13–40.
23. DOLINGER, P.M., EHRLICH, P.R., FITCH, W.L., BREEDLOVE, D.E. 1973. Alkaloid and predation patterns in Colorado lupine populations. Oecologia 13: 191–204.
24. WALLER, G.R., NOWACKI, E. K. 1978. Alkaloid biology and metabolism in plants. Plenum Press, New York, p. 246.
25. APPEL, H.M. AND MARTIN, M.M. 1992. Significance of metabolic load in the evolution of host specificity of *Manduca sexta*. Ecology 73: 216–228.
26. JUVICK, J.A. AND STEVENS, M.A. 1982. Inheritance of foliar tomatine content in tomatoes. J. Amer. Soc. Hort. Sci. 107: 1061–1065.

27. STURGEON, K.B. 1979. Monoterpene variation in ponderosa pine xylem resin related to western pine beetle predation. Evolution 33: 803–814.
28. NORRIS, D.M., KOGAN, M. 1980. Biochemical and morphological bases of resistance. In: Breeding Plants Resistant to Insects, (F.G. Maxwell and P.R. Jennings, eds.), John Wiley, New York, pp. 23–61.
29. CASWELL, H., REED, R.C. 1975. Indigestibility of C4 bundle sheath cells by the grasshopper, *Melanoplus confusus*. Ann. Entomol. Soc. Amer. 68: 686–688.
30. WALLACE, L.E., MCNEAL, F.H., BERG, M.A. 1973. Minimum stem solidness required in wheat for resistance to the wheat stem sawfly. J. Econ. Entomol. 66: 11221–1123.
31. GALLUN, R.L, ROBERTS, J.J., FINNY, R.E., PATTERSON, F.L. 1973. Leaf pubescence of field grown wheat: a deterrent to oviposition by the cereal leaf beetle. J. Environ. Qual. 2: 333–334.
32. SCHNEIDER, F. 1944. Eine Ursache der raschen Blattausvermehrunj an Bohnen. Fortschr. Ergebn. Gartenb. 5: 4.
33. MUMMA, R.O., CRAIG, R., COX-FOSTER, D., MEDFORD, J., GROSSMAN, H., GRAZZINI, R., HESK, D., WALTERS, D., YERGER, E. 1993. Biochemistry and Genetics of Small Arthropod Resistance in *Pelargonium*. In: Proceedings of the Third International. Geranium Conference, (R. Craig and J. Selchau, eds.), Ball Publishing, Geneva, IL, pp. 173–184.
34. GOODMAN, R.N., NOVACKI, A.J. 1994. The Hypersensitive Reaction in Plants to Pathogens, APS Press, St. Paul, p. 244.
35. PAXTON, J.D. 1981. Phytoalexins—a working redefinition. Phytopathol. Z. 101: 106–109.
36. KUĆ, J. 1995. Phytoalexins, stress metabolism and disease resistance in plants. Ann. Rev. Phytopathol. 33: 275–297.
37. SNYDER, B.A., LEITE, B., HIPSKIND, J., BUTLER, L.G., NICHOLSON, R.L. 1990. Accumulation of sorghum phytoalexins induced by *Colletotrichum graminicola* at the infection site. Physiol. Mol. Plant Pathol. 39: 463–470.
38. ESSENBERG, M., PIERCE, M.L., COVER, E.C., HAMILTON, B., RICHARDSON, P.E., SCHOLES, V.E. 1992. A method for determining phytoalexin concentrations in flurescent hypersenstively necrotic cells in cotton leaves. Physiol. Mol. Plant Pathol. 41: 101–109.
39. MAYAMA, S., TANI, T. 1982. Microspectrophotometric analysis of the location of avenalumin accumulation in response to fungal infection. Physiol. Plant Path. 21: 141–149.
40. VANETTEN, H.D., MATTHEWS, D.E., MATTHEWS, P.S. 1989. Phytoalexin detoxification: importance for pathogenicity and practical implications. Ann. Rev. Phytopathol. 27: 143–164
41. DESJARDINS, A.E. AND GARDNER, H.W. 1989. Genetic analysis in *Gibberella pulicaris*: Rishitin tolerance, rishitin metabolism, and virulence on potato tubers. Mol. Plant Microb. Interact. 2: 26–34.
42. TSUJI, J., JACKSON, E.P., GAGE, D.A., HAMMERSCHMIDT, R., SOMERVILLE, S.C. 1992. Phytoalexin accumulation in *Arabidopsis thaliana* during the hypersensitive response to *Pseudomonas syringae* pv. *syringae*. Plant Physiol. 98: 1304–1309.
43. GLAZEBROOK, J., AUSUBEL, F.M. 1994. Isolation of phytoalexin deficient mutants of *Arabidopsis thaliana* and characterization of their interactions with bacterial pathogens. Proc. Natl. Acad. Sci., USA 91: 8955–8959.
44. SCHAFER, W. 1994. Molecular mechanisms of fungal pathogenicity in plants. Ann. Rev. Phytopathol. 32: 461–477.
45. AIST, J.R. 1983. Structural responses as resistance mechanisms. In: The Dynamics of Host Defence, (J.A. Bailey and B.J. Deverall, eds.), Academic Press, Sydney, pp. 33–70.
46. RIDE, J.P. 1978. The role of cell wall alterations in resistance to fungi. Ann. Appl. Biol. 89: 302–306.
47. HIJWEGEN, T. 1963. Lignification, a possible mechanism of active disease resistance against pathogens. Neth. J. Plant Path. 69: 314–317.

48. KEEN, N.T., LITTLEFIELD, L.J. 1979. The possible association of phytoalexins with resistance gene expression in flax to *Melampsora lini*. Physiol. Plant Pathol. 14: 265–280.
49. PENG, M., KUĆ, J. 1992. Peroxidase-generated hydrogen peroxide as a source of antifungal activity in vitro and on tobacco leaf disks, Phytopathology 82: 696–699.
50. HAMMERSCHMIDT, R., KUĆ, J. 1982. Lignification as a mechanism for induced systemic resistance in cucumber, Physiol. Plant Pathol. 20: 61–71.
51. STEIN, B.D., KLOMPARENS, K. AND HAMMERSCHMIDT, R. 1993. Histochemistry and ultrastructure of the induced resistance response of cucumber plants to *Colletotrichum lagenarium*. J. Phytopath. 137: 177–188.
52. BAKER, C.J AND ORLANDI, E.W. 1995. Active oxygen in plant pathogenesis. Ann. Rev. Phytopathol. 33: 299–321.
53. MEHDY, M.C. 1994. Active oxygen species in plant defense against pathogens. Plant Physiol. 105: 467–472.
54. SUTHERLAND, M.W. 1991. The generation of oxygen radicals during host responses to infection. Physiol. Mol. Plant Path. 39: 79–94.
55. STERMER, B.A., HAMMERSCHMIDT, R. 1987. Association of heat shock induced resistance to disease with increased accumulation of insoluble extensin and ethylene synthesis. Physiol. Mol. Plant Path. 31: 453–461.
56. CHEN, Z., KLESSIG, D.F. 1991. Identification of a soluble salicylic-acid binding protein that may function in signal transduction in the plant disease resistance response. Proc. Natl. Acad. Sci. USA 88: 8179–8183.
57. CHEN, Z., SILVA, H., KLESSIG, D.F. 1993. Active oxygen species in the induction of plant systemic acquired resistance by salicylic acid. Science 262: 1883–1885.
58. SIEGEL, B.Z. 1993. Plant peroxidases—an organismic perspective. Plant Growth Reg. 12: 303–312.
59. MAYER,A.M. 1987. Polyphenol oxidase in plants—Recent Progress. Phytochemistry 26: 11–20.
60. LAZAROVITS, G., WARD, E.W.B. 1982. Polyphenoloxidase activity in soybean hypocotyls at sites incoulated with *Phytophthora megasperma* f.sp. *glycinea*. Physiol. Plant Pathol. 21: 227–236.
61. CUTT, J.R., KLESSIG, D.F. 1992. Pathogenesis-related proteins. In: Plant Gene Research, (F. Meins and T. Boller, eds.), Springer-Verlag, New York, pp. 181–216.
62. STERMER, B.A. 1995. Molecular regulation of induced systemic resistance. In: Induced Resistance to Disease in Plants, (R. Hammerschmidt and J. Kuć, eds.), Kluwer Academic Publishers, Amsterdam, pp. 111–140.
63. BROGLIE, K., CHET, I., HOLLIDAY, M., CRESSMAN, R., RIDDLE, P., KNOWLTON, S., MAUVIAS, C.J., BROGLIE, R. 1991. Transgenic plants with enhanced resistance to fungal pathogens. Science 254: 1194–1197.
64. KESSMANN, H., STAUB, T., HOFMANN, C., MAETZKE, T., HERZOG, J., WARD, E., UKNES, S., RYALS, J. 1994. Induction of systemic acquired disease resistance in plants by chemicals. Ann. Rev. Phytopath. 32: 439–459.
65. RYALS, J., UKNES, S., WARD, E. 1994. Systemic acquired resistance. Plant Physiol. 104: 1109–1112.
66. ALEXANDER, D., GOODMAN, R.M., GUT-RELLA, M., GLASCOCK, C., WEYMAN, K., FRIEDRICH, L., MADDOX, D., AHL-GOY, P., LUNTZ, T., WARD, E., RYALS, J. 1993. Increased tolerance to two oomycete pathogens in transgenic tobacco expressing pathogenesis-related protein 1a, Proc. Natl. Acad. Sci., USA 90: 7327–7331.
67. ENKERLI, J., GISI, U., MOSINGER, E. 1993. Systemic acquired resistance to *Phytophthora infestans* in tomato and the role of pathogenesis-related proteins. Physiol. Mol. Plant Pathol. 43: 161–171.

68. HAMMERSCHMIDT, R., SMITH BECKER, J. 1996. Acquired resistance to disease. Hort. Revs. 18: (in press)
69. HAMMERSCHMIDT, R. 1993. The nature and generation of systemic signals induced by pathogens, arthropod herbivores and wounds. Adv. Plant Pathol. 10: 307–337.
70. DANN, E.K., DEVERALL, B.J. 1995. Induced resistance in legumes, In: Induced Resistance to Disease in Plants (R. Hammerschmidt and J. Kuć, eds.) Kluwer Academic Publishers, Amsterdam, pp. 1–30.
71. ELLISTON, J., KUĆ, J., WILLIAMS, E.B., RAHE, J.E. 1977. Relationship of phytoalexin accumulation to local and systemic protection of bean against anthracnose, Phytopath. Z. 88: 114–130.
72. HAMMERSCHMIDT, R., YANG-CASHMAN, P. 1995. Induced resistance in cucurbits. In: Induced Resistance to Disease in Plants, (R. Hammerschmidt and J. Kuć, eds.), Kluwer Academic Publishers, Amsterdam, pp. 63–85.
73. HAMMERSCHMIDT, R., NUCKLES, E.M., KUĆ, J. 1982. Association of enhanced peroxidase activity with induced systemic resistance of cucumber to *Colletotrichum lagenarium*, Physiol. Plant Pathol. 20: 73–82.
74. RASMUSSEN, J.B., SMITH, J.A., WILLIAMS, S., BURKHARDT, W., WARD, E., SOMERVILLE, S., RYALS, J., HAMMERSCHMIDT, R. 1995. cDNA cloning and systemic expression of an acidic peroxidase associated with systemic acquired resistance in cucumber, Physiol. Mol. Plant Pathol. 46: 389–400.
75. MÉTRAUX, J.P., STREIT, L., STAUB, T. 1988. A pathogenesis related protein in cucumber is chitinase, Physiol. Mol. Plant Pathol. 33: 1–10.
76. SMITH, J.A., HAMMERSCHMIDT, R. 1988. Comparative study of acidic peroxidases associated with induced resistance in cucumber, muskmelon and watermelon. Physiol. Mol. Plant Path. 33: 255–261.
77. TUZUN, S., KUĆ, J. 1989. Induced systemic resistance to blue mold of tobacco, In: Blue Mold of Tobacco, (W.E. McKeen, ed.), American Phytopathological Society Press, St. Paul, MN, pp. 177–200.
78. CAMERON, R.K., DIXON, R.A., LAMB, C.J. 1994. Biologically induced systemic acquired resistance in *Arabidopsis thaliana*. Plant J. 5: 715–725.
79. MAUCH-MANI, B., SLUSARENKO, A. 1994. Systemic acquired resistance in *Arabidopsis thaliana* induced by a predisposing infection with a pathogenic isolate of *Fusarium oxysporum*. Mol. Plant-Microbe Int. 7: 378–383.
80. UKNES, S., MAUCH-MANI, B., MOYER, M., POTTER, S., WILLIAMS, S., DINCHER, S., CHANDLER, D., SLUSARENKO, A., WARD, E., RYALS, J. 1992. Acquired resistance in *Arabidopsis*. Plant Cell 4: 645–656.
81. UKNES, S., WINTER, A.M., DELANY, T., VERNOOIJ, B., MORSE, A., FRIEDRICH, L., NYE, G., POTTER, S., WARD, E., RYALS, J. 1993. Biological induction of systemic acquired resistance in *Arabidopsis*. Mol. Plant-Microbe Int. 6: 692–698.
82. KUĆ, J., SHOCKLEY G., KEARNEY, K. 1975. Protection of cucumber against *Colletotrichum lagenarium* by *Colletotrichum lagenarium*, Physiol. Plant Pathol. 7: 195–199.
83. RICHMOND, S., ELLISTON, J.E., KUĆ, J. 1979. Penetration of cucumber leaves by *Colletotrichum lagenarium* is reduced in plants systemically protected by previous infection with the pathogen. Physiol. Mol. Plant Pathol. 14: 329–338.
84. BASHAN, B., COHEN, Y. 1983. Tobacco necrosis virus induces systemic resistance in cucumber against *Sphaerotheca fuliginea*. Physiol. Plant Pathol. 23: 137–144.
85. CONTI, G., BASSI, M., CARMINUCCI, D., GATTI, L., BOCCI, A.M. 1990. Preinoculation with tobacco necrosis virus enahnces peroxidase activity and lignification in cucumber as a resistance response to *Sphaerotheca fuliginea*, J. Phytopath. 128: 191–202.
86. SCHMELE, I., KAUSS, H. 1990. Enhanced activity of the plasmamembrane localized callose synthase in cucumber leaves with induced resistance. Physiol. Mol. Plant Pathol. 37: 221–228.

87. MOERSBACHER, B.M., NOLL, U., GORRICHON, L., REISNER, H.J. 1990. Specific inhibition of lignification breaks hypersensitve resistance of wheat to stem rust. Plant Physiol. 93: 465–470.
88. AVDIUSHKO, S.A., YE, X.S., HILDEBRAND, D.F., KUĆ, J. 1993 Induction of lipoxygenase activity in immunized cucumber plants. Physiol. Mol. Plant Path. 42: 83–95.
89. CROFT, K.P.C., JUTTNER, F., SLUSARENKO, A.J. 1993. Volatile products of the liopxygenase pathway evolved from *Phaseolus vulgaris* L. leaves inoculated with *Pseudomonas syringae* pc. *phaseolicola*. Plant Physiol. 101: 13–24.
90. FARMER, E.E. 1994. Fatty acid signalling in plants and their associated microorganisms. Plant Mol. Biol. 26: 1423–1437.
91. CHAI, H.B., DOKE, N. 1987. Systemic activation of O_2^- generating reaction, superoxide dismutase, and peroxidase in potato plants in relation to induction of systemic resistance to *Phytophthora infestans*, Ann. Phytopath. Soc. Japan 53: 645–649.
92. BI, Y.-M., KENTON, P., MUR, L., DARBY, RL., DRAPER, K. 1995. Hydrogen peroxide does not function downstream of saliacylic acid in the induction of PR protein expression. Plant J. 8: 235–245.
93. NEUENSCHWANDER, U., VERNOOIJ, B., FRIEDRICH, L., UKNES, S., KESSMANN, H., RYALS, J. 1995. Is hydrogen peroxide a second messenger of salicylic acid in systemic acquired resistance. Plant J. 8: 227–233.
94. RUFFER, M., STEIPE, B., ZENK, M. 1995. Evidence against specific binding of salicylic acid to plant catalase. FEBS Lett. 377: 175–180.
95. YE, X.S., JARLFORS, J., TUZUN, S., PAN, S.Q., KUĆ, J. 1992. Biochemical changes in cell walls and cellular responses of tobacco leaves related to systemic resistance to blue mold (*Peronospora tabacina*) induced by tobacco mosaic virus. Can. J. Bot. 70: 49–57.
96. CONN, E.E. 1979. Cyanide and cyanogenic glycosides. In: Herbivores: Their Interactions with Secondary Plant Metabolites, (G.A. Rosenthal and D.H. Janzen, eds.), Acad. Press, New York, pp. 387–412.
97. CHEW, F.S. 1988. Biological effects of glucosinolates. In: Biologically Active Natural. Products: Potential. Use in Agriculture, (H.G. Cutler, ed.), Amer. Chem. Soc., Washington, DC, pp. 155–181.
98. TALLAMY, D.W., MCCLOUD, E.S. 1991. Squash beetles, cucumber beetles, and inducible cucurbit responses. In: Phytochemical Induction by Herbivores, (D.W. Tallamy and M.J. Raupp, eds.), John Wiley, New York, pp. 155–181.
99. CLAUSEN, T.P., REICHARDT, P.B., BRYANT J.P., WERNER, R.A. 1991. Long-term and short-term induction in quaking aspen: related phenomena? In: Phytochemical Induction by Herbivores, (D.W. Tallamy and M.J. Raupp, eds.), John Wiley, New York, pp. 71–84.
100. APPEL, H.M. 1993. Phenolics in ecological interactions: The importance of oxidation. J. Chem. Ecol. 19: 1521–1552.
101. FELTON, G.W., SUMMERS, C.B., MEULLER, A.J. 1994. Oxidative responses in soybean foliage to herbivory by bean leaf beetle and three-cornered alfalfa hopper. J. Chem. Ecol. 20: 639–650.
102. FOWLER, S.V., MACGARVIN, M. 1986. The effects of leaf damage on the performance of insesct herbivores on birch, *Betula pubescens*. J. Anim. Ecol. 55: 565 573.
103. HARTLEY, S.E., LAWTON, J.H. 1987. Effects of different types of damage on the chemistry of birch foliage, and the responses of birch feeding insects. Oecologia 74: 432–437.
104. WRATTEN, S.D., EDWARDS, P.J., DUNN, I. 1984. Wound-induced changes in the palatability of *Betula pubescens* and *Betula pendula*. Oecologia 61: 372–375.
105. MAURICIO, R., BOWERS, M.D., BAZZAZ, F.A. 1993. Pattern of leaf damage affects fitness of the annual plant *Raphanus sativus* (Brassicaceae). Ecology 74: 2066–2071.
106. EDWARDS, P.J., WRATTEN, S.D., GIBBERD, R.M. 1991. The impact of inducible phytochemicals on food selection by insect herbivores and its consequences for the distribution of

grazing damage. In: Phytochemical Induction by Herbivores, (D.W. Tallamy and M.J. Raupp, eds.), John Wiley, New York, pp. 204–221.
107. TUOMI, J., T. FAGERSTROM, AND P. NIEMELA. 1991. Carbon allocation, phenotypic plasticity, and induced defenses. In: Phytochemical Induction by Herbivores, (D.W. Tallamy and M.J. Raupp, eds.), John Wiley, New York. pp. 85–104
108. RHOADES, D.F. 1985. Offensive-defensive interactions between herbivores and plants: their relevance in herbivore population dynamics and ecological theory. Am. Nat. 125: 205–238.
109. FLOSS, H. 1986. The shikimate pathway—an overview. Rec. Adv. Phytochem. 20: 13–55.
110. SCHULTZ, J.C., BALDWIN, I.T. 1982. Oak leaf quality declines in response to defoliation by gypsy moth larvae. Science 217: 149–151.
111. BALDWIN, I.T. 1993. Chemical changes rapidly induced by folivory. In: Insect-Plant Interactions, (E.A. Bernays, ed.), CRC Press, Boca Raton, FL, pp. 1–23.
112. PARSONS, T.J., BRADSHAW, JR., H.D., GORDON, M.P. 1989. Systemic accumulation of specific mRNAs in response to wounding in poplar trees. Proc. Natl. Acad. Sci. USA 86: 7895–7899.
113. TURLINGS, T.C.J., TUMLINSON, J.H., HEATH, R.R., PROVEAUX, A.T., DOOLITTLE, R.E. 1991. Isolation and identification of allelochemicals that attract the larval parasitoid, *Cotesia marginiventris* (Cresson), to the microhabitat of one of its hosts. J. Chem. Ecol. 17: 2235–2251.
114. TURLINGS, T.C.J., MCCALL, P.J., ALBORN, H.T., TUMLINSON, J.H. 1993. An elicitor in caterpillar oral secretions that induces corn seedlings to limit chemical signlas attractive to parasitic wasps. J. Chem. Ecol. 19: 411–425.
115. TRULINGS, T.C.J., TUMLINSON, J.H., LEWIS, W.J. 1990. Exploitation of herbivore-induced plant odors by host-seeking parasitic wasps. Science 250: 1251–1253.
116. MATTIACCI, L., DICKE, M., POSTHUMUS, M.A. 1994. Beta-glucosidase: an elicitor of herbivore-induced plant odor that attracts host-searching parasitic wasps. Proc. Natl. Acad. Sci. USA 92: 2036–2040.
117. HARTLEY, S.E., LAWTON, J.H. 1991. Biochemical aspects and significance of the rapidly-induced accumulation of phenolics in birch foliage. In: Phytochemical. Induction by Herbivores, (D.W. Tallamy and M.J. Raupp, eds.), John Wiley, New York, pp. 105–132.
118. TALLAMY, D.W., RAUPP, M.J. 1991. Phytochemical Induction by Herbivores, John Wiley, New York. p. 431.
119. CARROL, C.R., HOFFMAN, C.A. 1980. Chemical feeding deterrent mobilized in response to insect herbivory and counteradaptation by *Epilachna tredecimnotata*. Science 209: 414–416.
120. TALLAMY, D.W. 1985. Squash beetle (*Epilachna borealis*) feeding behavior: an adaptation against induced cucurbit defenses. Ecology 66: 1545–1579.
121. HAYNES, R.L., JONES, C.M. 1975. Wilting and damage to cucumber by spotted and striped cucumber beetles. Hortscience 10: 265–266.
122. APRIYANTO, D., POTTER, D.A. 1990. Pathogen-activated induced resistance of cucumber: response of arthropod herbivores to systemically protected leaves, Oecologia 85: 25–31.
123. BAR-NUN, N., MAYER, A.M. 1990. Cucurbitacins protect cucumber tissue against infection by *Botrytis cinerea*. Phytochemistry 29: 787–791.
124. HUNTER, M.D., SCHULTZ, J.C. 1993. Inducible plant defenses breached? Phytochemical induction protects an herbivore from disease. Oecologia 94:195–203.
125. GUEDES, M.E.M., RICHMOND, S., KUĆ, J. 1980. Induced systemic resistance to anthracnose in cucumber as influenced by the location of the inducer inoculation with *Colletotrichum lagenarium* and the onset of lowering and fruiting. Physiol. Plant Path. 17: 229–233.
126. TUZUN, S., KUĆ, J. 1985. Movement of a factor in tobacco infected with *Peronospora tabacina* which systemically protects against blue mold. Physiol. Mol. Plant Path. 26: 321–330.
127. DEAN, R.A., KUĆ, J. 1986. Induced systemic protection in cucumber: The source of the "signal", Physiol. Mol. Plant Path. 28: 227–233.

128. DEAN, R.A., KUĆ, J. 1986. Induced systemic protection in cucumber: Time of production and movement of the signal, Phytopathology 76: 966–970.
129. DAVIS, J.M., GORDON, M.P., SMIT, B.A. 1991. Assimilate movement dictates remote sites of wound-induced gene expressin in poplar leaves. Proc. Natl. Acad. Sci. USA 88: 2392–2398.
130. WHITE, R.F. 1979. Acetylsalicylic acid (aspirin) induced resistance to tobacco maosaic virus in tobacco. Virology 99: 410–412.
131. MILLS, P.R., WOOD, R.K.S. 1984. The effects of polyacrylic acid, acetylSA and SA on resistance of cucumber to *Colletotrichum lagenarium*. Phytopathol. Z. 111: 209–216.
132. LINTHORST, H.J.M. 1991. Pathogenesis-related proteins in plants. Crit. Rev. Plant Sci. 10: 123–150.
133. PIERPONT, W.S. 1994. Salicylic acid and its derivatives in plants: Medicines, metabolism and messenger molecules. Adv. Bot. Res. 20: 164–235.
134. MÉTRAUX, J.P., SIGNER, H., RYALS, J., WARD, E., WYSS-BENZ, M., GAUDIN, J., RASCHDORF, K., SCHMID, E., BLUM, W., INVERARDI, W. 1990. Increase in salicylic acid at the onset of systemic acquired resistance in cucumber, Science 250: 1004–1006.
135. RASMUSSEN, J.B., HAMMERSCHMIDT, R., ZOOK, M. 1991. Systemic induction of salicylic acid accumulation in cucumber after inoculation with *Pseudomonas syringae* pv. *syringae*, Plant Physiol. 97: 342–1347.
136. YALPANI, N., SILVERMAN, P., WILSON, T.M.A., KLEIER, D.A., RASKIN, I. 1991. Salicylic acid is a systemic signal and inducer of pathogenesis-related proteins in virus infected tobacco. Plant Cell. 3: 809–818.
137. MALAMY, J., CARR, J.P., KLESSIG, D.F., RASKIN, I. 1990. Salicylic acid—a likely endogenous signal in the resistance response of tobacco to tobacco mosaic virus. Science 250: 1002–1004.
138. LAWTON, K., BECK, J., POTTER, S., WARD, E., RYALS, J. 1994. Regulation of cucumber class III chitinase gene expression. Molec. Plant-Microbe Interact. 7: 48–57.
139. GAFFNEY, T., FRIEDRICH, L., VERNOOIJ, B., NEGROTTO, D., NYE, G., UKNES, S., WARD, E., KESSMANN, H., RYALS, J. 1993. Requirement of SA for the induction of systemic acquired resistance. Science 261: 754–766.
140. DELANEY, T.P., UKNES, S., VERNOOIJ, B., FRIEDRICH, L., WEYMANN, K., NEGROTTO, D., GAFFNEY, T., GUT-RELLA, M., KESSMANN, H., WARD, E., RYALS, J. 1994. A central role of salicylic acid in plant disease resistance. Science 266: 1247–1250.
141. SMITH, J.A., FULBRIGHT, D.W., HAMMERSCHMIDT, R. 1991. Rapid induction of systemic resistance in cucumber by *Pseudomonas syringae* pv. *syringae*, Physiol. Mol. Plant Pathol. 38: 223–235.
142. VERNOOIJ, B., FRIEDRICH, L., MORSE, A., REIST, R., KOLDITZ-JAWHAR, R., WARD, E., UKNES, S., KESSMANN H., RYALS, J. 1994. Salicylic acid is not the translocated signal responsible for inducing systemic acquired resistance but is required in signal transduction. Plant Cell 6: 959–965.
143. SHULAEV, V., LEON, J., RASKIN, I. 1990. Is salicylic acid a tranlocated signal of systemic acquired resistance in tobacco? Plant Cell 7: 1691–1701.
144. SEMBDNER, G., PARTHIER, B. 1993. The biochemistry and the physiological and molecular actions of jasmonates. Ann. Rev. Plant Physiol. Plant Mol. Biol. 44: 569–589.
145. BALDWIN, I.T., SCHMELZ, E.A., OHNMEISS, T.E. 1994. Wound-induced changes in root and shoot jasmonic acid pools corrleate with induced nicotine synthesis in *Nicotiana sylvestris* Spegazzini and Comes. J. Chem. Ecol. 20: 2139–2157.
146. KAUSS, H., KRAUSE, K., JEBLICK, W. 1992. Methyl jasmonate conditions parsley suspension cells for increased elicititation of phenylpropanoid defense responses. Biochem. Biophys. Res. Commun. 189: 304–308.
147. PEARCE, G., STRYDROM, G., JOHNSON, S., RYAN, C.A. 1991. A polypeptide from tomato leaves induces the expression of proteinase inhibitor genes. Science 253: 895–897.

148. MCGURL, B., PEARCE, G., OROZOCO-CARDENAS, M., RYAN, C.A. 1992. Structure, expression and antisense inhibition of the systemin precursor gene. Science 255: 1570–1573.
149. RYAN, C.A. 1992. The search for the proteinase inhibitor induceing factor, PIIF. Plant. Mol. Biol. 19: 123–133.
150. RYAN, C.A., FARMER, E.E. 1991. Oligosaccharide signals in plants: A current assessment. Ann. Rev. Plant Physiol. Plant Mol. Biol. 42: 651–674.
151. BISHOP, P.D., PEARCE, G., BRYANT, J.E., RYAN, C.A. 1984. Isolation and characterization of the proteinase inhibitor inducing factor from tomato leaves. J. Biol. Chem. 259: 13172–13177.
152. BAYDOIUN, E.A.-H., FRY, S.C. 1985. The immobility of pectic substances in injured tomato leaves and its bearing on the identity of the wound hormone. Planta 165: 269–276.
153. MACDOUGALL, A.J., RIGBY, N.M., NEEDS, P.W., SELVENDRAN, R.R. 1992. Movement and metabolism of oligogalacturonide elicitors in tomato shoots. Planta 188: 566–574.
154. RIGBY, N.M., MACDOUGALL, A.J., NEEDS, P.W., SELVENDRAN, R.R. 1994. Phloem translocation of a reduced oligogalacturonide in *Ricinus communis* L. Planta 193: 536–541.
155. KARBAN, R., ADAMCHAK, R., SCHNATHORST, W. 1987. Induced resistance and interspecific competition between spider mites and a vascular wilt fungus in cotton plants. Science 235: 678–680.
156. KARBAN, R., CAREY, J.R. 1984. Induced resistance of cotton seedlings to mites. Science 225: 53–54.
157. BELL, A.A., STIPANOVIC, R.D., ELZEN, G.W., WILLIAMS, H.J. 1987. Structural and genetic variation of natural pesticides in pigment glands of cotton (*Gossypium*), In: Allelochemicals: Role in Agriculture and Forestry, ACS Symp. 330, (G.R. Waller, ed.), Amer. Chem. Soc., Washington, DC, pp. 477–490.
158. KOGAN, M., FISCHER, D.C. 1991. Inducible defenses in soybean against herbivorous insects. In: Phytochemical Induction by Herbivores, (D.W. Tallamy and M.J. Raupp, eds.), John Wiley, New York, pp. 347–378.
159. KOGAN, M., PAXTON, J. 1983. Natural inducers of plant resistance to insects. Symp. Amer. Chem. Soc. 208: 153–171.
160. MCINTYRE, J.L., DODDS, J.A., HARE, J.D. 1981. Efects of localized resistance against diverse pathogens and insects. Phytopathology 71: 297–301.
161. AJLAN, A.M., POTTER, D.A. 1992. Lack of effect of tobacco mosaic virus-induced systemic acquired resistance on arthropod herbivores in tobacco. Phytopathology 82: 647–651.
162. AJLAN, A.M., POTTER, D.A. 1991. Does immunization of cucumber against anthracnose by *Colletotrichum lagenarium* affect host suitability for arthropods, Ent. Exper. Appl. 58: 83–91.
163. PURCELL, A.H., NAULT, L.R. 1991. Interactions among plant pathogenic prokaryotes, plants, and insect vectors. In: Microbial. Mediation of Plant-Herbivore Interactions. (P. Barbosa, V.A. Krischik, and C.G. Jones, eds.), J. Wiley, New York, pp. 383–405.
164. BARBOSA, P. 1991. Plant pathogens and nonvector herbivores. In: Microbial. Mediation of Plant-Herbivore Interactions, (P. Barbosa, V.A. Krixhik, and C.G. Jones, eds.), J. Wiley, New York, pp. 341–382.
165. RAFFA, K.F. 1991. Induced defensive reactions in conifer -bark beetle systems. In: Phytochemical. Induction by Herbivores, (D.W. Tallamy and M.J. Raupp, eds.), John Wiley, New York, pp. 245–276.
166. KRISCHIK, V.A., GOTH, R.W., BARBOSA, P. 1991. Generalized plant defense: effects on multiple species. Oecologia 85: 562–571.

Chapter Six

PHYTOCHEMISTRY OF THE MELIACEAE
So Many Terpenoids, So Few Insecticides

Murray B. Isman,[1] Hideyuki Matsuura,[2] Shawna MacKinnon,[3] Tony Durst,[3] G. H. Neil Towers,[2] and John T. Arnason[4]

[1] Department of Plant Science
[2] Department of Botany
University of British Columbia
Vancouver, British Columbia, Canada V6T 1Z4
[3] Department of Chemistry
[4] Department of Biology
University of Ottawa
Ottawa, Ontario, Canada K1N 6N5

Introduction 156
    Neem Insecticides 156
    Nature versus the Chemical Industry 157
    Definitions 157
    Some Properties of Antifeedants 158
    Triterpenoid Chemistry of the Meliaceae 159
Structure-Activity Studies 160
    Azadirachtin 160
    Salannin 163
    Toosendanin 165
    Gedunin 166
Other Compounds Recently Isolated 168
    Limonoids 168
    Non-limonoid Insecticides 171
The Significance of Mixtures 171
    Melia toosendan Studies 172
    Neem (*Azadirachta indica*) Studies 172
Conclusion 174

Phytochemical Diversity and Redundancy in Ecological Interactions,
edited by Romeo et al., Plenum Press, New York, 1996

INTRODUCTION

Neem Insecticides

Utilization of, and scientific interest in, botanical insecticides diminished precipitously following the introduction of DDT, parathion, and numerous other synthetic insecticides in the late 1940's and early 1950's. However, concerns for human health and environmental impacts of synthetic insecticides (along with other pesticides) has provided the impetus for the search for alternative pest management products. Investigations into new botanical insecticides have increased substantially since the early 1980's. In fact, one plant species is almost entirely responsible for reigniting interest and credibility in the discovery, development and use of botanical insecticides: the Indian neem tree, *Azadirachta indica* A. Juss.

More than three decades of research on neem has spawned six international conferences and publication of at least six volumes compiling research findings on this species. A vast amount of scientific literature is available on the insect control properties of derivatives from neem seeds. More importantly, this research has culminated in successful commercialization and regulatory and industrial acceptance of neem-based pest control products, providing a new paradigm for the development of insect management products derived from plants. So much has been written on the subject of neem (the compendium edited by Schmutterer being the most recent[1]), that even a brief review here is unnecessary.

In nature, plant chemical defense against insect herbivory almost never depends on a single compound, but instead depends on mixtures of compounds. Often these mixtures are made up of one or more series of structurally closely related compounds, and crude plant extracts are frequently more bioactive than that of their most active constituent once isolated and purified. This phenomenon is well exemplified by neem insecticides. Commercial neem insecticides are derived from the seeds of this tree, from which over 100 triterpenoid compounds have been isolated and characterized. As we will discuss later, many of these have measurable bioactivity against insects, so that the natural blend that occurs in a neem-seed based insecticide may be more effective than the single most active constituent alone.

There are some other unique aspects to the neem story that are worth highlighting. The first is that the earliest research and the initial strategy for its use in pest management were all based on its antifeedant action in the desert locust *Schistocerca gregaria*. The more profound physiological effects of neem, namely its interference with molting and development in insects, were discovered subsequent to investigations of the antifeedant effect. The irony of this, according to most industry specialists, is that it is the insect growth regulatory properties of neem that are of operational importance in the field, not the antifeedant properties (at least for most pest species). The second unique aspect of neem, relative to other commercial botanical insecticides (e.g. pyrethrum, rotenone, ryania), is its slow action as an insecticide. Neem's abrupt antifeedant effect on many insects notwithstanding, its toxicity would have been totally overlooked in conventional bioassays geared toward acute toxins (neurotoxins, muscle poisons, respiratory inhibitors) such as those found in other botanicals and most synthetic insecticides.

Nature versus the Chemical Industry

The commercial success of neem insecticides, despite their lack of acute toxicity, bodes well for the search for additional pest management products derived from plants. Whereas the chemical industry has focussed primarily on acutely toxic chemicals over the past fifty years, chemical defenses of higher plants, shaped by 300 million years of evolutionary "experience", tend to be far more subtle in their actions on insects and other herbivores. Plants tend to discourage insect herbivory by deterring feeding and oviposition or by inhibiting growth, rather than by poisoning their enemies outright. Therefore, the search for botanical insecticides with bioactivity comparable to that of synthetic insecticides is a search for biological oddities and extremes. This generalization may help to explain why there appear to be far more plant natural products with antifeedant effects on insects than those which are toxic to the same insects, although recent studies suggest that the line between behavioral deterrence and toxicity may be less distinct than previously thought.[2,3]

In evolutionary terms, compounds with a high degree of toxicity may be counterproductive if they select for resistance in insect populations. Numerous laboratory selection experiments utilizing synthetic insecticides have demonstrated that resistance can arise rapidly, often within as few as ten insect generations when mortality rates are high.[4] Natural products are not immune to this phenomenon; witness the spreading resistance to the endotoxin of *Bacillus thuringiensis* in natural populations of the diamondback moth, *Plutella xylostella*.[5]

Definitions

Terms like antifeedant, growth inhibitor, and insecticide are sometimes used interchangeably in the literature, and are often misused. We think it worthwhile, therefore, to provide our own working definitions for these terms so that they can be used without ambiguity as descriptors for specific natural products to be discussed in this chapter.

- antifeedant: we favor a restrictive definition for this term - a peripherally-mediated *behavior*-modifying substance (i.e. acting directly on chemosensilla) resulting in feeding deterrence. This definition thus excludes compounds that suppress feeding through centrally-mediated effects. Some compounds produce a condition that can best be described as anorexia *following topical administration*; others produce general malaise in insects some time following ingestion. In either of these latter cases, the net result is that (over time), the insect consumes less food, but this effect is the result of a physiological response, rather than a behavioral one.

- growth inhibitor: a substance that suppresses larval growth of an insect. In our combined experience with over one hundred allelochemicals, very few actually kill treated insects, except at unrealistically high doses or concentrations. Far more often, a tested compound will inhibit larval growth, resulting in small individuals if growth is evaluated over a fixed time interval, or as a delay in the time to pupation. In either case, the effect can be dramatic, but without any mortality. One of the most common bioassays for evaluating plant extracts or pure compounds is the chronic growth bioassay wherein test materials are added to an artificial medium on which larvae are allowed to feed and grow. However, it must be kept in mind that growth inhibition in this type of bioassay can arise from either a behavioral response (feeding deterrence as defined above) or a physiological one (post-ingestive toxicity). An allelochemical can therefore only be unambiguously described as a growth inhibitor if the effect can also be produced following administration through an alternate route (eg. topically, or by injection).

- insecticide: a substance that kills insects at reasonable doses or concentrations. The vast majority of commercial insecticides (including botanicals like pyrethrum and rotenone) are acute toxins - treated insects succumb within 24 hours of exposure. On the other hand, some synthetic insect growth regulators (e.g. juvenile hormone mimics) are, like neem, slow acting, but they ultimately lead to death of the insect. Although many insecticides will also inhibit larval growth when administered at sublethal doses, the key distinction between an insecticide and a growth inhibitor remains: insecticides normally kill the target insect, growth inhibitors do not.

It is important to note that these terms are not mutually exclusive, and in fact, many compounds show an array of actions depending on the dose or concentration. In many cases, the bioactivity reported is defined only by the bioassay used by the investigator.

Some Properties of Antifeedants

The idea of using non-toxic feeding deterrents (=antifeedants) as crop protectants has attracted much attention in the past two decades, and has been the subject of many reviews.[6-8] Several authors have pointed out a number of shortcomings in using antifeedants as stand-alone crop protectants, and these criticisms have led to some innovative approaches to their deployment.[9-11] On the other hand, many if not most, antifeedants also have some post-ingestive physiological activity.[2] That could be important in the field as the insects' threshold for rejection diminishes with increasing hunger and they eventually feed.

A likely candidate for commercialization as an antifeedant among plant natural products is the abundant triterpene, limonin, from the seeds of grapefruit, *Citrus paradisi*. This compound is a strong antifeedant against the Colorado potato beetle, *Leptinotarsa decemlineata*,[11] the most economically important pest of potato in North America, Europe and Russia.

From the practical viewpoint, the most obvious problems for antifeedants are: large differences in susceptibility between species; and the potential for habituation (desensitization) in initially-deterred insects. Azadirachtin (1), an exceptional antifeedant for many insects, provides examples of both situations. Though it is the most potent antifeedant known for the desert locust (EC_{50} = 0.05 ppm), the North American migratory grasshopper, *Melanoplus sanguinipes*, is completely insensitive (EC_{50} > 1000 ppm).[12] In an evaluation of susceptibility of six species of noctuid caterpillars to azadirachtin, no significant inter-specific differences in the growth-inhibiting effects occurred.[13] In contrast, the antifeedant effect, measured in a leafdisc choice bioassay, differed significantly between species, with a more than 30-fold difference in EC_{50} values between the least and most sensitive species. Previous investigators have demonstrated that individual insects can habituate (become desensitized) to particular antifeedants upon repeated exposures.[14,15] We have observed the antifeedant effect of azadirachtin to be diminished by as much as 50% upon second exposure (24 hours after initial exposure) to *Spodoptera litura* larvae in leafdisc choice tests (M. Bomford and M. Isman, unpublished data). It may be noteworthy that *S. litura* was found to be the most susceptible species to the antifeedant effect of azadirachtin in the earlier study.[13] Therefore, if azadirachtin acted strictly as an antifeedant, its protective action in the field could be very short-lived.

Triterpenoid Chemistry of the Meliaceae

In terms of secondary plant chemistry, the Meliaceae (mahogany family) is best characterized by the production of limonoids, a group of modified triterpenes with/or derived from a precursor with, a 4,4,8-trimethyl-17-furanyl-steroid skeleton.[16] The neem tree, *Azadirachta indica*, unquestionably the most intensively studied species of the family, itself contains upwards of 100 different limonoids in its different tissues.[17] Because so many of the limonoids are biologically active to insects, at least as antifeedants, they have been the subject of recent reviews.[17,18] Although there have been exceptions, bioassay-driven isolation of Meliaceous plant extracts normally results in the isolation of limonoids, even when insects have not been the bioassay organism. The breadth of biological activities of limonoids, from microorganisms to vertebrates, is truly impressive, and suggests the hypothesis that the evolution of limonoids in the Meliaceae has not depended on insects as the sole, or even main, selecting agents.

A variety of oxidations and skeletal rearrangements occur among the limonoids. The A and D rings are often oxidized to lactones or epoxides. Where more than one ring is oxidized, extensive rearrangements can yield complex structures barely recognizable as limonoids.[18] Arguably the most biologically active limonoids are the C-*seco* types, exemplified by azadirachtin. These likely differ in biogenic origin from the other limonoid types. C-*seco* limonoids are restricted to the tribe Melieae, containing only two genera, *Azadirachta* and *Melia*.

STRUCTURE-ACTIVITY STUDIES

The impetus for structure-activity investigations is the optimization of biological activity through synthetic chemical derivatization of a bioactive natural product. An alternative goal, for natural products that are extremely complex and thus unlikely to be amenable to synthesis on a commercial scale, is the discovery of a simpler structure, maintaining sufficient bioactivity, but amenable to large scale synthesis. Both approaches have been widely used by the chemical industry, but success has been elusive with the exception of the synthetic pyrethroids.

Azadirachtin

The major active principle in neem seed kernels, and the basis for the commercial insect control products, is the C-*seco* limonoid, azadirachtin (**1**). Following the initial isolation of this fascinating compound by Butterworth and Morgan,[19] the structure of the complex molecule remained uncertain until extensive NMR analyses by Kraus and associates led to the currently accepted form.[20] For well over a decade, Steven Ley and colleagues have been proceeding toward a total synthesis of the azadirachtin molecule.[21] To date, they have succeeded in synthesizing the intact decalin fragment (the "western" half of the molecule) and the intact dihydrofuranacetal moiety (the "eastern" half), but have yet to link these together through the crucial C8-C14 bond to form the complete azadirachtin molecule. Nonetheless, their synthetic efforts have produced not only the above-mentioned fragments, but also a large number of synthetic derivatives of azadirachtin itself. Taken together, these compounds provide glimpses into structural requirements for the various bioactivities of azadirachtin against insects.

	R₁	R₂	R₃	R₄	
1	OTig	OAc	COOCH₃	OH	Azadirachtin
2	OH	OTig	COOCH₃	H	3-Tigloylazadirachtol
3	OTig	OAc	H	OAc	Marrangin

Kraus has also made significant contributions to this end, primarily through the isolation of a wide array of minor limonoid constituents both from *A. indica* and from a series of closely related species in the genus *Melia*.[22] Likewise, Lee et al.[23] and Hansen et al.[24] have compared azadirachtin with minor constituents from neem and *M. azedarach*, and with some synthetic derivatives of azadirachtin. Finally, Rembold has isolated nine azadirachtin analogues from neem and synthesized more than a dozen derivatives.[25] From this work he has established minimum structural requirements for the insect growth regulatory (IGR) effects of azadirachtin, and has applied a quantitative structure-activity relations (QSAR) model to a selected group of natural and synthetic azadirachtin analogues.[26] Unfortunately, differences between these investigating laboratories, with respect to bioassay methods and species, make it tenuous to draw strong conclusions, although certain inferences appear quite obvious.

Both Kraus and Rembold have used the Mexican bean beetle, *Epilachna varivestis*, as their main bioassay species, but while Rembold has used a bioassay with molting disruption as the endpoint, Kraus has focussed primarily on the antifeedant effects against this insect. Structure-activity studies conducted by Ley's colleagues[27,28] utilized four different caterpillar species: *Spodoptera littoralis*, *S. frugiperda*, *Heliothis virescens* and *Helicoverpa armigera*. Later antifeedant studies were based on bioassays with *S. littoralis* alone.[29] Lee[23] used neonate *H. virescens* for his studies; Blaney et al.[27] used last instar larvae.

The studies by Ley and colleagues indicate clearly that the complete carbon skeleton is required for bioactivity. Neither the decalin fragment nor the furanacetate fragment possess IGR or growth inhibiting activity.[29] With respect to antifeedant effects, Blaney et al.[29] found that a decalin fragment similar but not identical to that in azadirachtin has modest activity against *S. littoralis* (50% deterrence at 8×10^{-5}M), and a furanacetate fragment lacking the C14-C15 epoxide function was also active (7×10^{-6}M). Although these synthetic compounds are active in the 1–10 ppm range, the phenomenal antifeedant action of azadirachtin (50% deterrence estimated at 3.3×10^{-15}M ~ nine orders of magnitude more active) - makes reference to the synthetic fragments as antifeedants questionable.

As molt disruptors for the bean beetle, the naturally-occurring azadirachtins are all highly active, with LC_{50} values ranging from 0.3–2.8 ppm[26] (note - concentrations in this bioassay are not comparable to those used by Blaney et al.[27]). On the practical side, it should be mentioned that only three of the nine azadirachtin analogues are of commercial significance. In neem (*Azadirachta indica*), kernels contain an array of azadirachtins, but two of these, azadirachtin (**1**) and 3-tigloylazadirachtol (**2**), account for about 99% of the total.[30] They normally occur in a ratio of 3–5 to 1, with azadirachtin the dominant compound. In the closely related species *A. excelsa*, azadirachtin is also the main bioactive constituent, but a related compound, called marrangin (**3**) or azadirachtin L, is present in appreciable amounts.[17] According to Rembold's *Epilachna* bioassay, this latter compound is about four times more active as an IGR than azadirachtin.

From his structure-activity studies, Rembold[25] concludes that important centers on the azadirachtin molecule for binding to its hypothetical receptor include the oxygens at C1, 2 and 7 of the decalin moiety, and the acetal oxygens of ring D. Interpretation of QSAR analysis also reveals that bioactivity (to *Epilachna*) is positively correlated with polarity of ring A; removal of the ester functions on C-1 and 3 (e.g. via saponification) enhances activity. Bioactivity also appears related to steric factors of substituents on C-7 and 11; increasingly bulky substituents diminish activity. This last observation, together with the deshielding effect of removal of the hydroxyl at C-11 (as occurs in marrangin [aza L] and 3-tigolylazadirachtol [aza B]), suggests free rotation around the central C-8 to C-14 bond.

In addition to isolating the naturally-occurring azadirachtins (further divided into three groups, the azadirachtins, azadirachtols, and meliacarpins), Kraus[22] has isolated a large number of other limonoids ("meliacins", **4–10**) from neem and related species. Bioassay of these compounds for antifeedant and molt-disrupting effects in *Epilachna* provides further insight into the importance of the carbon skeleton for the full range of biological activities. These effects are summarized in Table 1.

4 Azadiradione   **5** Gedunin

6 Sendanin

7 1-Tigloyl-3-acetylvilasinin

8 Nimbin

9 Ochinolide B

10 Salannin

1 Azadirachtin

The most obvious conclusion from Kraus' studies is that the insect growth regulatory effects are exclusive to the azadirachtins; none of the other groups of limonoids possess this type of physiological activity in insects. Within the azadirachtin group, the azadirachtins, azadirachtols and meliacarpins all appear comparable as both IGRs and antifeedants.[25–27] On the other hand, many of the neem limonoids with less complex carbon skeletons are effective antifeedants against the bean beetle. The salannins, nimbins and nimbolinins are all comparable to azadirachtin as antifeedants (Table 1). However, as larval growth inhibitors against *Heliothis virescens*[31] and *Spodoptera litura*, the salannins and nimbins (major limonoid constitutents of neem kernels) are at least 200 times less active than the azadirachtins (Isman et al., unpublished data).

Salannin

Although salannin (**10**) is relatively ineffective as an insect growth inhibitor, and lacks molt-disrupting bioactivity, it does have pronounced antifeedant

Table 1. Meliacins from neem (*Azadirachta indica*): bioactivities of the major skeletal types of limonoids as antifeedants, growth inhibitors and growth regulators against insects

	Major feature	Example	Antifeedant EC_{50}*	Larval growth EC_{50}†	IGR‡
Skeletal type					
Azadirone	rings A-D intact	azadiradione (**4**)	320	560	no
Gedunin	ring D lactone	gedunin (**5**)	930	50	no
	C-19/28 oxygen bridge	sendanin (**6**)	**	45	no
Amoorastatin					
Vilasinin	C-6/28 oxygen bridge	1-tigloyl-3-acetylvilasinin (**7**)	10	—	no
C-seco type					
Nimbin	C-7/15 ether bridge	nimbin (**8**)	50	††	no
	C-12/15 ether bridge	ochinolide B (**9**)	20	1500	no
Nimbolinin					
Salannin	C-6/28, C-7/15 oxygen bridges	salannin (**10**)	14	170	no
Azadirachtin	rings C/D as furanacetate	azadirachtin (**1**)	13	0.7	YES

*bioassay with the Mexican bean beetle, *Epilachna varivestis*; values shown are in ppm of solution applied to bean leaf discs; data from ref. 17.
†bioassay with the cotton bollworm, *Helicoverpa zea*; values shown are in ppm, dietary concentration; data from refs. 17 and 31.
‡bioassay for molt disruption in *Epilachna varivestis* or *Oncopeltus fasciatus*.
**this and related compounds are relatively strong antifeedants for *Spodoptera litura*.
†† for *S. litura*, dietary $EC_{50} > 200$ ppm.

effects in beetles. Yamasaki and Klocke[32] prepared fourteen derivatives of salannin and tested these as antifeedants against the Colorado potato beetle, a monophagous insect known to be sensitive to antifeedants. While salannin itself was not particularly active, antifeedant activity was increased more than 40-fold by hydrogenation of the furan ring and the tigloyl group at C-1, along with replacement of the acetoxy function with a methoxy group at C-3. Derivatives prepared in this way were actually 7.5-fold more active than azadirachtin. In contrast, the derivatives were no more active than the parent compound as antifeedants against the fall armyworm, *Spodoptera litura*. Kraus has evaluated the four major natural salannins as antifeedants against the Mexican bean beetle.[22] EC_{50} values for these compounds range from 9–20 ppm, making them comparable to azadirachtin ($EC_{50} = 13$ ppm). As both the potato beetle and bean beetle are specialist feeders, extrapolation of antifeedant effects of compounds against these species to other insects (for example, generalists like noctuid caterpillars) is risky.

Toosendanin

The limonoid toosendanin (**11**), obtained in stembark extracts of the tree *Melia toosendan* (possibly a synonym of *M. azedarach*[33]), is the primary active ingredient of a botanical insecticide recently developed in China.[34] Toosendanin acts as a growth inhibitor, stomach poison and antifeedant to a number of pest insects. This compound, possessing a C-19/28 oxygen bridge (a member of the amoorastatin group by Kraus' nomenclature), is closely related to the meliatoxins from *M. azedarach* and the trichilins from *Trichilia* species.[18] In evaluating the bioactivity against insects of refined bark extracts, containing 60–75% toosendanin, we observed that on a weight-to-weight basis, the extracts were more inhibitory to the growth of the variegated cutworm, *Peridroma saucia*, than was pure toosendanin.[35] This suggested the presence of minor constituents with significant bioactivity and/or synergistic effects among related active constituents.

Bioassay driven fractionation by column chromatography over silica gel and Sephadex LH-20 followed by preparative HPLC led to the isolation of the 12- and 3-deacetyl analogues (**12, 13**) of toosendanin. The latter compound is a new natural product, whereas the 12-deacetyl derivative had previously been isolated from *M. azedarach* var. *japonica*[36] and *M. dubia*.[17] Because sendanin (**6**) is the C-28 acetate of toosendanin, the former compound was readily synthesized, as were the 1- and 7-acetate derivatives (**17, 18**) of sendanin.

When tested at equimolar concentrations against *Spodoptera litura* neonate larvae in a 10-day growth bioassay and against fourth instar larvae in a leaf-disc choice test (for antifeedant effects), toosendanin and sendanin were found to be equally effective in both bioassays (Table 2). The 12-deacetyl analogue of toosendanin (**12**) was as effective as the parent compound, whereas the 3-deacetyl analogue (**13**) was significantly less active in both bioassays. The 3,12-dideacetyl synthetic derivative (**14**) was less effective as a growth inhibitor, but equally effective as an antifeedant. Acetylation of sendanin resulted in a slight, but significant, reduction in growth inhibitory activity, and a substantial reduction in antifeedant activity.

Unlike sendanin, toosendanin exists as an equilibrium mixture of two compounds (C-28 epimers). Benzoyl substitution on C-28 of toosendanin resulted in reductions in both inhibitory and antifeedant activities when tested as a racemic mixture. Interestingly, while both epimers were equally active as modest growth inhibitors, the S-28-benzoyl epimer (**15**) was a significantly stronger antifeedant than the R-epimer (**16**) (Table 2).

Based on our results, we conclude that acetyl substitution plays only a minor role in the structure-activity relations of this group of compounds. Although 12-deacetyl-toosendanin is as active as its parent compound, it is no more so in our bioassays with *S. litura*. In a bioassay utilizing the related armyworm *S. eridania*, the most active antifeedant among seven limonoids isolated from the bark of *M. toosendan* was azedarachin A, the only compound of the seven with a hydroxyl substituent at C-12.[37] There may yet remain other minor constituents of the *M. toosendan* bark extract that are more active, but more likely, the minor constituents may synergize toosendanin (the major constituent) such that the extract is more effective than any of its constituents in isolation.

	R₁	R₂	
11	Ac	Ac	Toosendanin
12	H	Ac	12-Deacetyltoosendanin
13	Ac	H	3-Deacetyltoosendanin
14	H	H	3,12-Dideacetyltoosendanin

	R₁	R₂	R₃	R₄	
6	OAc	H	H	H	Sendanin
17	OAc	H	Ac	H	1-Acetylsendanin
18	OAc	H	H	Ac	7-Acetylsendanin
15	OCOPh	H	H	H	(28S)-Benzoyltoosendanin
16	H	OCOPh	H	H	(28R)-Benzoyltoosendanin

Gedunin

Some genera of the Meliaceae, such as *Swietenia, Khaya* and *Cedrela* are important timber species in tropical regions. Bark and woodwaste (sawdust) from these timber species might represent inexpensive, locally abundant natural resources from which large quantities of limonoids could be extracted for commercial use. We have been following this strategy with *Cedrela odorata*, commonly known as spanish

Table 2. Larval growth inhibitory and antifeedant effects of toosendanin and analogues on larvae of *Spodoptera litura*

Compound	Larval weight (day 10) (% of controls)*	Feeding deterrence (%)†
1st experiment		
toosendanin (**11**)	56.3 a	77.7 a
3,12-dideacetyl-toosendanin (**14**)	76.9 cd	70.1 a
28-benzoyl-toosendanin (racemic)	90.1 d	53.8 b
sendanin (**6**)	60.2 ab	75.1 a
1-acetyl-sendanin (**17**)	73.2 bc	27.3 c
7-acetyl-sendanin (**18**)	79.7 cd	45.9 b
2nd experiment		
toosendanin (**11**)	32.8 a	86.8 a
3-deacetyl-toosendanin (**13**)	82.0 b	69.5 b
12-deacetyl-toosendanin (**12**)	40.4 a	98.8 a
28R-benzoyl-toosendanin (**15**)	76.8 b	32.3 d
28S-benzoyl-toosendanin (**16**)	77.0 b	55.3 c

*All compounds tested at a dietary concentration of 0.35 µmol g^{-1} fwt (approx. 200 ppm).
†All compounds tested at a concentration of 0.35 nmol cm^{-2} (approx. 2 µg cm^{-2}). See ref. 47 for methods.
NB Means followed by the same letter within a column within each experiment do not differ significantly (P<0.05, Tukey's test).

cedar, a key timber species in the tropical Americas. Solvent extraction of *C. odorata* sawdust yields large quantities of the simple limonoid gedunin (**5**). Gedunin is a modest growth inhibitor to caterpillars,[22] but a potent natural antimalarial, based on *in vitro* bioassays with *Plasmodium falciparum*.[38]

In an effort to enhance the bioactivity of gedunin and to determine structure requirements for activity, simple derivatives (**19–23**) were prepared and bioassayed for growth inhibition of the European corn borer, *Ostrinia nubilalis*. Modifications included saturation of ring A, epoxidation of ring A, and reduction and acetylation of the keto group at C-1. Whereas gedunin is at best a weak antifeedant to the bean beetle *Epilachna*, the closely related limonoid, khivorin (**24**)(from *Khaya nyasica*), is almost 20-fold more active.[22] Khivorin differs from gedunin only in having a saturated ring A and acetoxy functions at C-1 and 3.

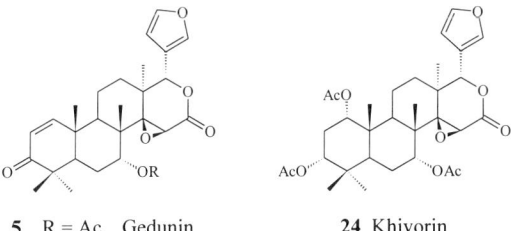

5 R = Ac Gedunin
22 R = H 7-Deacetylgedunin

24 Khivorin

Table 3. Larval growth inhibitory effects of gedunin and analogues on larvae of *Ostrinia nubilalis*

Compound	Larval weight (day 19) (% of controls)[*]
1,2-dihydro-3β-gedunol (**20**)	111.8 a
1,2-dihydrogedunin (**19**)	87.1 b
hexahydrogedunin	86.5 b
21-acetylgedunin	83.5 b
7-deacetylgedunin (**22**)	71.8 b
gedunin (**5**)	70.6 b
23-acetylgedunin	51.8 c
1,2-epoxygedunin (**23**)	25.9 d

[*]All compounds tested at a dietary concentration of 50 ppm.
NB Means followed by the same letter do not differ significantly ($P<0.05$, Tukey's test).

	R	
19	O=	1,2-Dihydrogedunin
20	OH	1,2-Dihydro-3β-gedunol
21	OAc	3β-Acetoxy-1,2-dihydrogedunin

23 1,2-Epoxygedunin

Most of the alterations to ring A and at C-7 of gedunin resulted in diminished bioactivity against both the corn borer (Table 3) and malaria parasites (J. Pezzutto, unpublished data). Conversion of the C-1/2 double bond into an epoxide increased growth inhibition about 2.5-fold, but reduced antimalarial activity two-fold. None of the derivatives were more active than gedunin itself when tested against *Spodoptera litura*, but we have not tested mixtures of these compounds.

OTHER COMPOUNDS RECENTLY ISOLATED

Limonoids

Native Meliaceae grown in plantations are frequently attacked as saplings by the mahogany shoot borer, *Hypsipyla*. Following attack apical dominance is lost and trees do not form a harvestable bole. The Asian species *Toona ciliata* var. *australis* appears immune from shoot borer attack, whereas *Swietenia mahogani* and *Cedrela odorata* are readily attacked. Kraus[22] has isolated a number of B-*seco* limonoids (toonacilins, toonacilids and toonaphyllins) from *T. ciliata*. As most of these compounds are effective antifeedants against *Epilachna*, it has been suggested that they may serve as resistance factors in *Toona* against the shoot borer. We recently bioassayed some B-*seco* limonoids isolated from different sources to determine if this type of limonoid is generally effective against insects. The compounds tested were: turraflorins A (**25**) and C, from *Turraea floribunda* (South Africa);[39] nymanin C (**26**) from *Dysoxylum malabaricum* (India)(T. Govindachari, unpublished data); and cedrelanolide I (**27**) from *Cedrela salvadorensis* (Costa Rica).[40] In our earlier screening of foliar extracts of meliaceaeous species, an extract of *Turraea holstii* was found to be very active as a growth inhibitor to the variegated cutworm, *Peridroma saucia* and as an antifeedant to the migratory grasshopper, *Melanoplus sanguinipes*.[41]

None of the B-*seco* limonoids were active as growth inhibitors when fed to neonate *Spodoptera litura* larvae at our screening concentration of 50 ppm, nor as molt-disruptants to the milkweed bug, *Oncopeltus fasciatus*, when topically applied to fifth instar nymphs at 50 µg per insect. Piscidinols C (**28**), E, and F, limonoids possessing a 7-membered lactone ring at C-17 (ring D), were isolated from *Walsura piscida*.[42] These compounds are also inactive against *S. litura*, even though an extract of this plant is reputed to have significant antifeedant activity against insects.[42] In contrast, sandoricin (**29**) and its 6-hydroxy analogue, which are B-*seco* limonoids isolated from the seeds of *Sandoricum koetjape*, are both significant antifeedants and growth inhibitors for the fall armyworm, *Spodoptera frugiperda* and the European corn borer, *Ostrinia nubilalis*.[43] Analogues of cedrelone, bearing hydroxyl groups on the butenolide ring, have recently been isolated from *Toona ciliata* growing in Brazil.[44] It has been suggested that these compounds might also be involved in resistance to the shoot borer.[44]

Several species in the genus *Trichilia* were found to be active when we screened extracts against the variegated cutworm.[45] Hirtin (**30**), isolated from the seeds of *T. hirta*,[46] is a strong growth inhibitor (EC_{50} = 11.5 ppm; cf. cedrelone from *T. ciliata*, EC_{50} = 53 ppm[47]). Species of *Chisocheton* also proved to be active in our systemic investigation of the Meliaceae. From *C. microcarpus* (Indonesia) we isolated two new limonoids (e.g. **31**) of the azadirone-type, but these compounds were inactive against the cutworm.[48] The active principles remain to be found in this species. Closely related limonoids have been isolated from *C. paniculatus*.[49] Finally, humilinolides, mexicanolide-type limonoids isolated from *Swietenia humilis*, appear to be moderately active growth inhibitors for the mealworm, *Tenebrio molitor*.[50]

25 Turraflorin A

26 Nymanin C

27 Cedrelanolide I

28 Piscidinol C

29 Sandoricin

30 Hirtin

31 *Chisocheton* Cmpd 1

32 Rocaglamide

Non-limonoid Insecticides

In screening meliaceous plants for insect bioactivity, one of the most active species found was *Aglaia odorata*.[41] Bioassay-driven fractionation of this species led to the isolation of a series of potent benzofurans, including the known anti-leukemic agent, rocaglamide (**32**).[51] Rocaglamide is the most potent plant natural product we have isolated, with an EC_{50} in the variegated cutworm of 0.91 ppm (cf. azadirachtin, EC_{50} = 0.19 ppm). Its activity as a growth inhibitor for the European corn borer is comparable to that of azadirachtin. However, the three other rocaglamide analogues isolated are also strikingly active, and the natural mixture of these, as occurs in the plant may be more active than the sum of these constituents alone.

As previously mentioned, extracts of several species of *Trichilia* were active in our investigations.[45] In particular, wood and bark extracts of *T. hirta* and *T. glabra* are potent growth inhibitors and antifeedants to both *Peridroma saucia* and *Spodoptera litura*. Given that sendanin was previously isolated from *T. roka*,[52] and hirtin and related limonoids have been isolated from *T. hirta* fruits,[46] we expected limonoids to be responsible for the bioactivity of wood and bark extracts of *T. hirta*. However, no peak corresponding to hirtin was found in HPLC chromatograms of our extracts, nor were any obvious signals indicative of limonoids seen in NMR analyses of the extracts. Instead, bioassay-driven fractionation has led us to the isolation of two compounds, $C_{21}H_{32}O_3$ and $C_{19}H_{24}O_3$, as the putative active principle in *T. hirta* wood extracts (D. Chauret et al., unpublished data). Their structures have been determined with the aid of a variety of NMR probe sequences. Toxicological characterization of this and a related compound in *S. litura* and other insects is underway.

THE SIGNIFICANCE OF MIXTURES

When a plant is investigated for biologically-active phytochemicals, invariably two or more related compounds are discovered, even if one particular compound is predominant in terms of bioactivity toward a certain organism. It can be said that individual plant species appear to synthesize "families" of closely related compounds. In the case of the Meliaceae, these families are usually limonoids differing by patterns of acetylation or other ester groups alone. In the extreme case of neem (*Azadirachta indica*), several skeletal types of limonoids are present in a single tissue (seeds).

What is the significance of these mixtures of related compounds? One possibility is that they equip the plant with a diverse arsenal capable of defense against a wide array of potential enemies - pathogens, herbivores, and competing plants. Another possibility is that the mixtures act in a synergistic fashion, i.e. on a weight-to-weight basis, a multicomponent mixture is more efficacious than a single compound. This latter hypothesis has been tested several times with respect to plant-insect chemical interactions, and was addressed over a decade ago.[53]

Melia toosendan Studies

As previously mentioned, we found that bark extracts of *M. toosendan*, containing as much as 75% toosendanin (**11**), were more inhibitory to larval growth of *Peridroma* than pure toosendanin itself.[35] Bioassay-driven fractionation of the semi-purified extracts led to the isolation of two analogues of toosendanin, and while these compounds were also active, they were minor constituents, by weight, and neither was more active than toosendanin. Since we could not find a minor constituent with exceptional bioactivity (contributing more bioactivity to the extract than its mass would dictate), the likely explanation for the enhanced bioactivity of the extract is a synergistic action among the limonoids present.

We have found that not only is the extract more active than pure toosendanin, but, confirming earlier work done in China, the extract is capable of synergizing commercial insecticides. For example, at 50 ppm in diet, the organophosphate insecticide malathion causes 40% mortality of fourth instar *Spodoptera litura*. At 800 ppm in diet, *M. toosendan* extract has no adverse effect on larvae. However, when these two treatments are combined at the concentrations shown above, mortality increases to more than 80%.[54] As malathion is detoxified primarily by carboxyesterase enzymes, we investigated the effects of toosendanin and the bark extract on midgut esterase activity in *S. litura*. The extract inhibited midgut esterase activity of both *S. litura* and the grasshopper *Melanoplus sanguinipes, in vitro*, with an IC_{50} of approximately 12 ppm.[54] However, pure toosendanin was not inhibitory, even at 24 ppm. Not surprisingly, neither of the deacetyl analogues we isolated were inhibitory, suggesting that inhibition of the enzyme activity also represents a synergistic action of the extract.

Neem (*Azadirachta indica*) Studies

While there is general agreement that for most insects azadirachtin (**1**) is the main active ingredient in neem, for several years scientists have debated the contribution of the other limonoids occurring in neem kernels at the same or greater concentrations than azadirachtin. In an attempt to address this question, we quantified the azadirachtin content of twelve samples of neem oil from India (ranging from <0.005 to >0.4% azadirachtin) by reverse-phase HPLC, and conducted a series of parallel bioassays. We determined that growth inhibition and antifeedant action in the variegated cutworm (*Peridroma saucia*) and molt disruption in the milkweed bug (*Oncopeltus fasciatus*) were all highly correlated with azadirachtin content. Azadirachtin accounted for 85–90% of the bioactivity in the growth inhibition and molt disruption bioassays, respectively, and 72% of the antifeedant action.[55]

In a sequel to that study, we conducted a similar one using refined neem seed concentrates (the technical grade material used to manufacture commercial neem insecticides) containing from 5–50% azadirachtin. Samples were analyzed by HPLC for azadirachtin (**1**), 3-tigloylazadirachtol (**2**)("aza B"), nimbin (**8**), deacetylnimbin, salannin (**10**) and deacetylsalannin (i.e. all the major limonoid constituents). Parallel bioassays for growth inhibition were conducted with *S. litura* larvae and for mortality (via molt disruption) with the green peach aphid, *Myzus persicae*. Azadirachtin, "aza B" and "total azadirachtins" were significantly correlated with larval growth inhibition, whereas the nimbins and salannins were not. In the case of IGR effects in the aphid, azadirachtin alone was significantly correlated with mortality, although correlations with "aza B" and "total azadirachtins" approached the level of statistical significance. As with *Spodoptera*, neither the nimbins nor the salannins were correlated with aphid mortality.

These results are not surprising when the bioactivities of the individual compounds are examined (Table 4). In *S. litura*, the dietary EC_{50} value for salannin is 75 times higher than that for azadrachtin, and the nimbins are at least two orders of magnitude less active than azadirachtin. For 2nd instar *M. persicae*, the LC_{50} values for salannin and nimbin are ca. 300 and 450 times higher, respectively, than that for azadirachtin. At least in these two species, azadirachtin is almost two orders of magnitude more effective than the other major limonoids from neem.

The results shown above could easily create the impression that neem insecticides are little more than a preparation of azadirachtin along with some "inert" limonoid contaminants! However, we have recently obtained evidence that the "other" constituents may have important consequences for the practical use of neem for insect control. It has been suggested that the development of resistance to neem insecticides in insects may be less probable than that to

Table 4. Efficacy of pure neem constituents to larvae of *Spodoptera litura* and nymphs of *Myzus persicae*

Compound	*S. litura* larval growth, dietary EC_{50} (ppm) (day 10)	*M. persicae* mortality, LC_{50} (ppm in solution)(day 6)
Azadirachtin (**1**)	0.21	1.3
3-Tigloyl-azadirachtol (aza B) (**2**)	0.09	—
Salannin (**10**)	15.7	383
Deacetyl-salannin	> 25	—
Nimbin (**8**)	> 25	> 500
Deacetyl-nimbin	> 25	> 500

synthetic insecticides, because neem contains numerous ingredients with different structures and actions. At the same time, we have speculated that under sufficient selection pressure, insects could evolve resistance to pure azadirachtin if exposed to that chemical alone.

To test these hypotheses, we conducted an eight-month selection experiment using the green peach aphid. From a parental line never exposed to neem, we selected two lines, one treated once a week with pure azadirachtin, and a second treated at the same frequency with a neem concentrate containing the same amount of azadirachtin. Each month the two selected lines and the parental line were tested for their susceptibility to pure azadirachtin. After two months, the aza-selected line demonstrated significantly decreased susceptibility to azadirachtin, and after eight months (approximately 40 generations), had developed nine-fold resistance to azadirachtin, compared to the parental line.[56] However, at no point in the experiment did the neem-selected line exhibit any resistance to pure azadirachtin. These results suggest that the blend of constituents in neem may somehow diffuse the selection process, mitigating resistance to azadirachtin.

We have also found that larvae of *S. litura* can rapidly habituate (i.e. become desensitized) to the antifeedant effects of azadirachtin upon repeated or continuous exposure (Bomford and Isman, unpublished data). In contrast, when exposed to neem producing the same concentration of azadirachtin, habituation does not take place and the caterpillars remain continuously sensitive to the antifeedant action. It is even conceivable that this situation is a paradigm, partly explaining the value of mixtures against a vast range of biotic selecting agents.

Therefore, the durability of neem as both an antifeedant and insecticide, relative to azadirachtin alone, argue in favor of the use of neem as a true botanical preparation, rather than the use of pure azadirachtin as a commercial insecticide, should it ever be possible to synthesize the chemical or produce it in on a commercial scale from tissue culture.

CONCLUSION

Many of the meliaceous limonoids isolated to date, at least those used in our bioassays, are relatively inactive against insects at biologically realistic concentrations. On the other hand, a majority of limonoids appear to have a reasonable level of antifeedant activity, at least against the Mexican bean beetle. More importantly, the vast majority of these compounds have not yet been widely screened against other organisms, for example plant pathogens and other microorganisms. Some of the limonoids, such as azadirachtin, attack unique endocrine targets in insects, and may not, therefore, be detected in common cytotoxicity bioassays (e.g. brine shrimp bioassay, human cell line bioassays). Potentially useful biological activities may have been overlooked. We also must acknowledge that our understanding of the role of these compounds in the natural context (i.e. as protective agents in the trees producing them) is almost nonexistent.

Neem provides an excellent opportunity to investigate structure-activity relations because phytochemical work on this species has revealed a large number of limonoids representing an array of skeletal types. Even among the neem limonoids, however, there are vast differences in efficacy against insect species, between bioassays with different endpoints (e.g. behavioral versus physiological), and between insects and other organisms. At our current state of knowledge, we cannot predict the bioactivity of individual, novel limonoids to insects or other target organisms with any degree of confidence.

Finally, with respect to the commercial development of insect control products based on plants of the Meliaceae, we conclude that: natural insecticides (by our definition given herein) are rare (limited to neem and its congeners); insect growth inhibitors are considerably more common; and antifeedants may even be considered widespread. Only time will tell if any of the plant derivatives under consideration for commercialization, apart from neem, will produce viable insect control products.

ACKNOWLEDGMENTS

We thank Drs. D. Mulholland (University of Natal, South Africa), R. Mata (UNAM, Mexico) and G. Suresh (SPIC Foundation, India) for providing limonoids for bioassay, Drs. P. Gunning and W. Chen for technical assistance, and Dr. Claus Passreiter for preparation of the figures. The authors acknowledge the generous support of their research from the Natural Sciences and Engineering Research Council (Canada) through research and strategic grants, as well as funding from the Integrated Forest Pest Management initiative under Natural Resources Canada's Green Plan.

REFERENCES

1. SCHMUTTERER, H. (ed.), 1995. The Neem Tree *Azadirachta indica* A. Juss. and Other Meliaceous Plants: Sources of Unique Natural Products for Integrated Pest Management, Medicine, Industry and Other Purposes. VCH, Weinheim, p. 696.
2. NAWROT, J., KOUL, O., ISMAN, M.B., HARMATHA, K., 1991. Naturally occurring antifeedants: effects on two polyphagous lepidopterans, J. Appl. Ent. 112: 194–201.
3. MULLIN, C.A., MASON, C.H., CHOU, J.C., LINDERMAN, J.R., 1992. Phytochemical antagonism of γ-aminobutyric acid based resistances in *Diabrotica*. In: Molecular Mechanisms of Insect Resistance (C.A. Mullin and J.G. Scott, eds.). American Chemical Society, Washington, D.C., pp. 288–308.
4. GOULD, F., 1991. Evolutionary potential of crop pests, Amer. Sci. 79: 496–507.
5. TABASHNIK, B.E., CUSHING, N.L., FINSON, N., JOHNSON, M.W., 1990. Field development of resistance to *Bacillus thuringiensis* in diamondback moth (Lepidoptera: Plutellidae), J. Econ. Entomol. 83: 1671–1676.

6. FRAZIER, J.L., 1986. The perception of plant allelochemicals that inhibit feeding. In: Molecular Aspects of Insect-Plant Associations (L.B. Brattsten and S. Ahmad, eds.). Plenum Press, New York, pp. 1–42.
7. JERMY, T., 1990. Prospects of antifeedant approaches to pest control - a critical review, J. Chem. Ecol. 16: 3151–3166.
8. SCHOONHOVEN, L.M., 1982. Biological aspects of antifeedants, Entomol. exp. appl. 31: 57–69.
9. GRIFFITHS, D.C., MANIAR, S.P., MERRIT, L.A., MUDD, A., PICKETT, J.A., PYE, B.J., SMART, L.E., WADHAMS, L.J., 1991. Laboratory evaluation of pest management strategies combining antifeedants with insect growth regulator insecticides, Crop Protection 10: 145–151.
10. MILLER, J.R., COWLES, R.S., 1990. Stimulo-deterrent diversion: a concept and its possible application to onion maggots, J. Chem. Ecol. 16: 3197–3212.
11. MURRAY, K.D., ALFORD, A.R., GRODEN, E., DRUMMOND, F.A., STORCH, R.H., BENTLEY, M.D., SUGATHAPALA, P.M., 1993. Interactive effects of an antifeedant used with *Bacillus thuringiensis* var. *san diego* delta endotoxin on Colorado potato beetle (Coleoptera: Chrysomelidae), J. Econ. Entomol. 86: 1793–1801.
12. CHAMPAGNE, D.E., ISMAN, M.B., TOWERS, G.H.N., 1989. Insecticidal activity of phytochemicals and extracts of the Meliaceae. In: Insecticides of Plant Origin (J.T. Arnason, B.J.R. Philogene and P. Morand, eds.). American Chemical Society, Washington, D.C., pp. 95–109.
13. ISMAN, M.B., 1993. Growth inhibitory and antifeedant effects of azadirachtin on six noctuids of regional economic importance, Pestic. Sci. 38: 57–63.
14. JERMY, T., BERNAYS, E.A., SZENTESI, A., 1982. The effect of repeated exposure to feeding deterrents on their acceptability to phytophagous insects. In: Insect-Plant Relationships (J.H. Visser and A.K. Minks, eds.), Pudoc, Wageningen, pp. 25–32.
15. RAFFA, K.F., FRAZIER, J., 1988. A generalized model for quantifying behavioral de-sensitization to antifeedants, Entomol. exp. appl. 46: 93–100.
16. CONNOLLY, J.D., 1983. Chemistry of the limonoids of the Meliaceae and Cneoraceae. In: Chemistry and Chemical Taxonomy of the Rutales (P.G. Waterman and M.F. Grundon, eds.), Academic Press, London, pp. 175–213.
17. KRAUS, W., 1995. Azadirachtin and other triterpenoids. In: The Neem Tree, *Azadirachta indica* A. Juss. and Other Meliaceous Plants: Sources of Unique Natural Products for Integrated Pest Management, Medicine, Industry and Other Purposes (H. Schmutterer, ed.), VCH, Weinheim, pp. 35–88.
18. CHAMPAGNE, D.E., KOUL, O., ISMAN, M.B., SCUDDER, G.G.E., TOWERS, G.H.N., 1992. Biological activity of limonoids from the Rutales, Phytochem. 31: 377–394.
19. BUTTERWORTH, J.H., MORGAN, E.D., 1968. Isolation of a substance that suppresses feeding in locusts, J. Chem. Soc., Chem. Commun., 23–24.
20. KRAUS, W., BOKEL, M., KLENK, A., PÖHNL, H., 1985. The structure of azadirachtin and 22,23-dihydro-23β-methoxyazadirachtin, Tetrahedron Lett. 26: 6435–6438.
21. LEY, S.V., DENHOLM, A.A., WOOD, A., 1993. The chemistry of azadirachtin, Nat. Prod. Rep., 100–157.
22. KRAUS, W., BOKEL, M., SCHWINGER, M., VOGLER, B., SOELLNER, R., WENDISCH, D., STEFFENS, R., WACHENDORFF, U., 1993. The chemistry of azadirachtin and other insecticidal constituents of Meliaceae. In: Phytochemistry and Agriculture (T.A. van Beek and H. Breteler, eds.), Clarendon Press, Oxford, pp. 18–39.
23. LEE, S.A., KLOCKE, J.A., BARNBY, M.A., YAMASAKI, R.B., BALANDRIN, M.F., 1991. Insecticidal constituents of *Azadirachta indica* and *Melia azedarach* (Meliaceae). In: Naturally Occurring Pest Bioregulators (P.A. Hedin, ed.), American Chemical Society, Washington, D.C., pp. 293–304.
24. HANSEN, D.J., CUOMO, J., KHAN, M., GALLAGHER, R.T., ELLENBERGER, W.P., 1994. Advances in neem and azadirachtin chemistry and bioactivity. In: Natural and Engineered Pest

Managment Agents (P.A. Hedin, J.J. Menn and R. M. Hollingworth, eds.), American Chemical Society, Washington, D.C., pp. 103–129.
25. REMBOLD, H., PUHLMANN, I., 1993. Phytochemistry and biological activity of metabolites from tropical Meliaceae. In: Phytochemical Potential of Tropical Plants (K.R. Downum, J.T. Romeo and H.A. Stafford, eds.), Plenum Press, New York, pp. 153–165.
26. REMBOLD, H., PUHLMANN, I., 1995. Azadirachtins: structure and activity relations in case of *Epilachna varivestis*. In: The Neem Tree *Azadirachta indica* A. Juss. and Other Meliaceous Plants: Sources of Unique Natural Products for Integrated Pest Management, Medicine, Industry and Other Purposes (H. Schmutterer, ed.), VCH, Weinheim, pp. 222–230.
27. BLANEY, W.M., SIMMONDS, M.S.J., LEY, S.V., ANDERSON, J.C., TOOGOOD, P.L., 1990. Antifeedant effects of azadirachtin and structurally related compounds on lepidopterous larvae, Entomol. exp. appl. 55: 149–160.
28. SIMMONDS, M.S.J., BLANEY, W.M., LEY, S.V., ANDERSON, J.C., TOOGOOD, P.L., 1990. Azadirachtin: structural requirements for reducing growth and increasing mortality in lepidopterous larvae, Entomol. exp. appl. 55: 169–181.
29. BLANEY, W.M., SIMMONDS, M.S.J., LEY, S.V., ANDERSON, J.C., SMITH, S.C., WOOD, A., 1994. Effect of azadirachtin-derived decalin (perhydronaphthalene) and dihydrofuranacetal (furo[2,3-*b*]pyran) fragments on the feeding behaviour of *Spodoptera littoralis*, Pestic. Sci. 40: 169–173.
30. REMBOLD, H., 1989. Isomeric azadirachtins and their mode of action. In: Focus on Phytochemical Pesticides. Volume 1, The Neem Tree (M. Jacobson, ed.), CRC Press, Boca Raton, pp. 47–67.
31. KLOCKE, J.A., 1987. Natural plant compounds useful in insect control. In: Allelochemicals: Role in Agriculture and Forestry (G.R. Waller, ed.), American Chemical Society, Washington, D.C., pp. 396–415.
32. YAMASAKI, R.B., KLOCKE, J.A., 1989. Structure-bioactivity relationships of salannin as an antifeedant against the Colorado potato beetle (*Leptinotarsa decemlineata*), J. Agric. Food Chem. 37: 1124–1130.
33. MABBERLEY, D.J., 1984. A monograph of *Melia* in Asia and the Pacific. The history of white cedar and Persian lilac, Garden's Bulletin 37: 49–64.
34. ISMAN, M.B., 1995. Leads and prospects for the development of new botanical insecticides, Rev. Pestic. Toxicol. 3: 1–20.
35. CHEN, W., ISMAN, M.B., CHIU, S.-F., 1995. Antifeedant and growth inhibitory effects of the limonoid toosendanin and *Melia toosendan* extractson the variegated cutworm, *Peridroma saucia* (Lep., Noctuidae), J. Appl. Ent. 119: 367–370.
36. AHN, J.-W., CHOI, S.-U., LEE, C.-O., 1994. Cytotoxic limonoids from *Melia azedarach* var. *japonica*, Phytochem. 36: 1493–1496.
37. ZHOU, J. B., OKAMURA, H., IWAGAWA, T., NAKATANI, M., 1995. Limonoid antifeedants from *Melia toosendan*, Phytochem. 41: 117–120.
38. KHALID, S.A., FAROUK, A., GEARY, T.G., JENSEN, J.B., 1986. Potential antimalarial candidates from African plants: an *in vitro* approach using *Plasmodium falciparum*, J. Ethnopharmacol. 15: 201–209.
39. FRASER, L.A., MULHOLLAND, D.A., NAIR, J.J., 1994. Limonoids from the seed of *Turraea floribunda*, Phytochem. 35: 455–458.
40. SEGURA, R., CALDERON, J., TOSCANO, R., GUTIERREZ, A., MATA, R., 1994. Cedrelanolide I, a new limonoid from *Cedrela salvadorensis*, Phytochem. Soc. N. Amer. Newslett. 4(1): 29.
41. CHAMPAGNE, D.E., ISMAN, M.B., DOWNUM, K.R., TOWERS, G.H.N., 1993. Insecticidal and growth-reducing activity of foliar extracts from Meliaceae, Chemoecology 4: 165–173.
42. GOVINDACHARI, T.R., KRISHNA KUMARI, G.N., SURESH, G., 1995. Triterpenoids from *Walsura piscida*, Phytochem. 39: 167–170.

43. POWELL, R.G., MIKOLAJCZAK, K.L., ZILKOWSKI, B.W., 1991. Limonoid antifeedants from seed of *Sandoricum koetjape*, J. Nat. Prod. 54: 241–246.
44. AGOSTINHO, S.M.M., SILVA, M.F. das G.F. da, FERNANDES, J.B., VIEIRA, P.C., PINHEIRO, A.L., VILELA, E.F., 1994. Limonoids from *Toona ciliata* and speculations on their chemosystematic and ecological significance, Biochem. Syst. Ecol. 22: 323–328.
45. XIE, Y.S., ISMAN, M.B., GUNNING, P., MACKINNON, S., ARNASON, J.T., TAYLOR, D.R., SANCHEZ, P., HASBUN, C., TOWERS, G.H.N., 1994. Biological activity of extracts of *Trichilia* species and the limonoid hirtin against lepidopteran larvae, Biochem. Syst. Ecol. 22: 129–136.
46. CHAN, W.R., TAYLOR, D.R., 1966. Hirtin and deacetylhirtin: new "limonoids" from *Trichilia hirta*, J. Chem. Soc., Chem. Commun., 206–207.
47. KOUL, O., ISMAN, M.B., 1992. Toxicity of the limonoid allelochemical cedrelone to noctuid larvae, Entomol. exp. appl. 64: 281–287.
48. GUNNING, P.J., JEFFS, L.B., ISMAN, M.B., TOWERS, G.H.N., 1994. Two limonoids from *Chisocheton microcarpus*, Phytochem. 36: 1245–1248.
49. BORDOLOI, M., SAIKIA, B., MATHUR, R.K., GOSWAMI, B.N., 1993. A meliacin from *Chisocheton paniculatus*, Phytochem. 34: 583–584.
50. SEGURA-CORREA, R., MATA, R., ANAYA, A.L., HERNANDEZ-BAUTISTA, B., VILLENA, R., SORIANO-GARCIA, M., BYE, R., LINARES, E., 1993. New tetranortriterpenoids from *Swietenia humilis*, J. Nat. Prod. 56: 1567–1574.
51. ISHIBASHI, F., SATASOOK, C., ISMAN, M.B., TOWERS, G.H.N., 1993. Insecticidal 1*H*-cyclopenta[*b*]benzofurans from *Aglaia odorata* (Lour.)(Meliaceae), Phytochem. 32: 307–310.
52. KUBO, I., KLOCKE, J.A., 1982. An insect growth inhibitor from *Trichilia roka* (Meliaceae), Experientia 38: 639–640.
53. BERENBAUM, M., 1985. Brementown revisited: interactions among allelochemicals in plants, Rec. Adv. Phytochem. 19: 139–169.
54. FENG, R., CHEN, W., ISMAN, M.B., 1995. Synergism of malathion and inhibition of midgut esterase activities by an extract from *Melia toosendan* (Meliaceae), Pestic. Biochem. Physiol. 53: 34–41.
55. ISMAN, M.B., KOUL, O., LUCZYNSKI, A., KAMINSKI, J., 1990. Insecticidal and antifeedant bioactivities of neem oils and their relationship to azadirachtin content, J. Agric. Food Chem. 38: 1406–1411.
56. FENG, R., ISMAN, M.B., 1995. Selection for resistance to azadirachtin in the green peach aphid, *Myzus persicae*, Experientia 51: 831–833.

Chapter Seven

THE ROLE OF MIXTURES AND VARIATION IN THE PRODUCTION OF TERPENOIDS IN CONIFER-INSECT-PATHOGEN INTERACTIONS

Rex G. Cates

Chemical Ecology Laboratory
Department of Botany and Range Science
Brigham Young University
Provo, Utah 84602-5181

Introduction 180
    Objectives 180
    Terpenoid Structures 181
Terpenoid Mixtures and Variation: Bark Beetles and Fungal Pathogens ... 181
    Defensive System of Conifers 181
    Oleoresin Effects on Tree-Killing Bark Beetles: The First Line of
        Defense 188
    Mixtures in the Wound-Induced Response against Bark Beetles
        and Fungi 189
    Contribution of Phenolics to Mixtures, Increasing Oleoresin
        Variation, and Resistance 195
Terpenoid Mixtures and Variation: Interactions with Defoliators 197
    Terpenoid Defensive Systems 197
    Terpenoid Interactions with Primary Nutrients 200
Ecological Consequences of Mixtures and Variation in Terpenoids and
    Nutrients 203
    Characteristics of Mixtures 204
    Ecological and Evolutionary Effects of Mixtures 206
Physiological Stress Effects on Mixtures and Variation in Their
    Production 208

Phytochemical Diversity and Redundancy in Ecological Interactions,
edited by Romeo et al., Plenum Press, New York, 1996

INTRODUCTION

The regulation of natural populations was debated intensively during the 1950s and 1960s, and continues to be an important topic of investigation. Although several treatises on this subject could be cited, one of the most important to the area of plant-herbivore-pathogen interactions was published by Hairston, Smith, and Slobodkin.[1] In this stimulating paper the authors put forward three tenants, the third of which states that "Herbivores are seldom food-limited, appear most often to be predator-limited, and therefore are not likely to compete for common resources." Contributing to this conclusion is the disparity between plants and herbivores in mobility, generation times, and the ability of insects and pathogens to adapt to host trees. Generation times of the myriad of insect herbivores that use woody plants are short, and the insects are highly mobile and depend upon host plants for most life processes.[2] Alternatively, woody perennials are long-lived, immobile, and seemingly at the mercy of their natural enemies. This explicit yet sweeping scenario provided a major impetus toward investigations that have led to our current knowledge of plant-herbivore-pathogen interactions at the population, community, and ecosystem levels. Five years later a seminal paper by Ehrlich and Raven[3] enhanced the field of plant-herbivore interactions and chemical ecology with a provocative discussion of the coevolution between butterflies and host plants. Now the field has advanced to investigations at the tritrophic level; and the linking thread among tritrophic interactions is to a large extent the natural products produced by plants.[4]

Objectives

The purpose of this paper is to expand on the ecological role of mixtures of natural plant products and variation in the production of these products as regulating factors in the interactions among plants and their associated herbivores and pathogens. This role emerged in the early 1980s.[5,6] Since then, information has accumulated defining the importance of mixtures as defensive compounds, their composition, variation in their production within and among host plants and populations, and their interaction with nutrients. In this paper, emphasis is placed on patterns in the production of terpenoids produced by conifers and tropical species (e.g. *Hymenaea*, *Copaifera*), and their interactions with tree-killing bark beetles (Scolytidae), fungal pathogens, and defoliators. The literature cited suggests that plants are well-defended against herbivores and pathogens, yet insect outbreaks are now a common occurence. Abiotic and biotic stress is suggested to be the major factor in reducing the effectiveness of mixtures and variation in their production. Several areas will be covered only briefly or not at all because recent reviews are available. These include terpenoid biosynthesis,[7–9] the metabolic costs of production of terpenoids,[10] and an overview of the ecological roles of higher plant terpenoids.[11] Host compounds as semiochemicals, in tritrophic interactions, and the overall significance of terpenoids in ecosystems are not discussed.[4,11–15] Throughout this paper the composition of a mixture is defined as the components and their relative proportion or percentages of the components making up the mixture.[11]

Terpenoid Structures

Terpenes are composed of multiples of five-carbon isoprene units originating in the mevalonate pathway. A great diversity of terpenoids, in the vicinity of 15,000 to 20,000 compounds, have been characterized.[11] In addition, terpenoid carbon skeletons contribute to the structures of other secondary metabolites,[11] and are the most expensive compounds to synthesize.[16] Terpenoids may be transported to flowers and rhizomes to be metabolized in these tissues and serve as a pool of recyclable carbon.[17,18] They are implicated in allelopathic interactions, influence litter decomposition and nutrient dynamics, are sources of energy to soil microbes, serve as defenses against herbivores and pathogens, and are involved in numerous other physiological and ecological interactions.[11,15,16] Categories of terpenoids based on polymers of the five-carbon isoprene unit are the C_{10} (monoterpenes), C_{15} (sesqui-), C_{20} (di-), C_{30} (tri-), C_{40} (tetra-), and $>C_{40}$ (poly) terpenoids.[11] Groups discussed here are mono-, sesqui-, and diterpenes; mono- and sesquiterpenes are referred to as "essential oils," and resins include all three groups. Because of their volatility mono- and sesquiterpenes are easily analyzed by capillary gas chromatography, and upon derivatization diterpenes also resolve well using this method.

The production of terpenoids is under strong genetic control.[14,16,19–21] Alternatively, as with classes of primary metabolites, natural products including terpenes can be strongly modified by environmental influences.[20,22] Stress due to abiotic and biotic factors diminish the inherent toxicity, deterrency, and inhibitory advantages provided by mixtures and variation in the production of these chemicals. Reduction in the effectiveness of these chemical adaptations is proposed to be a most significant cause of outbreaks of forest pests.[23]

TERPENOID MIXTURES AND VARIATION: BARK BEETLES AND FUNGAL PATHOGENS

Defensive System of Conifers

The viscous oleoresin system is the important mechanical and chemical component in the defensive repertoire of conifers against tree-killing bark

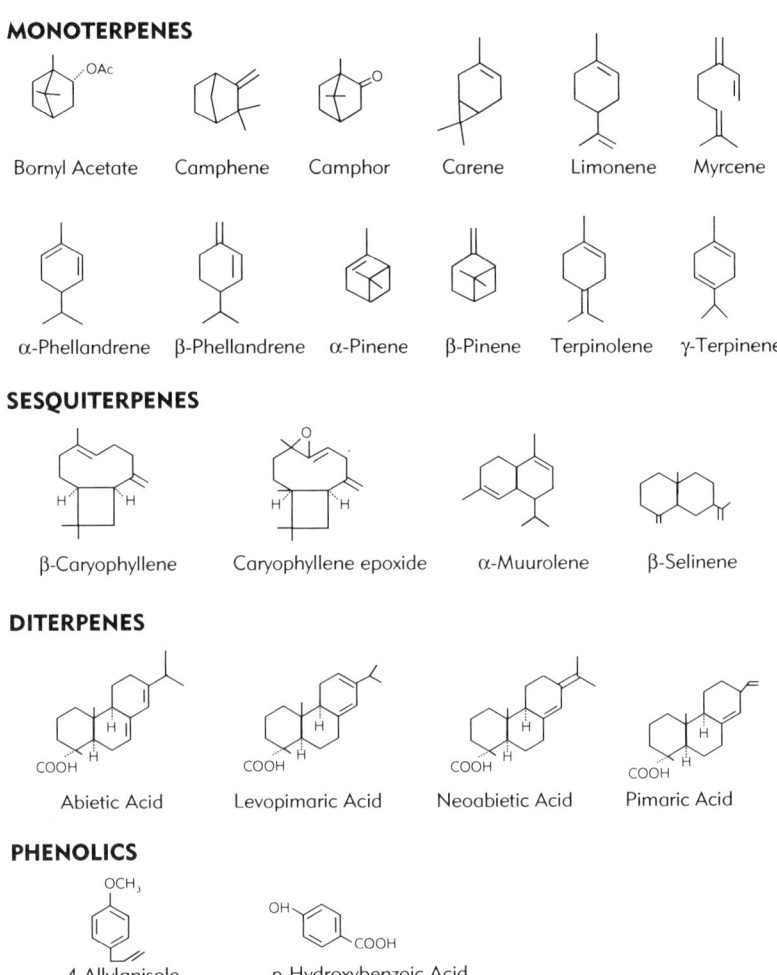

Figure 1. Structures of monoterpenes, sesquiterpenes, diterpenes, and phenolics found in the oleoresin and needles of conifers, and the leaves of some tropical woody species.

beetles and associated pathogenic fungi. Oleoresin in conifers is a mixture of about equal amounts of volatile cyclic monoterpenes (C_{10}) referred to as turpentine, and nonvolatile diterpene resin acids (C_{20}) or rosin.[24] Their qualitative and quantitative production is under genetic control.[14, 16, 19–21] Minor components within the oleoresin are oxygenated monoterpenes (e.g. bornyl acetate, α-terpeniol), sesquiterpenes (C_{15}) and phenolics (e.g. the phenylpropanoid 4-allylanisole = methyl chavicol, estragol, tarragon), and perhaps stilbenes, and lignans. Although conifers produce a variety of monomeric phenolics and condensed tannins in xylem or phloem, these compounds to date have been relegated as a minor factor in their defenses. However, there is increasing interest in phenolics as defensive mechanisms.

A description of the oleoresin system of conifers indicates why oleoresin is the major source of mixtures and variation in the production of individual components, and illustrates the similarities and differences among species in the Pinaceae. Lewinsohn et al.[24,25] detail the anatomical organization and biochemistry of the preformed (also referred to as constitutive or primary) oleoresin system and the wound-induced (hypersensitive, reaction, secondary, resinosus) system. Oleoresin is produced by specialized secretory structures in the needles, stems, and roots, and the organization of these structures varies among taxa in the Pinaceae. One end of the spectrum is represented by *Thuja plicata* which produces only resin cells; similarly, species in the genus *Abies* produce short-lived, resin-producing cells but also sac-like structures termed resin blisters that are lined with resin-producing epithelial cells. This system is not continuous throughout phloem and xylem. The other end of the spectrum is represented by species in the genus *Pinus* which have a more detailed anatomical organization consisting of resin ducts that form interconnected passages throughout the xylem. These ducts are lined with thin-walled, long-lived epithelial cells that produce oleoresin profusely in healthy trees.

In species characterized by a resin duct system (e.g. *Pinus*), large amounts of preformed or constitutive defenses are continuously in place in non-stressed trees. Attack by bark beetles and their associated fungi trigger an immediate massive accumulation of oleoresin at the site of the wound due to disruption of resin ducts. Simultaneously, newly synthesized oleoresin occurs, new anatomically indistinguishable resin ducts are formed, and resin continues to accumulate in the phloem and xylem at the site of the wound. Alternatively, species with resin cells and unconnected, constitutive resin blisters (e.g. *Abies*) respond to attack and infection by *de novo* biosynthesis of oleoresin which is deposited in traumatic resin ducts.[24,25] Regardless of which mechanism is deployed, the success of each in healthy trees depends on a rapid-acting, massive response. Monoterpene cyclase enzymes show activity that is consistent with the organization of these two basic types of resin-producing structures, and that activity, at least in grand fir, is proportional to the intensity of the injury incurred.[24,25]

Genera in the Pinaceae differ in their reliance on consitutive or wound-induced resin defenses.[24] Pine species store large quantities of constitutive oleoresin in extensive duct systems resulting in less of an elevated response to bark beetle attack and fungal infection. True firs (i.e. *Abies*), however, have preformed cortical blisters but lack the massive, interconnected consitutive defenses found in pines. Fir species rely instead on fast-acting and massive increases of oleoresin at the site of wounding. A ranking from the constitutive or preformed mechanism to the reduced constitutive but high-induction mechanism is *Pinus, Larix, Picea,*

Pseudotsuga, and *Abies*. Among the species of *Abies* analyzed, *A. grandis* shows the largest qualitative and quantitative difference between a weak constitutive oleoresin defense and the wound-inducible oleoresinosis.[24] In both types of mechanisms, synthesis of terpenes and the process of resinosus begin immediately upon attack or infection, and the chemical response reaches near maximum between 3 and 10 days in vigorous trees.

These two approaches differ significantly in their response to attack and/or infection (Table 1).[26] Both species differ qualitatively and quantitatively in the monoterpene chemistry of constitutive and reaction oleoresins (Table 1). In the constitutive resin, *P. contorta* produces three terpenes not found in *A. grandis*, and *A. grandis* produces two terpenes not found in *P. contorta* resin. β-Phellandrene makes up 78% of the terpenes in *P. contorta* while α-pinene and β-pinene make up 48.1% and 42%, respectively, of the *A. grandis* monoterpenes. The

Table 1. Qualitative and quantitative compositional differences in the constitutive monoterpene fraction between two conifer species, and changes in monoterpenes in reaction tissue between the same conifers following inoculation with fungi* (modified from Raffa and Berryman 1987[26])

Constitutive phloem tissue†	% Composition	Reaction phloem tissue	% Composition
Pinus contorta			
β-Phellandrene	78.0	β-Phellandrene	80.0
β-pinene	8.1	β-Pinene	7.2
δ-3-Carene	4.7	α-Pinene	4.1
α-Pinene	2.9	δ-3-Carene	2.7
Myrcene	1.7	Limonene	1.3
Terpinolene	1.6	Terpinolene	1.2
		Myrcene	1.0
Abies grandis			
α-Pinene	48.1	α-Pinene	33.9
β-Pinene	42.0	β-Pinene	24.4
Myrcene	4.5	Myrcene	14.5
Sabinene	2.2	Sabinene	8.7
Tricyclene	1.4	β-Phellandrene	3.3
		Unknown no. 2	3.2
		δ-3-Carene	2.9
		Terpinolene	2.6
		Camphene	2.3
		Unknown no. 5	1.5
		Limonene	1.3

Pinus contorta was inoculated with *Ceratocystis clavigera* and *Abies grandis* with *Trichosporium symbioticum*. Monoterpenes representing less than 1% of the total are not shown.
†From cortical resin blisters in *A. grandis*.

reaction resin of *P. contorta* is qualitatively and quantitatively about the same percent composition as that of the constitutive resin with the exception of the addition of limonene at >1% concentration. Alternatively, major qualitative and quantitative changes occur in the induced reaction in *A. grandis* (Table 1). The major components of α- and β-pinene decrease significantly, myrcene and sabinene more than triple in percent composition, tricyclene decreases to <1% composition, and seven compounds are added or increase to >1% composition. Consequently, bark beetles and fungi attacking vigorously growing *A. grandis* trees encounter monoterpenes not typical of the preformed resin-blister oleoresin. This production of a different mixture of terpenes, along with a three-fold increase in the toxic and inhibitory myrcene, presents a formidable barrier to these natural enemies.

Not shown in Table 1 is the quantitatively higher amounts of oleoresin in the lodgepole pine preformed system compared to grand fir. Even though grand fir has preformed defenses located in resin pockets, these are not connected and are easily avoided by attacking beetles. Lodgepole pine is typical of other pines in that the induced reaction is not qualitatively and quantitatively greatly different from preformed oleoresin. Limonene represents the major qualitative change in that it is nonexistent or in low concentration in the preformed resin. It increases about 95% in lodgepole pine and 474% in grand fir.[26] Although at best a minor constituent, it is one of the most toxic monoterpenes to bark beetles and their associated fungi; minor constituents being highly effective against natural enemies is a pattern not uncommon to monoterpenes.[26,27]

Monoterpene production by saplings of conifer species that inhabit and often coexist in forests in western North America is found in Table 2. Species are arranged from those with well developed resin duct systems to those with resin producing cells and blisters to *Thuja plicata*, which only has resin cells. Species characterized by well developed resin duct systems typical of pines produce more terpenes in greater amounts in the constitutive oleoresin than those characterized by resin cells and resin blisters (e.g. *Abies, Sequoia, Thuja*).[24-26] Quantitative and qualitative variation in the preformed oleoresin components exists among all species whether comparing different genera or species within a genus. This diversity of compounds illustrates the potential for the evolution of mixtures and compositional variability in their production (Table 2). The induced reaction discussed below further elaborates on the theme of quantitative variation in mixtures and its role in the defensive system of conifers.

The typical chemical response of healthy pine trees to mechanical wounding and inoculation with a pathogenic fungus is massive (Table 3).[28] Damage induced by bark beetles and their associated fungi show similar patterns to those found in the inoculated phloem, but some differences do occur apparently due to differences in elicitors and virulence of the associated fungi.[29,30] In pines, it is difficult to determine exactly the concentration of components in the oleoresin,[24] especially the minor components, because the phloem is in direct contact

Table 2. Monoterpene percent composition of the extracted oleoresin from a range of conifer species (modified from Lewinsohn et al. 1991[25]). Species are arranged from those with a high degree of anatomical development in the resin duct system (*Pinus*) followed by those with a constricted resin duct system and resin cells (*L. occidentalis*, *P. menziesii*, *P. pungens*) to those with resin cells and resin blisters (*Abies*, *Sequoia*) or just resin cells (*T. plicata*).

Monoterpene	*Pinus contorta*	*Pinus ponderosa*	*Larix occidentalis*	*Pseudotsuga menziesii*	*Picea pungens*	*Abies grandis*	*Abies lasiocarpa*	*Abies concolor*	*Sequoia sempervirens*	*Thuja plicata*
α-Pinene	10.6	12.7	37.3	15.0	41.0	23.3	18.4	18.9	77.8	48.8
Camphene	0.5	0.1	0.5	3.3	2.4	5.6	3.7	0.1	0.8	ND
β-Pinene	37.6	40.9	39.9	7.4	22.7	36.2	40.5	56.1	3.7	ND
Sabinene	1.6	1.6	3.2	16.5	2.5	0.8	0.1	0.5	10.5	31.2
3-Carene	10.7	42.7	15.0	46.4	24.7	17.6	2.3	20.7	ND	ND
α-Phellandrene	0.8	ND*	0.1	ND	ND	ND	ND	0.1	ND	ND
Myrcene	0.8	0.3	1.1	0.7	0.5	0.32	0.1	0.4	0.6	ND
α-Terpinene	0.1	ND	0.1	ND	ND	ND	ND	ND	ND	ND
Limonene	0.9	0.2	1.0	4.6	4.7	1.7	9.1	0.4	1.8	ND
β-Phellandrene	36.0	0.4	0.9	1.4	0.9	14.3	25.3	2.5	3.1	ND
γ-Terpinene	ND	0.1	0.5	0.4	ND	ND	ND	ND	ND	ND
p-Cymene	ND	ND	0.1	ND	ND	ND	ND	ND	ND	ND
Terpinolene	0.5	1.3	0.5	4.2	0.6	0.2	0.2	0.2	1.6	ND
Oxygenated derivative[†]	<1	<1	<1	<1	<1	<1	<1	<1	<1	20[‡]

* ND, not detected (<0.1%).
† All values <1% made up primarily of bornyl acetate and α-terpineol.
‡ 18% thujone plus 2% isothujone.

Table 3. Monoterpene content of red pine reaction resin* (modified from Klepzig et al. 1995[38])

Monoterpene	Control	Mechanically wounded	Inoculated	
			1 week	3 week
α-Pinene	8.0 (7.5)a	113.5 (1.8)b	306.1 (34.7)c	288.6 (47.9)c
Camphene	0.003 (0.0)a	1.1 (0.1)b	3.1 (0.3)c	2.9 (0.5)c
β-Pinene	0.1 (0.0)a	38.9 (7.6)b	102.4 (17.0)b	118.2 (24.2)b
3-Carene	0.01 (0.0)a	3.6 (1.0)b	9.3 (1.9)b	9.0 (2.3)b
Myrcene	0.001 (0.0)a	0.9 (0.1)b	2.5 (0.3)c	2.6 (0.5)c
Limonene	0.001 (0.0)a	1.5 (0.2)b	4.0 (0.5)c	4.0 (0.6)c
Total monoterpenes	8.2 (35.6)a	159.9 (24.2)b	427.7 (50.9)c	425.5 (71.0)c

*Mean (mg monoterpene/g dry weight) (se) of reaction resin in red pine unwounded (control), mechanically wounded, and wound-inoculated with *Leptographium terebrantis* (sampled after one week and after three weeks). Means followed by different letters within a row are significantly different ($P < 0.05$, protected least squares means, ANOVA).

with the xylem where the preformed resin exists.[24] Numerous studies have shown that mechanical wounding does not produce the complete response that occurs when bark beetles or fungi attack trees.[13,26,27,30] Unless specific hypotheses dictate the use of mechanical wounding, this treatment should be dispensed with. Our understanding may be better served in using the areas on a tree bole for testing a different hypothesis or to increase sample size.

In summary, the terpenoid composition (number of compounds and the relative proportion of each)[11] of conifer oleoresin varies significantly within and among species. The monoterpene component may contain up to 30 different structural types,[28] which vary qualitatively, quantitatively, and in their enantiomeric composition. Variation in enantiomers needs further investigation as to its contribution to adversely affecting host location, pheromone synthesis, and toxicity to beetles and fungal associates. The diterpene fraction, which provides a high level of viscosity to the oleoresin, is a mixture of resin acids such as pimaric, palustric, abietic, neoabietic, and sandaracopimaric acids (Fig. 1). Diterpenes have not been found to be significantly toxic but contribute to the defensive properties of the oleoresin by influencing the physical characteristics of the oleoresin (e.g. resin flow, quantity, viscosity, and rate of crystallization).[27,28,31,33] Once oleoresin is exposed through rupture of cells, blisters, or ducts, the monoterpene olefins evaporate and the diterpenes begin to crystallize. Other components found in the oleoresin, especially after attack, are oxygenated monoterpenes, sesquiterpenes, phenylpropanoids, phenolic acids, stilbenes, lignans, and tannins.[26–33]

Upon wounding the mobilization of natural products occurs, and at the site of lesion formation the favorable nutrient conditions in the phloem decrease. The induced reaction is dynamic, and mobilization and biosynthesis of the oleoresin requires extensive amounts of carbohydrate and nutrients.[10,34] As induction proceeds the lesion surrounds the invading beetle and fungi, the phloem tissue becomes resin-soaked, nutrients decrease, and digestibility of the tissue is greatly reduced. As this process continues phloem and epithelial cells die, and eventually in vigorously growing trees the necrotic tissue containing the bark beetle and fungi is sealed from the nonlesion area by a periderm. The beetles die in the resin impregnated lesions and fungal growth is arrested. The physical characteristics of the resin, toxic properties, fast rates of biosynthesis and deployment, increase in compositional diversity, variation in quantity, barren nature of nutrition, and the sealing of a resin-impregnated lesion by a periderm, taken together, represent a complex and effective defensive system.

Oleoresin Effects on Tree-Killing Bark Beetles: The First Line of Defense

The physical characteristics of the oleoresin, primarily determined by the resin acids or diterpenes (Fig. 1), represent the first line of defense against invading bark beetles and fungi. These characteristics (resin quantity, flow, viscosity, rate of crystallization) retard beetle progress, are important in "pitching out" beetles from the tree, and provide time for the induced response to deploy.[27,33] The copious flow of resin may interfere with beetle pheromones thereby reducing attack densities.[35] However, beetles and fungi are known to tolerate the host's preformed defenses even though several of the monoterpene components in the constitutive oleoresin are known to be toxic.[13]

The well developed oleoresin system in conifers is recognized as a dynamic and costly defensive system. However, the role that the physical aspects of the constitutive vs. the induced oleoresins play in host-beetle-fungal interactions among different pine host species and in the regulation of bark beetle populations, are sources of disagreement. For example, Raffa and Berryman[36] indicate that resin flow did not differ significantly between *P. contorta* trees that survived or died during an outbreak of *Dendroctonus ponderosae*. In addition, Raffa et al.[13] state, in regard to firs and pines, that "Host allelochemicals can reach adulticidal, ovicidal, and fungistatic concentrations within a few days". Alternatively, Nebeker et al.[32] suggest "that it (induced response) probably has a minor role in the defense of *Pinus* spp. against bark beetle-fungus invasion in the southern US, largely due to the rapid mass attack behavior of *D. frontalis*." Perhaps these statements are mutually exclusive, but the evidence is equivocal. Understanding the effect of oleoresin defensive systems and the physiological state of the tree on population dynamics of beetles are areas that need investigation.

Mixtures in the Wound-Induced Response against Bark Beetles and Fungi

Terpenoids assessed using laboratory and field studies have been shown to significantly adversely affect tree-killing bark beetles (Table 4). Compounds are ranked in the table within a study from greatest to least effectiveness. Included are conifer species and their associated beetle taxa across the western, northern, and southern United States. No study indicates that a single compound in the induced response confers resistance at any level against bark beetles. Limonene is found in every study and often ranked at the higher end of the scale in its adverse effects. The next most common occurrences are α-pinene and β-pinene. Based on the number of times a compound was found to be effective against bark beetles, the ranking from most deterrent to least is limonene, β-pinene, α-pinene, myrcene = camphene, δ-3-carene and β-phellandrene. These studies also suggest that concentration of individual terpenes within the mixture affects ranking in deterrency. Unfortunately whole preformed and induced oleoresins are not tested often, but Delorme and Lieutier[45] found both highly effective against *Tomicus piniperda*, a bark beetle less tolerant to resin components compared to *Dendroctonus* species. The effect of α-pinene in tree resistance presents an interesting dichotomy since it is an attractant and synergist for pheromones in some beetle species but does show adverse effects on adult beetles.[13,37,38] Part of this dichotomy may be somewhat resolved by a dose-response threshold, a factor discussed by Raffa and Berryman.[35]

Table 4. Mixtures of terpenoids implicated as mechanisms of resistance in plants against tree-killing bark beetles (Scolytidae). Compounds and oleoresin are listed from most to least inhibitory for each study

Terpenoids	Plant taxon	Insect taxon
α-pinene, β-phellandrene, limonene, myrcene, β-pinene, δ-3-carene[39]	*Abies grandis*	*Scolytis ventralis*
Camphene, β-pinene, myrcene, limonene[40]	*P. taeda P. echinata*	*Dendroctonus frontalis*
Limonene, myrcene, β-phellandrene[41]	*Picea glauc*	*D. rufipenn*
	Larix laricina	*D. simplex*
α-Pinene, β-pinene, limonene, myrcene[42]	*Pinus resinosa*	*Ips pini*
	P. banksiana	
Limonene, 3-carene[43,44]	*P. ponderosa*	*D. brevicomis*
Camphene = limonene, δ-3-carene = α-pinene, preformed resin[45]	*P. sylvestris*	*Tomicus piniperda*
Limonene, α-pinene = β- pinene, camphene[46]	*P. taedaa**	*D. frontalis*
	P. echinata*	
Camphene, β-pinene, limonene = α-pinene[46]	*P. taeda**	*I. calligraphus*

*Myrcene and terpinolene occur in these species but were not tested.

Mixtures of mono- and sesquiterpenes increase resistance against fungal pathogens in host trees (Table 5). Studies cited include a variety of tree and fungal taxa, both laboratory and field studies, and a variety of bioassays.[48,50,51] For each study, compounds are arranged in the order of importance from greatest adverse effect to least. In laboratory studies, many compounds were found to be inhibitory. Consequently, where large numbers of compounds were tested (e.g. Cates et al.,[50] Bridges et al.[51]), those inhibiting growth 50% or more compared to the control are included. In the study by Paine and Hanlon[48] only those compounds found in the tree species that the fungi infect are included.

Limonene, β-pinene, and myrcene in mixture and individually were found to affect adversely fungal growth across a wide range of hosts and fungi. No single monoterpene was inhibitory to any of the fungal taxa. However, in the hosts *Abies* and *Picea*, limonene, myrcene, δ-3-carene, camphene, β-pinene, and β-phellandrene were significantly inhibitory to fungi.[37,39,47] The concentration of individual monoterpenes in the mixture also appeared important in tree resistance.[37,45,53] Mixtures of three, but more commonly five, monoterpenes across all hosts typically share in either toxicity, inhibition, or deterrence. In laboratory studies by Paine and Hanlon,[48] Cates et al.,[50] and Bridges,[51] 4-allylanisole or methyl chavicol (Fig. 1), was tested and found to be the most inhibitory compound to several conifer-infecting fungi. This compound originates via the shikimic acid pathway and is common to many species of pine. The presence of 4-allylanisole adds another level of compositional variation to the mixture in that it is a phenylpropanoid. As more unknowns, particularly in the induced resin, are identified, more diversity within and among classes of compounds will likely be found to be important to these interactions.

Sequoia produces a complex mixture of highly inhibitory terpenoids (Table 5). The monoterpenes sabinene and γ-terpinene significantly inhibited growth either individually when presented at several doses, or in three different ratios of a mixture, to the fungal endophytes *Botrytis cinerea*, *Crytosporiopsis albietina*, *Pestalotiopsis funerea*, *Phomopsis occulta*, *Pleuroplaconema* sp., and *Seiridum juniperi*.[53] Langenheim and coworkers[53,54] in a series of papers show the defensive effect of mono- and sesquiterpene mixtures on pathogens. From the essential oils of *Hymenaea* and *Copaifera* have been isolated chemical phenotypes which when presented in mixture reduced the growth of fungi. One of those effective against the *Pestalotia* isolates tested was composed of the sesquiterpenes α-, β-selinene and caryophyllene. This chemical phenotype reduced the growth of four fungal isolates of *H. courbaril*, *Copaifera multijuga*, *H. courbaril* var. *stilbocarpa*, and *Cocos nucifera* from 26% to 34% of that of the control. In addition, Langenheim has reviewed the dosage-dependent effect of a single highly reactive compound caryophyllene oxide, and determined that it is an effective, broad-spectrum deterrent to fungi and the lepidopteran generalist *Spodoptera exigua*.[11] Furthermore, its deterrency is dosage dependent for

Table 5. Mixtures of terpenoids and 4-allylanisole implicated as mechanisms of resistance in plants against fungal pathogens. Terpenoids and oleoresin are ranked from most to least inhibitory for each study

Terpenoids	Plant taxon	Fungal taxon
Coniferae		
Myrcene, limonene, β-pinene, δ-3-carene, camphene[39]	*Abies grandis*	*Trichosporum symbioticum* (white mycelium)
Limonene, δ-3-carene, β-pinene, camphene, myrcene[39]	*A. grandis*	*T. symbioticum* (brown mycelium)
Myrcene, 3-carene[47]	*A. grandis*	*T. symbioticum*
Limonene, myrcene, δ-3- carene, β-phellandrene, camphene[37]	*Picea lutzii* *P. sitchensis* *P. glauca**	*Leptographium abietinum*
p-Cymene, terpinolene, limonene, β-pinene, 4- allylanisole[48]	*Pinus contorta*	*Ophiostoma clavigerum*
N-Heptane, limonene, myrcene, n-nonane, β- pinene[48]	*P. jeffreyi*	Mycangial fungus from *Dendroctonus jeffreyi*†
Limonene, β-pinene, myrcene, terpinolene, β-phellandrene, 4-allylanisole[48]	*P. ponderosa*	*O. clavigerum*†
Preformed = induced oleoresin, limonene, camphene, δ-3-carene, β-pinene, γ- terpinene, α-pinene,[45]	*P. sylvestris*	*O. brunneo-ciliatum* *Leptographium wingfieldii*
Limonene, γ-terpinene, β- pinene[45]	*P. sylvestris*	*O. ips*
Myrcene, limonene, β- pinene, 3-carene, α- pinene[49]	*P. sylvestris*	*Crumenulopsis sororia* isolates
Myrcene, limonene, terpinolene[50]	*P. taeda*	*O. minus*
α-Pinene, β-pinene, limonene[50]	*P. taeda*	*Ceratocystiopsis ranaculosis*
Myrcene, limonene, γ- terpinene, terpinolene[51]	*P. taeda*	*Ceratocystis minor* (= *Ophiostoma minus*)
Camphene, β-pinene, myrcene, limonene[52]	*P. taeda*	*Ceratosystis minor*
Sabinene, γ-terpinene (individually and in mixture)[53]	*Sequoia sempervirens*	Six redwood endophytes
Dicotyledonae		
α-, β-Selinene, caryophyllene[54]	*Copaifera panamensis* *Hymenaea courbaril*	*Pestalotia* isolates. *P. subcuticularis*
Caryophyllene epoxide[55]	*H. courbaril*	Attine fungus

*Terpene ranking is for *P. glauca* only.
†Saturated atmospheres of resin constituents used.

different fungi. Hubbell et al. also found caryophyllene epoxide to be highly inhibitory to an attine fungus.[55]

Raffa and Berryman compiled the data for conifer monoterpenes, for which activity against bark beetles and associated fungi were known, and their relative rates of increase in the wound response following inoculation.[26] Monoterpenes were ranked according to their level of activity and their relative rate of increase, and a highly significant relationship was found (Spearman's coefficient-of-rank, $p = 0.81$, $t = 7.57$, $p<0.001$). Table 6 shows these data as well as results from recent studies amenable to the same statistical analysis. The analysis of the data in Table 6 also resulted in a highly significant relationship ($t = 7.84$, $p<0.005$)

Table 6. Relationship between the level of activity of a monoterpene against bark beetles and fungi and the amount of compound increase during the induced response. Terpenes ranked one showed the highest activity and percent increase while those with higher numbers had less activity and percent increase (added to Raffa and Berryman 1987[26])

				Rank order*	
Host species	Inoculating organism	Defensive property	Monoterpene	Level of activity	% Increase during response
Pinus contorta	C. clavigera (Europhium clavigerum)	Ovicidal to D. ponderosae	Limonene	1^{56}	1^{57}
			δ-3-Carene	2	3
			α-Pinene	3	2
P. echinata	Ips calli-graphus	Toxicity to adults	Limonene	1^{46}	1^{58}
			α-Pinene	3	3
			β-Pinene	3	4
			Camphene	5	5
P. taeda	Ceratocystis minor (also C. m. bar-rasii)	Toxicity to Dendroctonus frontalis adults	Limonene	1^{40}	1^{58}
			Myrcene	2	2
			α-Pinene	3	4
			β-Pinene	4	5
			Camphene	5	3
P. taeda	Ophiostoma minus; Ceratocys-tiopsis ranaculosis	Toxicity to Dendroctonus frontalis;[59] Inhibition in growth of fungi in lesions[50]	Limonene	1^{40}	2^{50}
			Myrcene	2	3
			α-Pinene	3	1
			β-Pinene	4	1
			Camphene	5	5
P. taeda	Ophiostoma minus	Inhibition of hyphal growth[50]	α-Pinene	1^{50}	1^{50}
			β-Pinene	1	1
			Terpinolene	1	3
			Limonene	2	2
			Myrcene	3	3
			Camphene	4	4
			Tricyclene	5	5

ROLE OF MIXTURES AND VARIATION IN THE PRODUCTION OF TERPENOIDS 193

indicating that the effectiveness of these compounds as defenses is related to their activity as well as their rate of increase. Selection by bark beetles and pathogenic fungi have favored, at least in the monoterpene fraction, biochemical adaptations that significantly adversely affect these natural enemies.[26]

In addition, selection seems to have favored a mixture of both major and minor constituents that are toxic, inhibitory, or repellent (Tables 4–6). For

Table 6. *Continued*

Host species	Inoculating organism	Defensive property	Monoterpene	Rank order* Level of activity	Rank order* % Increase during response
P. taeda	Ceratocystiopsis ranaculosis	Inhibition of hyphal growth[50]	β-Pinene	1^{50}	1^{50}
			α-Pinene	2	1
			Limonene	3	2
			Terpinolene	4	3
			Myrcene	5	3
			Tricyclene	5	5
			Camphene	6	4
P. taeda	D. frontalis	Toxicity to adults	Limonene	1^{46}	1^{58}
			α-Pinene	3	3
			β-Pinene	3	4
			Camphene	5	5
Abies grandis	Trichosporium symbioticum	Repellency to Scolytus ventralis adults	Limonene	1^{60}	1^{61}
			δ-3-Carene	2	2
			α-Pinene	3	4
			Myrcene	3	3
			β-Pinene	5	5
			Camphene	6	6
			Tricyclene	7	7
A. grandis	T. symbioticum	Toxicity to S. ventralis adults	α-Pinene	1^{39}	5^{61}
			β-Phellandrene	2	1
			Limonene	3	2
			Myrcene	4	4
			β-Pinene	5	6
			δ-3-Carene	6	3
A. grandis	T. symbioticum	Inhibition of T. symbioticum (white mycelium)	Myrcene	1^{39}	3^{61}
			Limonene	2	1
			β-Pinene	3	4
			δ-3-Carene	4	2
			Camphene	5	5
A. grandis	T. symbioticum	Inhibition of T. symbioticum (brown mycelium)	Limonene	1^{39}	1^{61}
			δ-3-Carene	2	2
			β-Pinene	3	4
			Camphene	4	5
			Myrcene	5	3

conifers the monoterpenes that are quantitatively low, or not found in the phloem or preformed resin, e.g. limonene, myrcene, δ-3-carene, terpinolene, and camphene (Table 2), increase significantly in the wound-induced oleoresin at concentrations that adversely effect bark beetles and fungi. Limonene is one of the most effective monoterpenes even at low concentrations and short exposure times.[62] Pathogenic fungi appear to be adversely affected by more compounds than beetles perhaps because fungi cannot easily move away from the oleoresin.

Studies cited in Tables 4–6 demonstrate the response typical of host trees to attacking bark beetles and their associated fungi. Data in these tables verify that the monoterpene composition of the induced resin differs from that of the preformed resin. Furthermore, qualitative differences in the induced reaction were especially pronounced in *Abies* and *Picea* compared to pines. In both types of conifer response, i.e. fir-spruce vs. pine, massive amounts of oleoresin were deposited in the phloem which is the tissue attacked by bark beetles and fungi. In addition, significant variations in terpene composition occurred in both the induced resin cell-resin blister system and the well developed duct system. In assays with oleoresins, the preformed and induced resins were particularly effective against less aggressive vs. more aggressive fungi, e.g. *Leptographium wingfieldii* and *Ophiostoma brunneo-ciliatum* but significantly less so against *O. ips*. Typical of pines, an increase in resin quantity and total monoterpene quantity rather than qualitative changes had the greatest effect.[45] Also, in the vapor form that fungi would encounter in the tree, several monoterpenes were effective including limonene, camphene, myrcene, δ-3-carene, γ-terpinene, β-pinene, and α-pinene.

α-Pinene is enigmatic. It is a common, major component in the monoterpene fraction in both preformed and induced resin, can act synergistically with pheromones, e.g. in pine bark beetles, yet often is found to be deterrent or toxic (Tables 3–5). In the induced oleoresin, α-pinene either stays at about the same level as the preformed oleoresin,[26] increases,[38] or decreases.[26,37] In laboratory studies, α-pinene is often one of the most inhibitory monoterpenes to less aggressive fungi, shows some adverse activity against more virulent forms, and is active against some fungi and bark beetles but not others (Tables 4–6). Cates et al.[50] found α-pinene to completely inhibit growth of *Ophiostoma minus*, and at the higher concentration found in trees, to cause 53% inhibition to the hyphal growth of the mycangial fungus *Ceratocystiopsis ranaculosis*. β-Pinene and terpinolene also were 100% deterrent to *O. minus*. Bridges[51] found no significant difference in the growth of *C. minor* (= *O. minus*) and the mycangial basidiomycete when challenged with enantiomers of α-pinene, but racemic α-pinene inhibited *C. minor*. In two field studies involving loblolly pine, α-pinene was not found to be correlated with hyphal growth.[50] Its high content in oleoresin, deterrent properties in laboratory studies, and highly variable adverse effects in field studies is a problem needing resolution.

Contribution of Phenolics to Mixtures, Increasing Oleoresin Variation, and Resistance

Phenolics, especially monomeric phenolics as opposed to the condensed tannins, are receiving more attention as to their role in mixtures in the resistance of host trees to tree-killing bark beetles and pathogenic fungi.[50,64] To date, the evidence for adverse effects attributable to condensed tannins produced in pine is lacking.[37,65] Hemingway et al.,[66] for example, found that condensed tannins, catechin, and flavonols from the phloem and xylem of *P. taeda* were degraded by *O. minus*. Also, in this same study the concentration of three stilbenes declined. Alternatively, stilbene aglycones and phenolics have been shown to be effective defenses. Two aglycone stilbenes were found to deter growth of *Leptographium abietinum* and feeding by spruce beetle adults and larvae.[41] Further tests are needed with other conifer species and bark beetle and associated fungi before the tannin and phenolic issue can be resolved. For example, stilbenes and phenolics appear to be effective defenses in spruce species.

Laboratory evidence suggests monomeric phenolics (Fig. 1) may adversely affect beetle-associated fungi. *O. minus* and *C. ranaculosis* were reared on agar containing varying concentrations of monomeric phenolics typical of loblolly pine trees (Table 7,8). All compounds were identified from loblolly pine phloem and wound-induced tissue except for quercetin. For *O. minus*, the most

Table 7. Hyphal growth of *Ophiostoma minus* when grown in agar having different concentrations (µg/ml) of phenolics typical of loblolly pine phloem lesions (from Cates et al. 1996[50])

Compound	% Growth inhibition or enhancement[†]			
	50	100	200	300
4-Allylanisole	+1*a	+1*a	7	14
Benzoic acid	6	15	43	100
Catechin	+3*	6†a	9a	16
Chlorogenic acid	5a	6a	14	21
p-Coumaric acid	7	26	48	66
Epicatechin	4a	4a	10	22
Ferulic acid	25	40	59	59
p-Hydroxybenzoic	1*a	2*a	9	23
Naringin	2	6	14	21
Quercetin	5	9	19	36
Sinapic acid	16	40	48	56
Taxifolin	5	13	26	35
Vanillic acid	23	36	48	53

*Not significantly different from control.
†Means followed by same letter not significantly different from one another.
+ Indicates fungal growth greater on phenolic diet compared to control (ANOVA, Student Newman–Keuls test, p≤0.05).

inhibitory at the two lower concentrations were benzoic, *p*-coumaric, ferulic, sinapic, and vanillic acids. At the two higher concentrations benzoic, *p*-coumaric, ferulic, sinapic, and vanillic acids were most inhibitory. For *C. ranaculosis* benzoic and vanillic acids were most inhibitory at the low concentrations but inhibition was only 16% and 25%, respectively. At the higher concentrations benzoic acid, *p*-coumaric acid, ferulic acid, naringin, and vanillic acid were inhibitory at 24% or greater. The more specialized *C. ranaculosis*, in contrast to *O. minus*, showed stimulation in growth for many compounds even at the highest concentrations suggesting that several of these compounds may be used as an energy source.

Since benzoic acid was the most consistent phenolic in inhibiting both fungal taxa, it was combined with each of the other phenolics and these combinations presented to the two species of fungi (Table 9). *O. minus* hyphal growth was significantly inhibited by every compound when each was combined with benzoic acid. While *C. ranaculosis* generally utilized most of the phenolics when presented individually as energy sources even at the highest concentration, when these compounds were mixed with benzoic acid, all were significantly inhibitory. No data indicated that in combination these compounds were used as energy sources. Finally, when all compounds were mixed at a concentration typical of loblolly pine reaction phloem, *O. minus* hyphal growth was completely inhibited

Table 8. Hyphal growth of *Ceratocystiopsis ranaculosis* when grown in agar having different concentrations (μg/ml) of phenolics typical of loblolly pine phloem (from Cates et al. 1996[50])

Compound	% Growth inhibition or enhancement[†]			
	50	100	200	300
4-Allylanisole	4	8	11	11
Benzoic acid	+24	16	48	100
Catechin	+24[a]	+27[a]	+6	5
Chlorogenic acid	+18	+9	10	20
p-Coumaric acid	+21[b]	+32[a]	+30[a]	24[b]
Epicatechin	+1[*]	+5	+18	+13
Ferulic acid	+19	+4	11	45
p-Hydroxybenzoic acid	3[a]	5[a]	8[b]	11
Naringin	+18[a]	+15[a]	10	26
Quercetin	+50	+35[a]	+30[a]	+22
Sinapic acid	+31	+21	+15	+4
Taxifolin	+10[a]	+10[a]	1[*]	7
Vanillic acid	8	25	38	63

[*]Not significantly different from control.
[†]Means followed by same letter not significantly different from one another.
+ Indicates fungal growth greater than control (ANOVA, Student Newman–Keuls test, $p \leq 0.05$).

Table 9. Response of *Ophiostoma* and *Ceratocystiopsis* to loblolly pine phloem phenolics when each individual phenolic was combined with benzoic acid (from Cates et al. 1996[50]). The right-hand column shows the average amount of each phenolic in the lesions

	% Inhibition		
Benzoic acid[*] plus	*O. minus*	*C. ranaculosis*	Phenolic content[†]
Catechin	33	48	0.75 ± 0.5
Chlorogenic acid	43	54	0.01 ± 0.0
p-Coumaric acid	76	63	0.01 ± 0.0
Epicatechin	37	49	0.01 ± 0.0
Ferulic acid	60	54	0.01 ± 0.0
p-Hydroxybenzoic acid	52	50	0.08 ± 0.1
Naringin	45	42	0
Quercetin	49	57	—
Sinapic acid	57	53	0.01 ± 0.0
Taxifolin	57	52	0.11 ± 0.0
Vanillic acid	61	65	0.02 ± 0.0
Benzoic acid	—	—	0.02 ± 0.0
Phenolic mixture[‡]	100	64	—

[*] 200 µg/ml benzoic acid plus 200 µg/ml of phenolic. All comparisons within a taxon significantly different from control, $p<0.001$ (ANOVA).
[†] From lesions induced by inoculating trees with *O. minus* (mg/g fresh wt.).
[‡] A mixture of all phenolics.

and 64% inhibition occurred for *C. ranaculosis*. Consequently, mixtures of two at a time or of all phenolics increased inhibition of hyphal growth in the bluestain fungus as well as the more specialized *C. ranaculosis*.

The phenolic compounds in the phloem lesion from loblolly pine trees due to inoculation with *O. minus* are listed in Table 9. These, in addition to several unknown compounds, are present in the lesions but their level of increase, the qualitative and quantitative variation in their production compared to unwounded phloem, and their source of production is not well known for most conifer species.

TERPENOID MIXTURES AND VARIATION: INTERACTIONS WITH DEFOLIATORS

Terpenoid Defensive Systems

For needle- or leaf-chewing herbivores, physical and chemical characteristics of the tissues eaten have been investigated as to their importance in determining the degree of utilization of the tissue and the level of tree resistance or susceptibility.[67,72] Physical characteristics include primarily toughness and bud phenology, both of which have been implicated in reducing preferences among trees within a population to spruce budworms[72] and in outbreaks of budworm species.[67,77] But further investigation is needed to determine the effects of physical characteristics on defoliator population dynamics. These include the effects of host and nonhost phenology on herbivores, changes in primary and secondary metabolites during budbreak and needle phenology, and the effect of qualitative and quantitative decreases in defenses of late emerging smaller buds in subcanopy trees.[78] Budbreak is genetically controlled, but a greater understanding of genetics and environmental effects on bud phenology is needed.

Defoliators are influenced by qualitative and quantitative variations in primary and secondary metabolites, mixtures of these compounds, and their variations among branches, canopy levels, individuals in the population, populations, and species.[71] Patterns in resistance based on both laboratory and field studies show that occasionally one or two compounds may reduce herbivore fitness significantly, but more often deleterious effects on different life stages are due to several compounds (Table 10). Ikedo et al.[69] found that one resin acid was highly effective against two sawfly species, and Cates et al.[72] showed that bornyl acetate increased budworm larval mortality 88% compared to the control. In this study only two of 35 budworm larvae reached the adult stage. Preference by sawflies for mature needles of hosts is thought to be determined by high concentrations of tricyclic diterpene resin acids in the current year's needles. When one-year old foliage was treated with purified individual resin acids, sawfly species avoided their preferred tissue.[70] Higher concentrations of resin acids in the foliage are correlated with nonpreference in many sawfly larvae, a pattern supported by the work of Ohigashi et al.[68] and Ikedo et al.[69]

Macedo and Langenheim[76] found that seven terpenoids, as well as total resin yield, in the leaves of *Copaifera langsdorfii* significantly reduced microlepidopteran herbivory. In addition, these authors noted that oecophorid herbivory and outbreaks of herbivores at one site were lower compared to another site. The chemical characteristics of trees where herbivory and outbreaks were lower included significant variation in leaf sesquiterpenes among trees, greater concentrations of cyperene, caryophyllene, $\alpha + \beta$-selinenes, and variation in total resin yields among trees. More studies of this type are needed to determine the utility of chemical variation in the silvicultural management of forest defoliators.[23]

Other studies show that four to seven compounds significantly increase larval mortality or reduce male and female biomass production (Table 10). Not uncommon is that among the several compounds that adversely affect a defoliator, one is most toxic or inhibitory. In *P. glauca* camphor is highly inhibitory, and in *P. menziesii* bornyl acetate is an important deterrent. In addition, the important effect of a single compound within an inhibitory mixture can be further enhanced

Table 10. Mixtures of terpenoids implicated as mechanisms of resistance in plants against needle- or leaf-attacking insects. Compounds are ranked from most to least toxic or inhibitory

Terpenoids	Plant taxon	Insect taxon
Camphene, α-pinene = β-pinene, β-phellandrene, terpinolene, sum of terpenes[67]	Abies balsamea	Choristoneura fumiferana (♀ wt. gain)
Camphene, β-phellandrene, β-pinene, α-pinene, terpinolene, bornyl acetate, sum of terpenes[67]	A. balsamea	C. fumiferana (♂ wt. gain)
Abietic, dehydroabietic, 12-methoxyabietic, sandaracopimaric, isopimaric acids[68]	Larix laricina	Pristiphora erichsonii
Camphor, 4 sesquiterpenes and monoterpene alcohols[67]	Picea glauca	Choristoneura fumiferana (♀ weight gain)
13-Keto-8 (14)-podocarpen- 18-oic acid[69]	P. banksiana	Neodiprion rugifrons N. swainei
Palustric, levopimaric acids[70]	P. banksiana	N. dubiosus
Dehydroabietic, neoabietic, palustric acids[70]	P. banksiana	N. rugifrons
Bornyl acetate, camphene, citronellyl acetate, limonene, geranyl acetate[71]	Pseudotsuga menziesii	Choristoneura occidentalis (larvae, Boulder, MT)
Bornyl acetate, camphene, tricyclene, β-pinene, terpinolene, myrcene, limonene[71]	P. menziesii	C. occidentalis (♀, Boulder, MT)
Bornyl acetate, camphene, tricyclene, myrcene, terpinolene, α-pinene[71]	P. menziesii	C. occidentalis (♂, Boulder, MT)
Bornyl acetate, camphene, terpinolene, limonene[71]	P. menziesii	C. occidentalis (larvae, Island Park, ID)
Acetate fraction, myrcene, unk. terpene 10[72]	P. menziesii	C. occidentalis (larvae, Boulder, MT)
Bornyl acetate, terpinolene, geranyl acetate[72]	P. menziesii	C. occidentalis (♀, Island Park, ID)
Bornyl acetate[71]	P. menziesii	C. occidentalis (♂, Island Park, ID)
Unk. terpene 5, β-pinene, myrcene[73]	P. menziesii	C. occidentalis (♀, Barley Canyon, NM)
Terpinolene, citronellyl acetate, α-pinene, bornyl acetate, myrcene, unk. terpene 8[73]	P. menziesii	C. occidentalis (♂, Barley Canyon, NM)
Sesquiterpene fraction (5 sesquiterpenes)[74]	P. menziesii	C. occidentalis (colony larvae)
Phellandrene, 3-carene, camphene[75]	Pinus nigra	Rhyacionia buoliana
Dicotyledonae		
Cyperene, caryophyllene, β-humulene, α- and β-selinene, γ-cadinene, γ-muurolene, total resin yield[76]	Copaifera langsdorfii	Stenoma aff. assignata (oecophorid), gelechids

through the interaction with nutrients. For example, the mixture in the foliage of white spruce is associated with a nitrogen level that is about 20% less than that of balsam fir. Based on total quantities of terpenoids, white spruce might be the preferred host. In reality, white spruce is not preferred, perhaps due to camphor and its inhibitory mixture interacting with lower foliage nitrogen.[67]

Terpenoid Interactions with Primary Nutrients

Plant tissues contain a myriad of nutritive and nonnutritive components which are correlated with bud and photosynthetic tissue phenology. Nutritive components occur in various forms and include at a minimum the essential elements, carbohydrates, proteins, fatty acids, and a host of minor components necessary for metabolic activity.[67,79] Since most of these in one form or another are essential, are qualitatively and quantitatively variable in plant tissues, and interact with natural products, the suggestion that nutrients may contribute to preference and resistance or susceptibility is attractive. The question quickly emerges as to which of the nutritive components should be investigated. Mattson and Scriber[80] suggested the best candidates are those that are essential to key metabolic processes and those that are most variable. Clues might also emerge through investigations of nutritive quality in nonhosts.[79]

Mattson et al.[67] investigated the effect of total nitrogen, individual mineral elements, individual monoterpenes, and total phenolics from balsam fir, white spruce, and black spruce on the growth and survival of the eastern spruce budworm *Choristoneura fumiferana*. In regressing generation survival against foliar elements, only about one-third of the variation was accounted for, and none accounted for more than 50% of the variation. They also found little consistency between years and between species, all of which pointed out the high level of variation in nutritional components. These results lent little support to the importance of nutrients to preference, resistance, or susceptibility. In comparing the variation in foliage nutrients among *Pseudotsuga menziesii*, *Picea engelmannii*, and *Abies concolor* and the relationship this variation might have to the western spruce budworm *C. occidentalis*, Clancy et al.[79] found that N,P, K, Ca, Fe and Zn, plus the Cu/Zn ratio were distinctly different among host species. In evaluating the overall results of this study the authors concluded that to resolve the issues of which hosts, stands, or individuals were "better or poorer" hosts, the budworm's dietary requirements for various nutrients first needed to be established. Subsequently, comparisons between putatively susceptible and resistant trees as determined by defoliation suggested that susceptible trees had lower amounts of N and sugars compared to resistant trees; this conclusion was in agreement with predictions from artificial diet bioassays.[81,82] No detectable differences were found in monoterpene concentrations between the putatively resistant and susceptible trees,[83] but the sample size was small, the monoterpenes showed high variation typical of natural populations, and defoliation is not always reliable in predicting resistance-susceptibility or budworm performance.[72] For example, from a natural population western spruce budworm larvae were reared on 12 putatively resistant trees (<25% defoliation) and 12 putatively susceptible trees (>85% defoliation) near Greenough, Montana in 1987.[71] No significant differences were found between male adult dry weight (18.3 ± 4.4mg vs 18.7 ± 4.4mg, respectively) or female adult dry weight (36.3 ± 8.2mg vs 38.4 ± 8.7mg, respectively -- R. G. Cates, unpubl. data). Another study following Clancy,[72] and also using colony western spruce budworm in a three-generation bioassay, determined that the ratio of Zn to N was the best predictor of budworm fitness, and not the actual N content of the diet.[84]

Sugar concentrations in Douglas-fir are highly variable[82,85] thus meeting the criteria of Mattson and Scriber.[80] Using colony budworm in the three-generation bioassay with two different levels of N, the optimum level of sucrose has been suggested to be 6%.[82] The upper sugar concentrations found in the current year's growth of Douglas-fir (18.4%) may be detrimental to budworm performance. Results from diet containing 1.2% N showed that budworm did best on diets with sugar concentrations near the lower limit observed for host foliage (5.7%), implying that an inferior host might have higher foliar sugar concentrations.[82]

In a field study evaluating the seasonal variation in monoterpenes, sesquiterpenes, phenolics, tannins and carbohydrates, Zou and Cates[85] found that galactose was the major carbohydrate in the current year's growth of Douglas-fir. This carbohydrate amounted to almost 80% of the total sugar concentration throughout the growing season. When testing the effects of carbohydrates on western spruce budworm, total carbohydrates at the level of 5.61% did not significantly increase colony larval mortality compared to the control, but 9.45% and 15.0% diets resulted in significantly higher mortality. When total carbohydrates at 5.61%, 9.45%, and 15.0% were incorporated with 0.08% total terpene, all carbohydrate levels adversely affected budworm mortality (Table 11).[86] To determine if a similar effect from galactose could be achieved, galactose, fructose, and glucose were placed in an agar diet at 6%. Galactose significantly increased larval mortality compare to the control, 6% glucose, and 6% fructose (Table 12).[84] Galactose at 6% in combination with 0.07% total monoterpene also significantly increased larval mortality compared to the control and the 6% fructose - 0.07% total monoterpene diet. In this set of studies, galactose was significantly active at levels much less than those found in the current year's growth of Douglas-fir, and in mixtures with other carbohydrates. In addition, in the presence of average tissue amounts of total monoterpenes, the adverse effect of galactose was significantly enhanced; however no interactions between galactose and terpenes were noted.[84]

The effects of high levels of sucrose and galactose on budworm need to be verified with larvae from natural populations. Differences between colony and

Table 11. The effect of total carbohydrate, and total carbohydrate plus total monoterpenes, on budworm larval mortality (from Zou and Cates 1994[84])

Terpenes (%)[*]	Carbohydrates (%)[*]	Mortality (%)[†]
0	0	44.0^a
0	5.61	43.4^a
0	9.45	64.0^b
0	15.0	96.1^d
0.08	0	78.9^b
0.08	5.61	82.5^{bc}
0.08	9.45	97.2^{cd}
0.08	15.0	100.0^d

[*]Terpene and carbohydrate diet composition in Zou and Cates.[84]
[†]Mortality values followed by different letters, significantly different p<0.05 (chi-square test).

natural budworm due to genetic variability, physiological tolerances, and lack of exposure to a number of biotic and abiotic selection pressures that may affect physiology and metabolism need to be accounted for.[87,88] Significant differences between natural populations and colony budworm do exist (Table 13). Larval, pupal, and overall mortality for female and male western spruce budworm obtained from a colony, when compared to mortality for natural populations from Montana and Idaho, was significantly different when all budworm sources were reared on a three-terpene/galactose diet.

Table 12. The effect of individual carbohydrates, and individual carbohydrates plus total monoterpenes, on budworm larval mortality (from Zou and Cates 1994[84])

Terpenes (%)[*]	Carbohydrates (%)[*]	Mortality (%)[†]
0	0	36^{bc}
0	glucose	20^{ab}
0	galactose	68^e
0	fructose	8^a
0.07	0	48^{bcd}
0.07	glucose	56^{de}
0.07	galactose	68^e
0.07	fructose	36^b

[*]Except for the control (0), individual carbohydrates were added at 6%; terpene composition in Zou and Cates.[84]
[†]Mortality values followed by different letters, significantly different, p<0.05 (chi-square test).

Table 13. Comparisons among Montana, Idaho and colony-reared budworm for stage specific larval and pupal mortality, and overall mortality for larvae reared on a 3-terpene/galactose diet. (Cates, Nay, Whitehead unpubl. results)

Mortality*	Montana	Idaho	Colony
Females (%)			
Larval mortality	7.1^a	13.7^a	63.3^b
Pupal mortality	19.2^a	15.9^a	97.0^b
Overall mortality	25.0^a	27.4^a	98.9^{b†}
Males (%)			
Larval mortality	11.4^a	14.5^a	70.6^b
Pupal mortality	34.4^a	27.7^a	96.7^b
Overall mortality	41.9^a	38.2^a	99.0^{b†}

*Different letters in a row indicate significant differences between means (P<0.05, chi-square tests).
†Only one pupa survived to the adult stage.

ECOLOGICAL CONSEQUENCES OF MIXTURES AND VARIATION IN TERPENOIDS AND NUTRIENTS

The above discussion has shown that considerable data on the composition of mixtures and the effect of composition on herbivores and pathogens has accrued over the past 20 years. However, superimposed on compositional effects of primary and secondary metabolites is another level of variation that has been implicated in the mediation of host tree-herbivore-pathogen interactions. This is the variation in the production of primary metabolites and natural products among tissues, plants, populations and the community of coexisting species.[71,79,89,90] Most types of variation found in terpenes and primary nutrients in conifer and tropical trees are depicted in Table 14. Discovering types of variation has been difficult since experimental designs usually are developed to reduce variation, and because in many cases the origin of the variation is unknown. For example, the production of primary nutrients, terpenes, and compositional profiles[76] is under genetic control, but the quality and quantity of these metabolites also are under environmental influence.

Investigating the level of genetic control for other parameters (e.g. changes in compound production within the same plant,[89] variation among crown levels and branches within a tree,[6,71] variation in compositional ratios[11]) is needed so that the significance of this variation to herbivore and pathogen population dynamics can be assessed and applied.[23] Some information exists about coevolutionary interactions among natural products and herbivores. For example Stephan[91] indicates that Douglas fir provenances show differential resistance to the woolly aphid *Gilletteella cooleyi*. Christiansen and Berryman[92] point out wide genetic variation in the Norway spruce *Picea abies* in preformed and

Table 14. Types of variation in terpenes, acetates, sesquiterpenes, other volatiles, and primary nutrients in current year's foliage of conifers that are hosts to *Choristoneura occidentalis* and *C. fumiferana*, and in some tropical plant species

Types of variation	Chemicals		
	Terpenes	Nitrogen	Soluble carbohydrates
Geographical variation within a species	X	X	X
Among coexisting host species	X	X	X
Across successional stages	X	X	X
Among populations within a species	X	X	X
Variation in same population among years	X	X	X
Tissue development (or within-season) variation within a population	X	X*	X
Among crown levels within a tree	X	—	—
Among branches within a crown level	X	—	—
Between sun and shaded branches	X	—	—
Due to thinning	X	X	X
Due to controlled burns	X	X	X
Within a population due to aspect	X	X	—
Within a population due to water stress	X	X	—

— = not determined.
*Also differences between putatively resistant and susceptible trees within a population in phosphorus and potassium (after Clancy 1991[83]).

wound-induced oleoresin chemistry, and that this variation has led to host-herbivore race formation.[5,6,71,73,86,87,90,93–95]

Characteristics of Mixtures

The defensive characteristics of complex mixtures increase heterogeneity and present a highly varying mosaic of tissue quality to the natural enemies of plants (Table 15). The diversity of herbivores and pathogens, and their frequency and intensity of interactions with potential hosts, provide selection pressures for the evolution of these defensive characteristics. These same selective agents may influence the diversity of primary metabolites, but plant physiological, developmental, and genetic controls constrain the qualitative and quantitative diversity that can be achieved among these classes of compounds as compared to secondary metabolites.[11,16,86,90] With regard to conifer-herbivore-pathogen interactions, terpenoids and low molecular weight phenolics show promise in fulfilling some of the defensive criteria (Table 15), but for phenolics more investigation is needed.[96] Conifer buds, leaves, phloem, and xylem produce phenolics, but conifers do not demonstrate the level of induction in phenolics that is known for deciduous trees.[64,89,96] In addition, phenolics may be metabolized by fungi,[66]

Table 15. Summary of the defensive characteristics of complex mixtures

1. Mixtures may be composed of three to many compounds within a class (e.g. monoterpenes or sesquiterpenes) with major and minor constituents active in the mixture (Tables 4–6,7,9,10).
2. Mixtures with defensive properties may be composed of two or more classes of compounds (e.g. monoterpenes, sesquiterpenes, resin acids, phenylpropanoids) with activity enhanced due to a mixture of compounds formed across classes (Tables 4–6).
3. Minor constituents of a mixture can have greater or equal activity compared to major components (Tables 4–6,7,9,10).
4. Compounds in mixtures vary in their effectiveness against different herbivore-pathogen taxa (including host races, isolates) as compositional ratios of the mixture vary.[89,101]
5. Effects of individual compounds, combined compounds, or total quantity often are concentration or dosage dependent.[11] Mixtures may magnify dosage dependence effects at lower concentrations.[45,76,84,86]
6. Mixtures provide a variety of mechanisms to enhance toxicity, deterrency, or inhibition.[11,86]
7. Mixtures that deviate qualitatively and quantitatively from the "population average," or from the more common chemical phenotype, confer greater resistance or deterrent properties.[88–90]
8. Individual plants undergoing the greatest enhanced chemical change when challenged by herbivores and pathogens are the more resistant or deterrent.[26,90,97,98]
9. Mixtures enhance additive and synergistic effects and complexity to defensive interactions involving primary and secondary metabolites.[67,86]
10. The more rapid the response in induction or production of defensive compounds during tissue development, the greater the toxicity, inhibition, and deterrency.[99,100]
11. Some mixtures can provide, depending on their chemical makeup, a physical barrier (e.g. those incorporating resin acids) as a defense as well as toxicity, deterrency, and inhibition.[27,32,33]

decrease in concentration during seasonal development,[85] or decrease during attack.[85]

The effect of variation in compositional ratios within an individual and among individuals within a host population is suggested to be an important factor in chemical defenses, and to be effective in reducing outbreaks of herbivores.[76,89,90] For example, four different chemical phenotypes were found to reduce fungal isolate growth (Table 16). In addition, chemical phenotypes II and IV were most effective against all five isolates, while phenotypes I and III had the greatest effect on certain isolates from certain individual trees.[101] Investigating such variation in mixtures will increase our understanding of what the diversity of effective mixtures are in a population of host trees, the potential that this variation has in regulating herbivore and pathogen populations, and the utility of maintaining hosts that vary in chemistry and phenology in silvicultural management.

Understanding the mechanism of action of natural products and how their complexity relates to additive and synergistic interactions are areas lacking information (Table 15). This understanding is important to the field of chemical ecology and the use of these compounds in applied management. Defining the number of individuals in a population that show differences from the "population

Table 16. Average linear growth (mm) of *Pleuroplaconema* sp. exposed to volatiles from four redwood phenotypes* (modified from Espinosa-Garcia and Langenheim 1991[101])

Chemical phenotype	Isolate group origin				
	Tree 1	Tree 2	Tree 3†	Tree 4	Tree 5
Control	9.36a‡	9.18a‡	8.00a	11.75a	8.87a
Phenotype I	3.92b	6.08b‡	3.25b	5.08b	4.67b
Phenotype II	1.83c	3.45c‡	2.37b,c	3.21c	2.75c
Phenotype III	2.73c‡	5.71b	2.87b,c	4.62b	2.96c
Phenotype IV	2.00c	1.50d	1.44c	2.92c	2.07c

*Means in each category come from 12 measurements belonging to three isolates. Means in columns with the same letter are not significantly different at $\alpha = 0.05$ according to Duncan's multiple-range test.
†Averages from eight measurements in each category belonging to two isolates.
‡Averages of 11 measurements belonging to three isolates.

average" and describing the types of chemical changes that individuals undergo during attack need resolution.

Finally, the rapidity of response in the induced reaction in xylem and phloem, and its effect on bark beetles and their associated fungi, is well documented.[25,26,38,99,100] However, the rapidity with which defenses can be deployed in the current year's growth of foliage is highly affected by physiological, developmental, and structural constraints in the tissue. Table 17 shows the changes in monoterpenes and sesquiterpenes in the current year's growth of Douglas-fir.[85] The major terpenoid defenses listed for this population from most effective to least are bornyl acetate, camphene, terpinolene, and limonene (Table 10).[71] Camphene and bornyl acetate triple in concentration in the first 10 days and then double in the next 17 days.[85] Although this population did not show a large increase in sesquiterpenes (unknowns 3–6), a population in Utah showed increases similar to the monoterpenes in Table 17. These sesquiterpenes are known to be highly toxic to western spruce budworm.[74] More information on the rate of deployment of primary nutrients such as galactose, as well as terpenoids in foliage, and how that varies among individuals within a population, are needed.[86]

Ecological and Evolutionary Effects of Mixtures

The ecological interactions leading to the evolution of effective complex mixtures are proposed to have significant effects on plant community structure and ecosystem sustainability (Table 18). The results of these interactions at the community level are a mosaic of tissue quality and host resources, the complexity this presents to the location of suitable resources needed to complete a life cycle,

Table 17. Variation in terpene concentration within a growing season in Douglas-fir current year's growth[*] (from Zou and Cates 1994[85])

	Sampling date				
Volatiles	June 11 $\bar{x} \pm se$	June 22 $\bar{x} \pm se$	July 9 $\bar{x} \pm se$	August 3 $\bar{x} \pm se$	September 20 $\bar{x} \pm se$	P
α-Pinene	0.43 ± 0.03	0.98 ± 0.10	2.01 ± 0.24	2.32 ± 0.26	1.42 ± 0.022	0.022
β-Pinene	0.39 ± 0.03	0.98 ± 0.10	2.01 ± 0.24	0.54 ± 0.10	0.38 ± 0.08	0.144
Camphene	0.49 ± 0.05	1.45 ± 0.18	3.31 ± 0.42	3.88 ± 0.45	2.62 ± 0.16	0.006
Limonene	0.19 ± 0.06	0.22 ± 0.06	0.34 ± 0.09	0.38 ± 0.10	0.27 ± 0.07	0.025
Myrcene	0.093 ± 0.02	0.103 ± 0.02	0.149 ± 0.02	0.166 ± 0.02	0.112 ± 0.01	0.012
Terpinolene	0.005 ± 0.00	0.028 ± 0.01	0.008 ± 0.00	0.004 ± 0.00	0.044 ± 0.01	0.030
Tricyclene	0.077 ± 0.01	0.221 ± 0.03	0.495 ± 0.06	0.589 ± 0.07	0.399 ± 0.02	0.006
Bornyl acetate	0.35 ± 0.04	1.14 ± 0.16	2.67 ± 0.41	2.82 ± 0.34	1.92 ± 0.12	0.001
α-Humulene	0.02 ± 0.01	0.01 ± 0.001	0.09 ± 0.03	0.11 ± 0.03	0.07 ± 0.02	0.079
Longibornyl	0.001 ± 0.00	0.008 ± 0.00	0.026 ± 0.01	0.044 ± 0.01	0.024 ± 0.01	0.313
Unknown 3	0.004 ± 0.00	0.004 ± 0.00	0.005 ± 0.00	0.006 ± 0.00	0.004 ± 0.00	0.207
Unknown 4	0.021 ± 0.00	0.023 ± 0.00	0.028 ± 0.00	0.029 ± 0.01	0.02 ± 0.004	0.185
Unknown 5	0.021 ± 0.00	0.042 ± 0.01	0.048 ± 0.02	0.054 ± 0.01	0.037 ± 0.008	0.150
Unknown 6	0.062 ± 0.01	0.033 ± 0.00	0.03 ± 0.00	0.028 ± 0.00	0.019 ± 0.003	0.005
Total	2.367 ± 0.21	4.922 ± 0.54	10.086 ± 1.30	11.372 ± 1.27	7.609 ± 0.49	0.006

[*]Values are mg/g fresh weight. Repeated measures was used to determine if a significant change occurred in the concentration of each compound across all sampling dates.

Table 18. Summary of the ecological and evolutionary consequences
of complex mixtures

---
1. Mixtures of primary and secondary metabolites among tissues, individuals within a species, and among species increase variation and heterogeneity in the quality of utilized and potential food resources.
2. Mixtures decrease host apparency to herbivores and pathogens (both the predictability and availability of suitable food resources for completion of the life cycle).
3. Mixtures and variation in the composition of mixtures increase the difficulty to evolve physiological and genetic resistance against defenses.
4. Mixtures increase competitive interactions among herbivores and among pathogens for suitable resources.
5. Mixtures increase herbivore and pathogen predictability and availability (apparency) to natural enemies by channeling herbivore and pathogens to trees with less effective defenses.
6. Components of mixtures that are plant defenses may be used by natural enemies to locate their host herbivores and plant pathogens in tritrophic interactions.
7. Maintaining the characteristics and mosaics of mixtures that lead to antiherbivore and antipathogen effectiveness is important to silvicultural management protocol.
8. Mixtures and variation in their components and production, are important to maintaining biological diversity and sustainable ecosystems.
---

and the constraints mixtures present to the evolution of resistance in herbivores and pathogens. Mixtures present barriers to resistance beyond the physiological or metabolic mechanisms of detoxification used for defensive compounds. Determining the diversity and kinds of effective mixtures among host plants in a community will increase our understanding of the role that natural products play in maintaining "healthy" ecosystems, and in applying this information to forest and agriculture management of pests and pathogens.

PHYSIOLOGICAL STRESS EFFECTS ON MIXTURES AND VARIATION IN THEIR PRODUCTION

I return to the statement by Hairston, Smith, and Slobodkin[1]: "Herbivores are not food-limited, appear most often to be predator-limited, and therefore are not likely to compete for common resources." The data and ideas presented here, and in other chapters in this volume e.g. Berenbaum, Hammerschmidt and Schultz, and Lindroth, suggest the contrary. Not all individuals within a species nor all species are available to any given herbivore or pathogen. In addition, significant biochemical, ecological, and evolutionary interactions leading to adaptations among host plants, their natural products, and herbivores and pathogens have occurred. The result of these interactions has been an elaboration of diversity and variety within a terpenoid defensive framework. Furthermore, these higher-order interactions contribute to the interrelationships, structure, and the dynamic nature of sustainable ecosystems.[102,103]

Table 19. Differences in *Picea glauca* chemistry between late successional upland sites where outbreaks of budworm defoliators and bark beetles initiated vs. late successional floodplain sites (Bonanza Creek Long-Term Ecological Research sites, Fairbanks, AK)

	Floodplain	Upland
Primary nutrients:[*]		
% Nitrogen	0.98	1.06[a]
% Phosphorus	0.11	0.19[a]
Secondary metabolites:[*]		
Total Terpenes[†]	3.8	2.2[a]
Tannin (mg/g d.w.)	24.4	27.4
Total Phenolics[‡]	738.9	674.6[b]

[*]Yarie and Van Cleve[105] for N, P; Cates et al. for secondary metabolites (in prep.).
[a]Significantly different between sites, $p<0.05$; [b]significantly different between sites, $p<0.10$.
[†]Terpenes expressed as mg/g fresh weight. [‡]Phenolics tentatively identified as hydroxyacetophenone glycosides, expressed as mean peak height fresh weight divided by 1,000.

In reality, however, massive outbreaks of insects and their associated pathogens have occurred regularly since the 1900s.[23,104] The reasons cited for these outbreaks have been well-documented and the "common denominator" is physiological stress generated by abiotic and biotic factors.[23] Associated with these stresses are often increases in primary nutrients and decreases in defensive compounds such that the salient characteristics of mixtures are reduced in effectiveness. For example, white spruce trees found in later successional stages at the upland sites at the Bonanza Creek LTER program have experienced heavy defoliation by the budworm species *Choristoneura occidentalis* and *C. orae* followed by attack from the bark beetle species *Ips perturbatus* and *Dendroctonus rufipennis*. Directly associated with these patterns of defoliation is the higher level of nitrogen and phosphorus in the upland site trees, but also, decreases in monoterpenes and phenolics in the foliage of mature to overmature trees at these sites (Table 19).[105]

Other stresses that ecosystems encounter include drought, lightning strikes, soil waterlogging, stress associated with stand characteristics such as high stocking levels of coexisting species, stand densities leading to crown closure, monocultures of mature to overmature stands, and human-related activities such as air pollution. The nature of these types of stress is to reduce the effects that individual compounds, variation in their production, and complex mixtures afford plants against herbivores and pathogens (Table 15,18). Consequently, forest and agriculture management protocol needs to favor regimes that maintain or enhance the advantages of the defensive mechanisms among cohabiting species.

ACKNOWLEDGMENTS

Thanks is extended to Annalise Carlquist for expert typing of the manuscript. This project was supported by National Science Foundation Project No. BSR 8702629.

REFERENCES

1. HAIRSTON, N.G., SMITH, F.E., SLOBODKIN, L.B. 1960. Community structure, population control, and competition. Am. Nat. 94: 421–425.
2. EMDEN, H.F. VAN (ed.). 1973. Insect/Plant Relationships. Symposia of the Royal Entomological Society of London: No. 6. John Wiley & Sons, New York, p. 215.
3. EHRLICH, P.T., RAVEN, P.H. 1964. Butterflies and plants: A study in coevolution. Evolution 18: 586–608.
4. RAFFA, K.F. 1995. Differential responses among natural enemies and prey to bark beetle pheromones: Implications of chemical, temporal, and spatial disparities to evolutionary theory and pest management. In: Behavior, Population Dynamics and Control of Forest Insects. (F.P. Hain, S.M. Salom, W.F. Ravlin, T.L. Payne, K.F. Raffa, eds.), Ohio State Univ., Wooster, Ohio, pp. 208–225.
5. WHITHAM, T.G. 1983. Host manipulation of parasites: Within-plant variation as a defense against rapidly evolving pests. In: Variable Plants and Herbivores in Natural and Managed Systems. (R.F. Denno, M.S. McClure, eds.), Academic Press, New York, pp. 15–41.
6. WHITHAM, T.G., SLOBODCHIKOFF, C.N. 1981. Evolution by individuals, plant-herbivore interactions, and mosaics of genetic variability: The adaptive significance of somatic mutations in plants. Oecologia 49: 287–292.
7. FUNK, C., LEWINSOHN, E., STOFER VOGEL, B., STEELE, C.L., CROTEAU, R. 1994. Regulation of oleoresins in grand fir (*Abies grandis*). Plant Physiol. 106: 999–1005.
8. LAFEVER, R.E., STOFER VOGEL, B., CROTEAU, R. 1994. Diterpenoid resin acid biosynthesis in conifers: Enzymatic cyclization of geranylgeranyl pyrophosphate to abietadiene, the precursor of abietic acid. Arch. Biochem. Biophys. 313: 139–149.
9. SAVAGE, T.J., HATCH, M.W., CROTEAU, R. 1994. Monoterpene synthases of *Pinus contorta* and related conifers: A new class of terpenoid cyclase. J. Biol. Chem. 269: 4012–4020.
10. GERSHENZON, J. 1994. Metabolic costs of terpenoid accumulation in higher plants. J. Chem. Ecol. 20: 1281–1328.
11. LANGENHEIM, J.H. 1994. Higher plant terpenoids: A phytocentric overview of their ecological roles. J. Chem. Ecol. 20: 1223–1280.
12. HOBSON, K.R., PARMETER, J.R. Jr., WOOD, D.L. 1994. The role of fungi vectored by *Dendroctonus brevicomis* Leconte (Coleoptera: Scolytidae) in occlusion of ponderosa pine xylem. Can. Entomol. 126: 277–282.
13. RAFFA, K.F., PHILLIPS, T.W., SALOM, S.M. 1993. Strategies and mechanisms of host colonization by bark beetles. In: Beetle-Pathogen Interactions in Conifer Forests. (T.D. Schowalter, G.M. Filip, eds.), Academic Press, London, pp. 103–128.
14. GERSHENZON, J., CROTEAU, R. 1991. Terpenoids. In: Herbivores, Their Interactions with Secondary Metabolites, Vol. 1., The Chemical Participants. (G.A. Rosenthal, M.R. Berenbaum, eds.), Academic Press, New York, pp. 165–219.
15. HARBORNE, J.B. 1991. Recent advances in the ecological chemistry of plant terpenoids. In: Ecological Chemistry and Biochemistry of Plant Terpenoids. (J.B. Harborne, F.A. Tomes-Barberan, eds.), Clarendon Press, Oxford, pp. 399–426.

16. GERSHENZON, J. 1993. The cost of plant chemical defenses against herbivory: A biochemical perspective. In: Insect-Plant Interactions, Vol. V. (E.A. Bernays, ed.), CRC Press, Boca Raton, Florida, pp. 105–173.
17. CROTEAU, R., SOOD, V.K. 1985. Metabolism of monoterpenes. Evidence for the function of monoterpenes and catabolism in peppermint (*Mentha piperita*). Plant Physiol. 77: 801–806.
18. CROTEAU, R., MARTINKUS, C. 1979. Metabolism of monoterpenes. Demonstration of (+)-neomenthyl-β-D-glucoside as a major metabolite of (-)-menthone in peppermint (*Mentha piperita*). Plant Physiol. 65: 169–175.
19. GERSHENZON, J., CROTEAU, R. 1990. Regulation of monoterpene biosynthesis in higher plants. In: Biochemistry of the Mevalonic Acid Pathway to Terpenoids. (G.H.N. Towers, H.A. Stafford, eds.), Plenum Press, New York, pp. 99–160.
20. GERSHENZON, J. 1984. Changes in the level of plant secondary metabolite production under water and nutrient stress. In: Phytochemical Adaptations to Stress, Recent Advances in Phytochemistry, Vol. 24. (B.N. Timmermann, C. Steelink, F.A. Loewus, eds.), Plenum Press, New York, pp. 273–320.
21. CROTEAU, R., GERSHENZON, J. 1994. Genetic control of monoterpene biosynthesis in mints (*Mentha*: Lamiaceae). In: Genetic Engineering of Plant Secondary Metabolism, Recent Advances in Phytochemistry, Vol. 28. (B.E. Ellis, G.W. Kuroki, H.A. Stafford, eds.), Plenum Press, New York, pp. 193–229.
22. CATES, R.G., REDAK, R.A., HENDERSON, C.B. 1983. Patterns in defensive natural products chemistry: Douglas-fir and western spruce budworm interactions. In: Plant Resistance to Insects. ACS Symposium Series 208. (P.A. Hedin, ed.), American Chemical Society, Washington, D.C., pp. 3–20.
23. WULF, W., CATES, R.G. 1987. Site and stand characteristics. In: Western Spruce Budworm. Technical Bull. 1694. (M. Brookes, R. Campbell, J. Colbert, R. Mitchell, R. Start, eds.), Washington, D.C., pp. 89–115.
24. LEWINSOHN, E., GIJZEN, M., CROTEAU, R. 1991. Defense mechanisms of conifers. Differences in constitutive and wound-induced monoterpene biosynthesis among species. Plant Physiol. 96:44–49.
25. LEWINSOHN, E., GIJZEN, M., SAVAGE, T.J., CROTEAU, R. 1991. Defense mechanisms of conifers. Relationship of monoterpene cyclase activity to anatomical specialization and oleoresin monoterpene content. Plant Physiol. 96:38–43.
26. RAFFA, K.F., BERRYMAN, A.A. 1987. Interacting selective pressures in conifer-bark beetle systems: A basis for reciprocal adaptations? Am. Nat. 129: 234–262.
27. CATES, R.G., ALEXANDER, H. 1982. Host resistance and susceptibility. In: Bark Beetles in North American Conifers: Evolution and Ecology. (J. Mitton, K. Sturgeon, eds.), Univ. of Texas Press, Austin, Texas, pp. 212–263.
28. LEWINSOHN, E., SAVAGE, T.J., GIJZEN, M., CROTEAU, R. 1993. Simultaneous analysis of monoterpenes and diterpenoids of conifer oleoresin. Phytochem. Anal. 4: 220–225.
29. MILLER, R.H., BERRYMAN, A.A., RYAN, C.A. 1986. Biotic elicitors of defense reactions in lodgepole pine. Phytochemistry 25: 611–612.
30. LEWINSOHN, E., WORDEN, E., CROTEAU, R. 1994. Monoterpene cyclases in grand fir callus cultures: Modulation by elicitors and growth regulators. Phytochemistry 36: 651–656.
31. RAFFA, K.F. 1991. Induced defensive reactions in conifer-bark beetle systems. In: Phytochemistry Induction by Herbivores. (D.W. Tallamy, M.J. Raupp, eds.), John Wiley & Sons, New York, pp. 245–276.
32. NEBEKER, T.E., HODGES, J.D., BLANCHE, C.A. 1993. Host response to bark beetle and pathogen colonization. In: Beetle-Pathogen Interactions in Conifer Forests. (T.D. Schowalter, G.M. Filip, eds.), Academic Press, London, pp. 157–173.
33. HODGES, J.D., ELAM, W.W., WATSON, W.F., NEBEKER, T.E. 1979. Oleoresin characteristics and susceptibility of four southern pines to southern pine beetle (Coleoptera: Scolytidae) attacks. Can. Ent. 111: 889–896.

34. BERRYMAN, A.A. 1972. Resistance of conifers to invasion by bark beetle-fungus associations. BioScience 22: 598–602.
35. RAFFA, K.F., BERRYMAN, A.A. 1983. The role of host plant resistance in the colonization behavior and ecology of bark beetles (Coleoptera: Scolytidae). Ecol. Monogr. 53: 27–49.
36. RAFFA, K.F., BERRYMAN, A.A. 1982. Physiological differences between lodgepole pines resistant and susceptible to the mountain pine beetle and associated microorganisms. Environ. Entomol. 11: 486–492.
37. WERNER, R.A., ILLMAN, B.L. 1994. Reponse of Lutz, Sitka, and white spruce to attack by *Dendroctonus rufipennis* (Coleoptera: Scolytidae) and blue stain fungi. Environ. Entomol. 23: 472–478.
38. HAIN, F.P., COOK, S.P., MATSON, P.A., WILSON, K.G. 1985. Factors contributing to southern pine beetle host resistance. In: Integrated Pest Management Research Symposium: The Proceedings. (S.J. Branham, R.C. Thatcher, eds.), USDA Forest Serv. Gen. Tech. Rep. SO-56. USDA Forest Serv., Southern Forest Exp. Stn., New Orleans, LA., pp. 154–160.
39. RAFFA, K.F., BERRYMAN, A.A., SIMASKO, J., TEAL, W., WONG, B.L. 1985. Effects of grand fir monoterpenes on the fir engraver, *Scolytus ventralis* (Coleoptera: Scolytidae), and its symbiotic fungus. Environ. Entomol. 14: 552–556.
40. COYNE, J.F., LOTT, L.H. 1976. Toxicity of substances in pine oleoresin to southern pine beetles. J. Georgia Entomol. Soc. 11: 297–301.
41. WERNER, R.A. 1995. Toxicity and repellency of 4-allylanisole and monoterpenes from white spruce and tamarack to the spruce beetle and eastern larch beetle (Coleoptera: Scolytidae). Environ. Entomol. 24: 372–379.
42. RAFFA, K.F., SMALLEY, E.B. 1995. Interaction of pre-attack and induced monoterpene concentrations in host conifer defense against bark beetle-fungal complexes. Oecologia 102: 285–296.
43. SMITH, R.H. 1965. Effect of monoterpene vapors on the western pine beetle. J. Econ. Entomol. 58: 509–510.
44. SMITH, R.H. 1966. Resin quality as a factor in the resistance of pines to bark beetles. In: Breeding Pest-Resistant Trees. (H.D. Gerhold, R.E. McDermott, E.H. Schreiner, J.A. Winioski, eds.), Permagon, Oxford, pp. 189–96.
45. DELORME, L., LIEUTIER, F. 1990. Monoterpene composition of the preformed and induced resins of Scots pine, and their effect on bark beetles and associated fungi. Eur. J. For. Path. 20: 304–316.
46. COOK, S.B., HAIN, F.P. 1988. Toxicity of host monoterpenes to *Dendroctonus frontalis* and *Ips calligraphus* (Coleoptera: Scolytidae). J. Entomol. Sci. 23: 287–292.
47. RUSSELL, C.E., BERRYMAN, A.A. 1976. Host resistance to the fir engraver beetle. 1. Monoterpene composition of *Abies grandis* pitch blisters and fungus-infected wounds. Can. J. Bot. 54: 14–18.
48. PAINE, T.D., HANLON, C.C. 1994. Influence of oleoresin constituents from *Pinus ponderosa* and *Pinus jeffreyi* on growth of mycangial fungi from *Dendroctonus ponderosae* and *Dendroctonus jeffreyi*. J. Chem. Ecol. 20: 2551–2563.
49. ENNOS, R.A., SWALES, K.W. 1991. Genetic variation in a fungal pathogen: Response to host defensive chemicals. Evolution 45: 190–204.
50. CATES, R.G., PAINE, T.D. 1996. Effects of terpenoid and phenolic compounds from *Pinus taeda* on growth of fungi associated with *Dendroctonus frontalis*. J. Chem. Ecol. submitted.
51. BRIDGES, J.R. 1987. Effects of terpenoid compounds on growth of symbiotic fungi associated with the southern pine beetle. Phytopathology 77: 83–85.
52. COOK, S.P., HAIN, F.P. 1985. Qualitative examination of the hypersensitive response of loblolly pine, *Pinus taeda* L., inoculated with two fungal associates of the southern pine beetle, *Dendroctonus frontalis* Zimmerman (Coleoptera: Scolytidae). Environ. Entomol. 14: 396–400.

53. ESPINOSA-GARCIA, F.J., LANGENHEIM, J.H. 1991. Effects of sabinene and γ-terpinene from coastal redwood leaves acting singly or in mixtures on the growth of some of their fungus endophytes. Biochem. Syst. Ecol. 19: 643–650.
54. ARRHENIUS, S.P., LANGENHEIM, J.H. 1983. Inhibitory effects of *Hymenaea* and *Copaifera* leaf resins on the leaf fungus *Pestalotia subcuticularis*. Biochem. Syst. Ecol. 11: 361–366.
55. HUBBELL, S.P., WIEMER, D.F., ADEJARE, A. 1983. An antifungal terpenoid defends a neotropical tree (*Hymenaea*) against attack by fungus-growing ants (*Atta*). Oecologia 60: 321–327.
56. RAFFA, K.F., BERRYMAN, A.A. 1983. Physiological aspects of lodgepole pine wound responses to a fungal symbiont of the mountain pine beetle, *Dendroctonus ponderosae* (Coleoptera: Scolytidae). Can. Ent. 115: 723–734.
57. RAFFA, K.F., BERRYMAN, A.A. 1982. Physiological differences between lodgepole pines resistant and susceptible to the mountain pine beetle and associated microorganisms. Environ. Entomol. 11: 486–492.
58. HAIN, F.P., MAWBY, W.D., COOK, S.P., ARTHUR, F.H. 1983. Host conifer reaction to stem invasion. Z. ang. Ent. 96: 247–256.
59. STEPHEN, F.M., LIH, M.P., PAINE, T.D., WALLIS, G.W. 1988. Using acute stress to modify tree resistance: Impact on within-tree southern pine beetle populations. In: Integrated Control of Scolytid Bark Beetles. (T.L. Payne, H. Saarenmaa, eds.), Virginia Polytechnic Inst. and State Univ., Blacksburg, Virginia, pp. 105–119.
60. BORDASCH, R.P., BERRYMAN, A.A. 1977. Host resistance to the engraver beetle, *Scolytus ventralis* (Coleoptera: Scolytidae). 2. Repellency of *Abies grandis* resins and some monoterpenes. Can. Entomol. 109: 95–100.
61. RAFFA, K.F., BERRYMAN, A.A. 1982. Accumulation of monoterpenes and associated volatiles following fungal inoculation of grand fir with a fungus vectored by the fir engraver *Scolytus ventralis* (Coleoptera: Scolytidae). Can. Entomol. 114: 797–810.
62. WERNER, R.A., ILLMAN, B.L. 1995. The role of stilbene-like compounds in host tree resistance of sitka spruce to the spruce beetle, *Dendroctonus rufipennis*. In: Behavior, Population Dynamics and Control of Forest Insects. Proc. IUFRO Joint Conf. (F.P. Hain, S.M. Salom, W.F. Ravlin, T.L. Payne, K.F. Raffa, eds.), Maui, Hawaii, pp. 123–133.
63. LIEUTIER, F. 1995. Associated fungi, induced reaction and attack strategy of *Tomicus piniperda* (Coleoptera: Scolytidae) in Scots pine. In: Behavior, Population Dynamics and Control of Forest Insects. Proc. IUFRO Joint Conf. (F.P. Hain, S.M. Salom, W.F. Ravlin, T.L. Payne, K.F. Raffa, eds.), Maui, Hawaii, pp. 139–151.
64. LIEUTIER, F., YART, A., JAY-ALLEMAND, C., DELORME, L. 1991. Preliminary investigations on phenolics as a response of Scots pine phloem to attacks by bark beetles and associated fungi. Eur. J. For. Pathol. 21: 354–64.
65. GAMBLIEL, H.A., CATES, R.G., CAFFEY-MOQUIN, M.K., PAINE, T.D. 1985. Variation in the chemistry of loblolly pine in relation to infection by the blue-stain fungus. In: Integrated Pest Management Research Symposium: The Proceedings. (S.J. Branham, R.C. Thatcher, eds.), USDA Forest Serv. Gen. Tech. Rep. SO-56. USDA Forest Serv., Southern Forest Exp. Stn., New Orleans, LA., pp. 177–184.
66. HEMINGWAY, R.W., MCGRAW, G.W., BARRAS, S. 1977. Polyphenols in *Ceratocystis minus* infected *Pinus taeda*: Fungal metabolites, phloem and xylem phenols. Agric. Food Chem. 25: 717–722.
67. MATTSON, W.J., SLOCUM, S.S., KOLLER, C.N. 1983. Spruce budworm (*Choristoneura fumiferana*) performance in relation to foliar chemistry of its host plants. In: Forest Defoliator-Host Interactions: A Comparison between Gypsy Moth and Spruce Budworms. USDA Forest Serv. Gen. Tech. Rep. NE-85. USDA Forest Serv., N.E. For. Exp. Stn., Broomhall, PA, pp. 55–65.

68. OHIGASHI, H., WAGNER, M.R., MATSUMURA, F., BENJAMIN, D.M. 1981. Chemical basis of differential feeding behavior of the larch sawfly, *Pristiphora erichsonii* (Hartig). J. Chem. Ecol. 7: 599–614.
69. IKEDA, T., MATSUMURA, F., BENJAMIN, D.M. 1977. Mechanism of feeding discrimination between matured and juvenile foliage by two species of pine sawflies. J. Chem. Ecol. 3: 677–694.
70. SCHUH, B.A., BENJAMIN, D.M. 1984. The chemical feeding ecology of *Neodiprion dubiosus* Schedl, *N. rugifrons* Midd., and *N. lecontei* (Fitch) on Jack pine (*Pinus banksiana* Lamb.) J. Chem. Ecol. 10: 1071–1079.
71. CATES, R.G., ZOU, J., CARLSON, C. 1991. The role of variation in Douglas-fir foliage quality in the silvicultural management of western spruce budworm. In: Interior Douglas Fir: The Species and its Management. (D. Baumgartner, J. Lotan, eds.), Washington State Univ., Pullman, WA, pp. 115–128.
72. CATES, R.G., REDAK, R., HENDERSON, C. 1983. Patterns in defensive natural product chemistry: Douglas-fir and western spruce budworm interactions. In: Mechanisms of Plant Resistance to Insects. (P. Hedin ed.), American Chemical Society, Washington, D.C., pp. 3–19.
73. REDAK, R., CATES, R.G. 1984. Douglas-fir—spruce budworm interactions. The effect of nutrition, chemical defenses, tissue phenology, and tree physical parameters on budworm success. Oecologia 62: 61–67.
74. ZOU, J., CATES, R.G. 1996. Effects of terpenes and phenolic and flavonoid glycosides in the current year's needles of Douglas-fir on western spruce budworm larval growth and pupal weight. Can. J. For. Res. submitted.
75. CHARLES, P.J., DELPLANQUE, A., MARPEAU, A., BERNARD-DAGAN, C., ARBEZ, M. 1982. Susceptibility of European black pine (*Pinus nigra*) to the European pine shoot moth (*Rhyacionia builiana*): variations of susceptibility at the provenance and individual level of the pine and effect of terpene composition. In: Resistance to Diseases and Pests in Forest Trees. Proc. 3rd Int. Workshop on the Genetics of Host-Parasite Interactions in Forestry. (H.M. Heybrock, B.R. Stephan, K. von Weissenberg, eds.), Center for Agricultural Publishing and Documentation, Wageningen, The Netherlands, pp. 206–212.
76. MACEDO, C.A., LANGENHEIM, J.H. 1989. A further investigation of leaf sesquiterpene variation in relation to herbivory in two Brazilian populations of *Copaifera langsdorfii*. Biochem. System. Ecol. 17: 207–216.
77. SHEPHERD, R.F. 1992. Relationships between attack rates and survival of western spruce budworm, *Choristoneura occidentalis* Freeman (Lepidoptera: Tortricidae) and bud development of Douglas-fir, *Pseudotsuga menziesii* (Mirb.) Franco. Canad. Entomol. 124: 347–358.
78. GAMBLIEL, H.A., CATES, R.G. 1995. Terpene changes due to maturation and canopy level in Douglas-fir (*Pseudotsuga menziesii*) flush needle oil. Biochem. Syst. Ecol. 23: 469–476.
79. CLANCY, K.M., WAGNER, M.R., TINUS, R.W. 1988. Variation in host foliage nutrient concentrations in relation to western spruce budworm herbivory. Can. J. For. Res. 18: 530–539.
80. MATTSON, W.J., SCRIBER, J.M. 1987. Nutritional ecology of insect folivores of woody plants: water, nitrogen, fiber, and mineral considerations. In: Nutritional Ecology of Insects, Mites, and Spiders (F. Slansky and J. Rodriguez, eds.), John Wiley and Sons, New York, pp. 105–146.
81. CLANCY, K.M., ITAMI, J.K., HUEBNER, D.P. 1993. Douglas-fir nutrients and terpenes: potential resistance factors to western spruce budworm defoliation. For. Science 39: 78–94.
82. CLANCY, K.M. 1992. The role of sugars in western spruce budworm nutritional ecology. Ecolog. Entomol. 17: 189–197.
83. CLANCY, K.M. 1991. Douglas-fir nutrients and terpenes as potential factors influencing western spruce budworm defoliation. In: Forest Insect Guilds: Patterns of Interaction with Host Trees (Y.N. Baranchikov, W.J. Mattson, F. Hain and T.L. Payne, eds.), United States Department of Agriculture Forest Service General Technical Report ME-153. Washington, D.C.

84. CLANCY, K.M. 1992. Response of western spruce budworm (Lepidoptera: Tortricidae) to increased nitrogen in artificial diets. Environ. Entomol. 21: 331–344.
85. ZOU, J., CATES, R.G. 1995. Foliage constituents of Douglas fir (*Pseudotsuga menziesii* (Mirb.) Franco (Pinaceae)): their seasonal variation and potential role in Douglas fir resistance and silviculture management. J. Chem. Ecol. 21: 387–402.
86. ZOU, J., CATES, R.G. 1994. The role of Douglas fir (*Pseudotsuga menziesii*) carbohydrates in resistance to budworm (*Choristoneura occidentalis*). J. Chem. Ecol. 20: 395–405.
87. SLANSKY, F.S. Jr., WHEELER, G.S. 1992. Feeding and growth responses of laboratory and field strains of velvetbean caterpillars (Lepidoptera: Noctuidae) to food nutrient and allelochemicals. J. Econ. Entomol. 85: 1717–1730.
88. MASON, L.J., PASHLEY, D.P., JOHNSON, S.J. 1987. The laboratory as an altered habitat: phenotypic and genetic consequences of colonization. Fla. Entomol. 70: 49–58.
89. CATES, R.G., REDAK, R. 1988. Variation in the terpene chemistry of Douglas-fir and its relationship to western spruce budworm success. In: Chemical Mediation of Coevolution. (K.C. Spencer ed.), Pergmon Press, N.Y., pp. 317–344.
90. CATES, R.G., ZOU, J. 1990. Douglas-fir (*Pseudotsuga menziesii*) population variation in terpene chemistry and its role in budworm (*Choristoneura occidentalis* Freeman) dynamics. In: Population Dynamics of Forest Insects. (A. Watt, S. Leather, M. Hunter, N. Kidd eds.), Intercept Ltd, London, England, pp. 169–182.
91. STEPHAN, B.R. 1987. Differences in the resistance of Douglas fir provenances to the woolly aphid *Gilletteella cooleyi*. Silv. Genet. 36: 76–79.
92. CHRISTIANSEN, E., BERRYMAN, A.A. 1994. Norway spruce clones vary widely in their susceptibility to a bark beetle-transmitted blue-stain fungus. In: Behavior, Population Dynamics and Control of Forest Insects. Proc. IUFRO Joint conf. (F.P. Hain, S.M. Salom, W.F. Ravlin, T.L. Payne, K.F. Raffa, eds.), Maui, Hawaii, pp. 152–153.
93. WILLHITE, E., STOCK, M. 1983. Genetic variation among western spruce budworm (*Choristoneura occidentalis*) outbreaks in Idaho and Montana. Can. Entomol. 11: 41–54.
94. STURGEON, K.B. 1979. Monoterpene variation in ponderosa pine xylem resin related to western pine beetle predation. Evolution 33: 803–814.
95. ENNOS, R.A., SWALES, K.W. 1991. Genetic variation in a fungal pathogen: Response to host defensive chemicals. Evolution 45: 190–204.
96. BRIGNOLAS, F., LIEUTIER, F., SAUVARD, D., YART, A., DROUET, A., CLAUDOT, A.C. 1995. Changes in soluble-phenol of Norway-spruce (*Picea abies*) phloem in response to wounding and inoculation with *Ophiostoma polonicum*. Eur. J. For. Path. 25: 253–265.
97. SCHULTZ, J.C. 1983. Habitat selection and foraging tactics of caterpillars in heterogeneous trees. In: Variable Plants and Herbivores in Natural and Managed Systems. (R.F. Denno, M.S. McClure, eds.), Academic Press, New York, pp. 61–90.
98. MATTSON, W.J., LORIMER, N., LEARY, R.A. 1982. Role of plant variability (trait vector dynamics and diversity) in plant/herbivore interactions. In: Resistance to Diseases and Pests in Forest Trees. Proc. 3rd Int. Workshop on the Genetics of Host-Parasite Interactions in Forestry. (H.M. Heybrock, B.R. Stephan, K. von Weissenberg, eds.), Center for Agricultural Publishing and Documentation, Wageningen, The Netherlands, pp. 295–303.
99. PAINE, T.D., STEPHEN, F.M., CATES, R.G. 1993. Within- and among-tree variation of the induced response of loblolly pine to a fungus associated with *Dendroctonus frontalis* Zimmerman (Coleoptera: Scolytidae) and sterile wounding. Can. Entomol. 125: 65–71.
100. PAINE, T., STEPHEN, F., CATES, R.G. 1991. Host defense reactions in response to inoculation with *Ophiostoma* species. In: Taxonomy and Biology of the Ophiostomatales. (K.A. Seifert, M. Wingfield, J. Webber eds.), APS Press, St. Paul, Minnesota. pp. 200–205.
101. ESPINOSA-GARCIA, F.J., LANGENHEIM, J.H. 1991. Effect of some leaf essential oil phenotypes from coastal redwood on growth of its predominant endophytic fungus, *Pleuroplaconema* sp. J. Chem. Ecol. 17: 1837–1857.

102. BERENBAUM, M.R. 1995. The chemistry of defense: Theory and practice. Proc. Natl. Acad. Sci. USA 92: 2–8.
103. MOLE, S. 1994. Trade-offs and constraints in plant-herbivore defense theory: A life-history perspective. Oikos 71: 3–12.
104. BERRYMAN, A.A. 1986. Forest Insects. Principles and Practice of Population Management. Plenum Press, New York. p. 279.
105. YARIE, J., VAN CLEVE, K. 1996. Effects of carbon, fertilizer and drought on foliar nutrient concentrations of taiga tree species in interior Alaska. Ecol. Applica. (in press).

Chapter Eight

RELATIONSHIPS BETWEEN THE DEFENSE SYSTEMS OF PLANTS AND INSECTS

The Cyanogenic System of the Moth *Zygaena trifolii*

Adolf Nahrstedt

Institut für Pharmazeutische Biologie und Phytochemie der Westf.
Wilhelms-Universität
Hittorfstr. 56, D-48149 Münster, Germany

Introduction 217
Origin of Linamarin and Lotaustralin in *Zygaena trifolii* 219
Localization of the Biogenetic Pathway and Its Specificity 222
Degradation of Linamarin and Lotaustralin 223
Conclusion 226

INTRODUCTION

Several secondary constituents which are produced by higher plants are used both by plants and also by insects for defense. For example, pyrrolizidine alkaloids accumulate in many Asteraeae and are taken up from the plant tissues by Nymphaelidae or Arctiidae and stored in their body tissues, thus becoming toxic to insectivores.[1] The same is true for: cardiac glycosides, produced by Asclepidaceae and used by the monarch butterfly;[2,3] mustard oils, produced by Brassicaceae and used by Pieridae;[4] quinolizidines, produced by Fabaceae and used by Pyralidae;[5,6] the azoxymethanol glucoside cycasin, produced by Cycadaceae and used by Lycaenidae.[7] Additionally, iridoids,[3,8] alkaloidal glycosidase inhibitors,[9] tropane alkaloids,[10] butenolides,[11] aristolochic acids,[12] and cyanogenic glycosides[13,14] also are all exploited to some degree by insects.

Cyanogenesis is a widespread phenomenon in higher plants, occurring in more than 2000 species.[15] Cyanogenesis depends on the accumulation of cyanogenic glycosides and degrading enzymes such as β-glucosidases and hydroxyni-

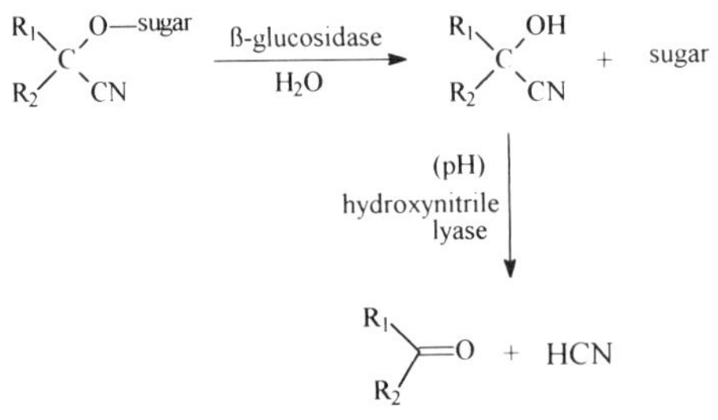

Figure 1. Cyanogenesis. The first step is catalyzed by a more or less specific β-glucosidase, the second occurs spontaneously at the cellular pH of 4–6 or is catalyzed by a hydroxynitrile lyase.

trile lyases.[16] Highly toxic hydrogen cyanide is liberated by the process.[17] Production of HCN usually occurs rapidly. The process is analagous to the so-called mustard oil bomb which results in the rapid liberation of mustard oils from glucosinolates.[18] One might use the term *cyanide bomb* for the release of cyanogenesis. Although cyanogenesis usually has been regarded as a defense system typical for plants,[19] some arthropods also contain cyanogenic compounds. Several Diplopoda, Chilopoda and Hexapoda species accumulate mandelonitrile for defense.[13,20] In addition to mandelonitrile the cyanogenic glucoside, prunasin, occurs in low concentration in *Paropsis atomaria* (Coleoptera);[21] the cyclopentenylcyanohydrin glucoside, gynocardin, appears to be sequestered by *Acraea horta* (Acraeinae) from its host *Kiggelaria africana*;[22,23] and the plant derived cyanogenic glucoside, cardiospermin, is accumulated by the true bug *Leptocoris isolata*.[24]

The cyanogenic glucosides linamarin and lotaustralin (Fig. 2) are widely distributed in the plant kingdom.[15] During the past 15 years, we have shown that both compounds occur also in several species of the Lepidoptera.[13,14] The moth family Zygaenidae, especially *Zygaena trifolii*, has been well investigated. This

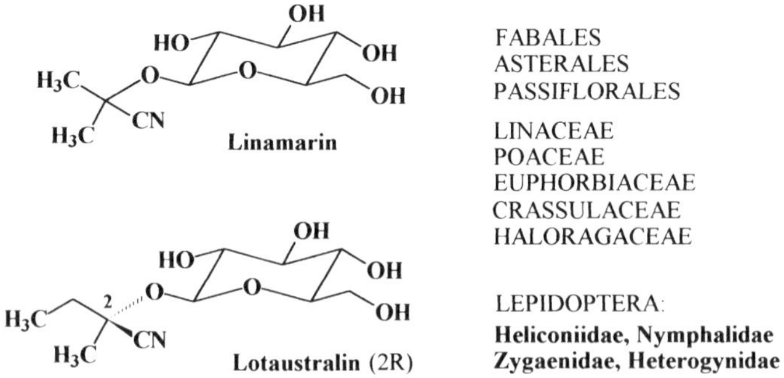

Figure 2. Linamarin and lotaustralin, both of which occur in the indicated families and orders of the plants as well as in the families of the Lepidoptera.

FABALES
ASTERALES
PASSIFLORALES

LINACEAE
POACEAE
EUPHORBIACEAE
CRASSULACEAE
HALORAGACEAE

LEPIDOPTERA:
Heliconiidae, Nymphalidae
Zygaenidae, Heterogynidae

species occurs in south and central Europe from Spain to Poland and is characterized by its typical aposematic coloration that is black with red spots for the adults and yellow with black spots for the larvae. At the beginning of this century, it was noticed that *Zygaena* species are resistant to HCN.[25] In 1962, Jones et al.[26] discovered that cyanogenesis occurred in *Zygaena filipendulae* when the insects were wounded. Later the source of HCN was shown to be linamarin and lotaustralin.[27,28] Since almost all species hitherto investigated contain both glucosides, cyanogenesis seems to be a monophyletic and, thus, an old feature of the Zygaenidae.[29,13,14] The following work reviews our present knowledge of the cyanogenic system of the larvae of *Z. trifolii*,[30] and compares it with what is known for higher plants. The example illustrates an interesting aspect of chemical redundancy between plants and insects and seems to be unique in the relationships between plants and insects.

ORIGIN OF LINAMARIN AND LOTAUSTRALIN IN *ZYGAENA TRIFOLII*

The distribution of linamarin and lotaustralin in the larval body is restricted to certain tissues (Table 1). Almost one third of the entire glucoside set is located in the hemolymph, the other two thirds in the integument. Thus a reasonable

Table 1. Distribution of linamarin and lotaustralin in different tissues of a larva of *Zygaena trifolii*

Tissue	amount [d.m.]	CN [µmol] per indiv.	% of total
hemolymph (serum)	50 µl	2.5	33.3
hemocytes		0.13	
fat body	8 mg	0.02	0.25
malpighii	2.5 mg	0.006	0.08
gut	3.4 mg	0.003	0.04
silk gland	n.d.	0.01	0.13
integument	14.2 mg	4.28	54
secretion	3 µl	0.93	11.8

amount of the cyanoglucosides is concentrated in the outer part of the larvae, the part which a potential predator first comes into contact with during an attack on its prey. Within the integument, cuticular cavities[31] are filled with a secretion that contains ca. 10% of the total cyanoglucosides.[32,33] The colorless fluid is secreted from the cavities in small sticky droplets when the larvae are attacked by birds, lizards or ants. Experiments with naive birds indicate the deterrent activity of the secretion.[34]

Larvae of *Z. trifolii* are monophagous and feed on *Lotus corniculatus* (Fabaceae), a plant that contains linamarin and lotaustralin. It has been shown, by feeding experiments of "cold" and ^{14}C-labelled linamarin and lotaustralin, that the larvae are able to take up both glucosides.[35] The radioactive glucosides used for these experiments contained the radioactive label only in the aglycone part; therefore it is not yet clear whether or not the glucosides are absorbed before degradation with subsequent re-glucosylation of the absorbed aglycone. This question is under investigation. By exhibiting this behavior the larvae are similar to many other insects that sequester toxic compounds from higher plants (see Introduction). More interestingly, however, it also was observed early on that many species of the Zygaenidae do not feed on plants that contain linamarin and lotaustralin, but nevertheless accumulate both glucosides[13] (Table 2). Nothing is known about the distribution pattern of linamarin and lotaustralin in such Zygaenidae.

This observation prompted us to investigate the possibility of *de novo* synthesis of linamarin and lotaustralin by *Z. trifolii* larvae. The biogenetic pattern of cyanogenic glycosides in plants is well known.[36,37] The current knowledge of the biosynthesis in *Zygaena*, obtained by following the protocol of Koch et al.,[37] is summarized (Fig. 3). The amino acids valine and isoleucine, well known from plants as precursors for linamarin and lotaustralin, were incorporated into both glucosides at rates from 1.5 to 3%. When valine and isoleucine enriched with the ^{13}C isotope were fed to the larvae, the ^{13}C-atoms of the amino acids were found at the predicted positions, as proven by ^{13}C-NMR spectroscopy.[38] The

Table 2. Foodplants (family level) of the Zygaenidae

Larvae of the Zygaenidae feed on

* Fabaceae (e.g. *Lotus* spec);
  cyanogenic with linamarin and lotaustralin

* Rosaceae;
  cyanogenic with mainly prunasin

* Apiaceae; Celastraceae; Polygonaceae; Lamiaceae;
  Ericaceae; Vitaceae; Aquifoliaceae;
  non-cyanogenic

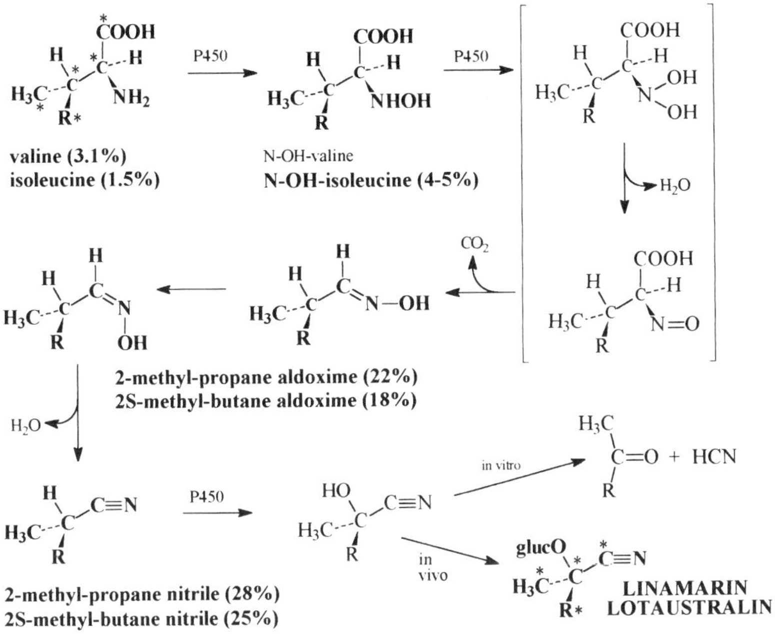

RELATIONSHIPS BETWEEN THE DEFENSE SYSTEMS OF PLANTS AND INSECTS 221

Figure 3. Incorporation rates of ^{14}C-labelled amino acids and intermediates of the biogenetic pathway of cyanogenic glycosides into linamarin and lotaustralin in the larvae of *Z. trifolii*. ^{13}C-Isotopes are marked with an asterisk. Protocol is that of Koch et al.[37]

nitriles corresponding to valine and isoleucine showed high incorporation rates.[39] The corresponding synthetically available N-hydroxyamino acids and aldoximes were shown in recent similar studies to have intermediate incorporation rates, as would be expected for a biogenetic pathway with competing side reactions on going from the precursors to the products.[40]

These results provide clear evidence for the *de novo* synthesis of linamarin and lotaustralin by *Zygaena* larvae, and show that the insects use the same established precursors and intermediates as higher plants to produce the corresponding cyanogenic glycosides. The results show, on the other hand, that, in contrast to higher plants, the pathway does not appear to be channelled,[35] since it is possible for intermediates to enter the pathway. Thus, we have discovered a unique example of apparent chemical redundancy in which an insect synthesizes the same compounds *de novo* that it is able to sequester from its host. These two independent, yet complementary, ways of obtaining cyanogenic glycosides may explain why those *Zygaena* species that feed on *Lotus* species contain exceptionally high concentrations of linamarin and lotaustralin. How much and in what ways uptake influences biosynthesis is an interesting question which is currently under investigation.

LOCALIZATION OF THE BIOGENETIC PATHWAY AND ITS SPECIFICITY

In plants, the production of cyanogenic compounds is associated with the microsomal fraction. This has been shown for a small number of species.[36,41-44] In each case, the microsomes do not produce the glucosides, but rather the corresponding cyanohydrins as shown by the liberation of HCN from incubated samples. Cyanogenic glycosides have never been detected in the microsomal assays. In order to establish the site of biosynthesis for *Z. trifolii* larvae, we incubated several tissues obtained after dissection (gut tissue, malphigii, integument, hemolymph, fat body) with late intermediates of the biosynthetic pathway, *i.e* the nitriles. Only the fat body produced HCN from the nitriles.[45] Interestingly, as in plants, no glucoslytransferase was active in the preparations used for incubation. Linamarin and lotaustralin could not be detected in the incubation mixture run at pH 6.3 (which is the pH of the hemolymph of *Z. trifolii*). The isolated fat body, however, was a suitable model for studying the specificity of the biogenetic pathway. The rate of HCN liberation following incubation with the different pathway intermediates, and the precursors, valine and isoleucine was measured[45] (Fig. 4).

The nitriles are metabolized at a significant rate (Fig. 4). In addition to the "endogeneous" substrates, related aliphatic nitriles with a linear or a branched carbon chain are also metabolized to HCN. Indeed, two of these non-naturally occurring compounds were metabolized better than isobutyronitrile, the precursor of linamarin. The aromatic phenylacetonitrile (an intermediate in the biosynthesis of the cyanogenic glucoside prunasin) also is converted to the corresponding hydroxynitrile (HCN) at a rate of 75% that of isobutyronitrile. The observed production of HCN in these cases could represent a non-specific hydroxylation that occurs in order to detoxify xenobiotics such as the exogeneously applied nitriles. This possibility was eliminated, however, when the isolated fat body also produced HCN when incubated with earlier intermediates of the pathway such as the aldoximes and the N-hydroxyamino acids, the latter although at very low rates.

Conversion of valine and isoleucine could only be detected when ^{14}C-labelled, radioactive amino acids were used. In contrast to the nitriles, only ca. 1/400 of the amount of valine and isoleucine incubated was metabolized. In spite of the low conversion of valine and isoleucine, we showed that other amino acids, such as [U-^{14}C]-leucine, [U-^{14}C]-phenylalanine and [U-^{14}C]tyrosine did not yield H^{14}CN. This finding suggests specificity of the pathway at its earliest point, the hydroxylation of the amino acids. A similar suggestion has been made for the higher plants, *Turnera angustifolia* and *Passiflora morifolia*.[46]

When a cytochrome P450-inhibitor, tetcyclasis, was incubated together with the nitriles in the isolated fat body, no HCN production was observed. Thus,

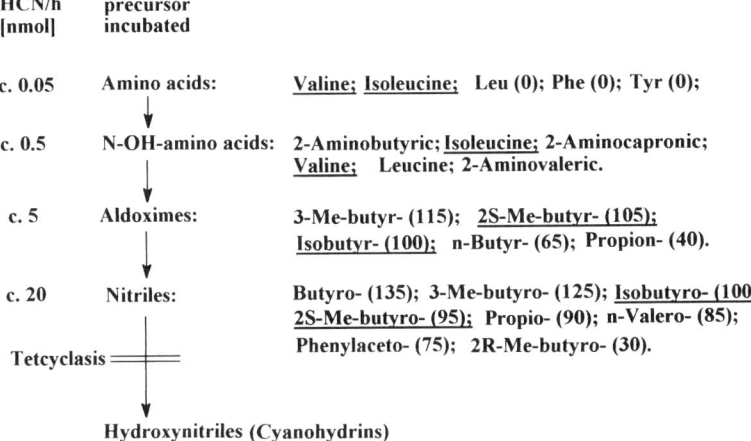

Figure 4. Metabolism of precursors and intermediates by the fat body of *Z. trifolii* larvae. Measured via HCN that is liberated from the hydroxynitrile produced. Endogenous precursors and intermediates are underlined. In brackets, the percentage of metabolization of the intermediate (that leading to linamarin was set to 100%). Tetcyclasis, a cytochrome P450 inhibitor, prevented production of HCN when nitriles were incubated. The left column shows the approximate amount of HCN produced by the fat bodies of 2 larvae per hour; experiments were run for 15 hours during which the production of HCN was linear.

the last step of the pathway in the insect appears to be catalyzed by a P450 dependent oxidase, as was shown earlier in plants for the microsomal fraction of the cereal *Sorghum bicolor*.[47] In summary, the fat body contains the entire enzymatic activity necessary to produce acetone cyanohydrin and methylethylketone cyanohydrin, the direct precursors of linamarin and lotaustralin. The localization of the glucosyltransferase, however, is still an open question. As the cyanohydrins (hydroxynitriles) are not stable, this enzyme ought to be present close to the site where the cyanohydrins are produced. This also is currently under study.

DEGRADATION OF LINAMARIN AND LOTAUSTRALIN

There are both similarities and differences in cyanogenic glycoside degradation in *Zygaena* and in plants. In cyanogenic plants, a β-glucosidase and, usually, a hydroxynitrile lyase catalyse the rapid liberation of hydrogen cyanide from crushed tissues in a two-step process (Fig. 1). When inspecting different

Table 3. Distribution of cyanogenic glucosides (linamarin and lotaustralin), linamarase and hydroxynitrile lyase (HNL) in tissues of a larva of *Z. trifolii*

organ	linamarase [%]Act.	hydroxynitrile lyase [%]Act.	CN-glucosides [%]Act.
hemolymph	82	84	33
integument	4	11	66
secretion	0	0	(11)
fat body	9	2	tr
gut epithel	2	1	tr
malpighii	<1	0	0
silk gland	0	0	0

tissues of the larvae of *Z. trifolii*, high β-glucosidase activity was shown to be present in the hemolymph (Table 3). The activity was purified up to ca. 90-fold[48,49] with DEAE-, CM- and concanavalin-columns run with McIlvaine buffer systems which contain citrate as a chelating agent (see below for the inhibition of the β-glucosidase). The enzyme is a dimer with a native mw of ca. 132 kD which is in the range of many plant β-glucosidases. Its pH optimum with linamarin as substrate is between 4.5 and 5, thus slightly more acidic than for many plant β-glucosidases. The β-glucosidase has a clear preference for the substrates linamarin and lotaustralin[48] and, thus, can be considered to be a linamarase. This finding is in direct contrast to many "linamarases" isolated from plants which show a comparatively broad substrate specificity.[50–53] Hydroxynitrile lyase activity was also isolated from the hemolymph (Table 3) and purified up to ca. 200-fold.[49,54] This enzyme is a dimer with a native mw of 143 kD with a pH optimum of 5.5 (acetone cyanohydrin) and wider substrate specificity. This is in accord with the hydroxynitrile lyase isolated from *Hevea brasiliensis*[55] but in contrast to that isolated from *Linum usitatissimum*.[56] Both plants contain linamarin and lotaustralin. The UV spectrum of the hydroxynitrile lyase isolated from *Z. trifolii* indicates the presence of flavin units associated with the protein (one per subunit). Hydroxynitrile lyases associated with a flavin unit have only been reported in plants from the seeds of rosaceous species.[16,57] Their function is not known.

The hydrolysis of cyanogenic glucosides in plants is prevented under most circumstances by compartmentation of the substrate and degrading enzymes. The β-glucosidases of some cyanogenic plants are associated with the apoplastic space[58,59] whereas the substrates are intracellular. In at least one other example the substrates and enzymes are in different tissues: dhurrin occurs in the epidermis and the degrading enzymes in the mesophyll of the leaves of *Sorghum*

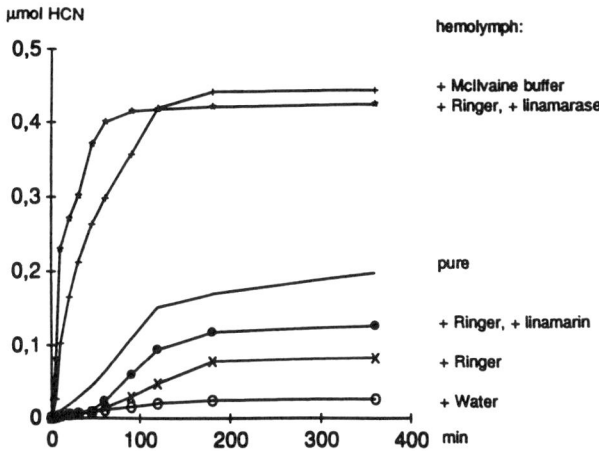

Figure 5. HCN production by hemolymph of the larvae of *Z. trifolii* under different conditions.

bicolor.[7,60] In *Z. trifolii* larvae, however, 1/3 of the entire glucoside content and more than 80% of the activity of the degrading enzymes is located in the hemolymph (Table 3) without any obvious compartmentation. Despite this, cyanogenesis seems to be inhibited in the freshly obtained hemolymph (Fig. 5).

Undiluted hemolymph shows a slow and relatively weak liberation of HCN. When diluted with Ringer-solution or water, the rate of HCN liberation is even less. When linamarin is added to the hemolymph diluted with Ringer-solution, cyanogenesis is again less than with pure hemolymph. This latter observation shows that substrate availablity is not the reason for slow cyanogenesis. When linamarase (obtained from *Hevea brasiliensis*[50]) is added to the hemolymph diluted with Ringer-solution, a rapid and effective reaction occurs and liberation of total HCN is finished after ca. 1 hr. Almost the same thing is observed after addition of McIlvaine buffer, a chelating agent, (or EDTA solution, not shown in Fig. 5) to the hemolymph instead of *Hevea* linamarase. This suggests that the insect linamarase is normally inhibited by certain metals, since chelating agents which mask the ions, are able to activate it.[49] The hemolymph contains 18.3 mM Mg^{++} and 7.5 mM Ca^{++}, as the likely candidates for chelation[49,61](all other metal ions were less than 0.1 mM). When isolated *Zygaena* linamarase at 50 mM was incubated with different amounts of Ca/Mg ions in the ratio that occurs naturally in the hemolymph, hydrolytic activity was inhibited dose dependently to approx.10% of the original activity without metal ions (Fig. 6). We thus conclude that these two alkaline earth metals ions are the natural inhibitors that prevent continous HCN production in the hemolymph of *Z. trifolii* larvae.[61] It is, however, still an open question how the linamarase and, thus, the cyanogenic system is activated after the larvae are wounded.

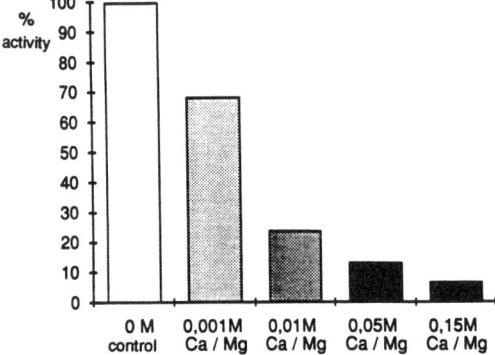

Figure 6. Inhibition of pure linamarase obtained from the hemolymph of the larvae of Z. trifolii by different concentration of Ca^{++} and Mg^{++} ions in the naturally occurring ratio of ca. 1:2.

CONCLUSION

The larvae of Zygaena trifolii possess a cyanogenic system that is, in some aspects, similar to those of plants. The moth, Z. trifolii, is obviously quite successful when it settles on a food plant that provides linamarin and lotaustralin, both of which are also synthesized de novo by the insect. One may speculate that the insects save metabolic energy by sequestering those cyanogenic glucosides from their host which they usually synthezise de novo. Slowly, we are gaining insight into the biochemistry of the insect system. But there are still many unanswered questions concerning origins, regulation of the system, and ecological importance. There is little information about how Lepidoptera are able to sequester other cyanogenic glycosides such as gynocardin.[23] There are recent reports that the non-cyanogenic nitrile glucoside, sarmentosin, occurs in the magpie moth and the Apollo butterfly.[62,63] This suggests that sequestration of nitrile compounds may occur more widely than hitherto believed. It will be of special interest to determine whether the cyanogenic system of Heliconius species[38,39,64] (Nymphalidae) is biochemically related to that of Zygaena. In conclusion, cyanogenic and also non-cyanogenic nitrile glucosides, with their occurrence in both plants and insects, and with both similarities and differences in their synthesis, compartmentalization, and degradation in these organisms, may provide future insight or at least fuel for the redundancy/functional diversity debate.

ACKNOWLEDGMENT

I acknowledge the cooperation, excellent experimentation and several discussions with Drs. I. Ackermann, R. Davis, S. Franzl, G. Holzkamp, E. Mueller, Prof. C. M. Naumann and V. Wray. I thank B. Quandt for valuable technical assistance and Profs. D. S. Seigler and J. Romeo for linguistic help.

REFERENCES

1. BOPPRE, M. 1986. Insects pharmacophagously utilizing defensive plant chemicals (Pyrrolizidine alkaloids). Naturwissenschaften 73: 17–26.
2. MACOLM, S. B. 1990. Chemical defense in chewing and sucking insect herbivores: plant-derived cardenolides in the monarch butterfly and oleander aphid. Chemoecology 1: 12–21.
3. PASTEELS, J. M., BRAEKMAN, J.-C., DALOZE, D. 1988. Chemical defense in the Chrysomelidae. In: (P. Jolivet, E. Petitpierre, T. H. Hsiao, eds.), Biology of Chrysomelidae. Kluwer Academic Pub. pp. 233–252.
4. APLIN, R. T., d'ARCY WARD, R., ROTHSCHILD, M., 1975. Examination of the large white and small white butterflies (Pieris spp) for the presence of mustard oils and mustard oil glycosides. J. Entomol. (A) 50: 73–78.
5. WINK, M. 1985. Chemische Verteidigung der Lupinen: Zur biologischen Bedeutung der Chinolizidinalkaloide. Pl. Syst. Evol. 150: 65–81.
6. MONTLLOR, C. B., BERNAYS, E. A., BARBEHENN, R. V. 1990. Importance of quinolizidine alkaloids in the relationship between larvae of *Uresiphita reversalis* (Lepidoptera: Pyralidae) and a host plant, *Genista monspessulana*. J. Chem. Ecol. 16: 1853–1865.
7. NASH, R. J., BELL, E. A., ACKERY, P. R., 1992. The protective role of cycasin in cycad-feeding Lepidoptera. Phytochemistry 31: 1955–1957.
8. STERMITZ, F. R. 1988. Iridoid glycosides and aglykones as chiral synthons, bioactive compounds and lepidopteran defenses. ACS Symp. Ser. 380: 397–402.
9. FELLOWS, L. E., KITE, G. C., NASH, R. J., SIMMONDS, M. S. J., SCOFIELD, A. 1992. Distribution and biological activity of alkaloidal glycosidase inhibitors from plants. In: (K. Mengel, D. J. Philbeam, eds.), Nitrogen metabolism of plants. Oxford Science Publ., Oxford, pp. 271–82.
10. ROTHSCHILD, M., APHIN, R., BAKER, J., MARSH, N. 1979. Toxicity induced in the tobacco hornworm (*Manduca sexta* L., Sphingidae, Lepidoptera). Nature 280: 487–488.
11. FUNG, S. Y., HERREBOUT, W. M., VERPOORTE, R., FISCHER, F. C. 1988. Butenolides in small ermine moths, *Yponomeuta* spp. (Lepidoptera : Yponomeutidae), and spindle tree (*Euonymus europaeus*, Celastraceae). J. Chem. Ecol. 14: 1099–1111.
12. NISHIDA, R., FUKAMI, H. 1989. Ecological adaptation of an Aristolochiaceae-feeding swallowtail butterfly, *Atrophaneura alcinous*, to aristolochic acids. J. Chem. Ecol. 15: 2549–2563.
13. DAVIS, R. H., NAHRSTEDT, A. 1985. Cyanogenesis in insects. In: (G. A. Kerkut, L. I. Gilbert, eds), Comprehensive Insect Physiology, Biochemistry and Pharmacology, Vol 11. Pergamon Press, Oxford, pp. 635–654.
14. NAHRSTEDT, A. 1988. Cyanogenesis and the role of cyanogenic compounds in insects. In: (CIBA Foundation, ed.), Cyanide Compounds in Biology, Vol 140. Wiley, Chichester, pp. 131–145.
15. HEGNAUER, R. (1964–1992). Chemotaxonomie der Pflanzen, Vol I-X. Birkhäuser Verlag, Basel.
16. POULTON, J. E. 1990. Cyanogenesis in plants. Plant Physiol. 94: 401–405.
17. NAHRSTEDT, A. 1993. Cyanogenesis and food plants. In: (T. van Beek, H. Breteler, eds), Proc. Phytochem. Soc. Europe. Phytochemistry and Agriculture, Vol 34. Oxford University Press, Oxford, pp. 107–129.
18. LUETHY, B., MATILE, P. 1984. The mustard oil bomb: Rectified analysis of the subcellular organisation of the myrosinase system. Biochem. Physiol. Pflanzen 179: 5–12.
19. NAHRSTEDT, A. 1985. Cyanogenic compounds as protecting agents for organisms. Plant Syst. Evol. 150: 35–47.
20. DUFFEY, S. S. 1981. Cyanide and arthropods. In: (B. Vennesland, E. E. Conn, C. J. Knowles, J. Westley, F. Wissing, eds.), Cyanide in Biology, Academic Press, London, pp. 385–414.

21. DAVIS, R. H., NAHRSTEDT, A. 1986. (R)Mandelonitrile and prunasin, the sources of hydrogen cyanide in all stages of *Paropsis atomaria* (Coleoptera:Chrysomelidae). Z. Naturforsch. 41c: 928–934.
22. RAUBENHEIMER, D. 1987. *Kiggelaria africana*, *Acraea horta* and the use of cyanide as a chemical defense. Veld and Flora p. 27.
23. RAUBENHEIMER, D. 1989. Cyanoglucoside gynocardin from *Acraea horta* (L) (Lepidoptera: Acraeinae): Possible implications for evolution of Acraeinae host choice. J. Chem. Ecol. 15: 2177–2189.
24. BRAEKMAN, J. C., DALOZE, D., PASTEELS, J. M. 1982. Cyanogenic and other glucosides in a Neo-Guinean bug *Leptocoris isolata*: possible precursors in its host plant. Biochem. Syst. Ecol. 10: 355–364.
25. BURGEFF, H. 1915. Zur Frage des Tötens der Zygaenen. Entomol. Z. 29: 49–50.
26. JAROSZEWSKI, J. W., JENSEN, P. S., CORNETT, C., BYBERG, J. R. 1988. Occurrence of lotaustralin in *Berberidopsis beckleri* and its relation to the chemical evolution of Flacourtiaceae. Biochem. Syst. Ecol. 16: 23–28.
27. DAVIS, R. H., NAHRSTEDT, A. 1979. Linamarin and lotaustralin as the source of cyanide in *Zygaena filipendulae* L (Lepidoptera). Comp. Biochem. Physiol. 64B: 395–397.
28. DAVIS, R. H., AND NAHRSTEDT, A. 1982. Occurrence and variation of the cyanogenic glucosides linamarin and lotaustralin in species of the Zygaenidae (Insecta:Lepidoptera). Comp. Biochem. Physiol. 71B: 329–332.
29. WITTHOHN, K., NAUMANN, C. M. 1987. Cyanogenesis - a general phenomenon in the Lepidoptera? J. Chem. Ecol. 13: 1789–1809.
30. SEIPEL, H. 1980. *Zygaena trifolii* ssp *barcelonensis* f. loc. *saleria* Burgeff, Nachr. Ent. Ver. Apollo, N. F. 1: 2–4.
31. FRANZL, S., NAUMANN, C. M., and NAHRSTEDT, A. 1988. Cyanoglucoside storing cuticle of Zygaena larvae (Insecta, Lepidoptera). Zoomorphology 108: 183–190.
32. WITTHOHN, K., NAUMANN, C. M. 1984. Qualitative and quantitative studies on the compounds of the larval defensive secretion of *Zygaena trifolii* (Esper, 1783) (Insecta, Lepidoptera, Zygaenidae). Comp. Biochem. Physiol. 79c: 103–106.
33. FRANZL, S., NAHRSTEDT, A., NAUMANN, C. M. 1986. Evidence for site of biosynthesis and transport of the cyanoglucosides linamarin and lotaustralin in larvae of *Zygaena trifolii* (Insecta:Lepidoptera). J. Insect Physiol. 32: 705–709.
34. RAMMERT, U. 1992. The reaction of birds to the larval defense system of *Zygaena trifolii*. In (C. Dutreix, C. M. Naumann, W. G. Tremewan, eds), Recent Advances in the Burnet Moth Research 1987. Koelz Scient. Koenigstein, pp. 38–52.
35. NAHRSTEDT, A., DAVIS, R. H. 1986. Uptake of linamarin and lotaustralin from their foodplant by larvae of *Zygaena trifolii*. Phytochemistry 25: 2299–2302.
36. CONN, E. E. 1991. The metabolism of a natural product: lessons learned from cyanogenic glycosides. Planta Med. 57: S1-S9.
37. KOCH, B., NIELSEN, V. S., HALKIER, B. A., OLSEN, C. E., MØLLER, B. L. 1992. The biosynthesis of cyanogenic glucosides in seedlings of cassava (*Manihot esculenta* Crantz). Arch. Biochem. Biophys. 292: 141–150.
38. WRAY, V., DAVIS, R. H., NAHRSTEDT, A. 1983. Biosynthesis of cyanogenic glycosides in butterflies and moths: Incorporation of valine and isoleucine into linamarin and lotaustralin by *Zygaena* and *Heliconius* species. Z. Naturforsch. 38c: 583–588.
39. DAVIS, R. H., NAHRSTEDT, A. 1987. Biosynthesis of cyanogenic glucosides in butterflies and moths Effective incorporation of 2-methylpropanenitrile and 2-methylbutanenitrile into linamarin and lotaustralin by *Zygaena* and *Heliconius* species. Insect Biochem. 17: 689–693.
40. HOLZKAMP, G., NAHRSTEDT, A. 1994. Biosynthesis of cyanogenic glucosides in the Lepidoptera. Incorporation of [U-14C]-2-methylpropanealdoxime, 2S-[U-14C]-methylbutanealdoxime and D,L-[U-14C]-N-hydroxyisoleucine into linamarin and lotaustralin by the larvae of *Zygaena trifolii*. Insect Biochem. Molec. Biol. 24: 161–165.

41. CUTLER, A. J., STERNBERG, M., CONN, E. E. 1985. Properties of a microsomal enzyme system from *Linum usitatissimum* which oxidases valine to acetone cyanohydrin and isoleucine to 2-methylbutanone cyanohydrin. Arch. Biochem. Biophys. 238: 272–279.
42. HOESEL, W., NAHRSTEDT, A. 1980. In vitro biosynthesis of the cyanogenic glucoside taxiphyllin in *Triglochin maritima*. Arch. Biochem. Biophys. 203: 753–757.
43. COLLINGE, D., HUGHES, M. A. 1982. In vitro characterization of the AC locus in white clover (*Trifolium repens* L.). Arch. Biochem. Biophys. 218: 38–45.
44. COLLINGE, D. B., HUGHES, M. A. 1984. Evidence that linamarin and lotaustralin, the two cyanogenic glucosides of *Trifolium repens* L, are synthesized by a single set of microsomal enzymes controlled by the Ac/ac locus. Plant Sci. Lett. 34: 119–125.
45. HOLZKAMP, G. 1994. Untersuchungen zur Biosynthesesequenz cyanogener Glukoside an lebenden Larven sowie am isolierten Fettkörper von *Zygaena trifolii* (Lepidoptera: Zygaenidae). PhD-thesis, Muenster, pp. 108–159.
46. OLAFSDOTTIR, E. S., JORGENSEN, L. B., JAROSZEWSKI, J. W. 1992. Substrate specificity in the biosynthesis of cyclopentanoid cyanohydrin glucosides. Phytochemistry 31: 4129–4134.
47. HALKIER, B. A., MØLLER, B. L. 1990. The biosynthesis of cyanogenic glucosides in higher plants - Identification of three hydroxylation steps in the biosynthesis of dhurrin in *Sorghum bicolor* (L) Moench and the involvement of 1-aci-nitro-2-(p-hydroxyphenyl)ethane as an intermediate. J. Biol. Chem. 265: 21114–21121.
48. FRANZL, S., ACKERMANN, I., AND NAHRSTEDT, A. 1989. Purification and characterization of a β-glucosidase (linamarase) from the haemolymph of *Zygaena trifolii* Esper, 1783 (Insecta, Lepidoptera). Experientia 45: 712–718.
49. MUELLER, E. 1992. Untersuchungen zur Enzymologie der Cyanogenese und zum Metabolismus von Blausäure in den Larven von *Zygaena trifolii* (Esper, 1783) (Insecta, Lepidoptera). PhD-thesis, Univ. Muenster, pp. 13–17 (HNL), pp. 59–64 (linamarase).
50. SELMAR, D., LIEBEREI, R., BIEHL, B. 1987. *Hevea* linamarase - a nonspecific β-glycosidase. Plant Physiol. 83: 557–563.
51. FAN, T. W. M., CONN, E. E. 1985. Isolation and characterization of two cyanogenic β-glucosidases from flax seeds. Arch. Biochem. Biophys. 243: 361–373.
52. MKPONG, O. E., YAN, H., CHISM, G., SAYRE, R. T. 1990. Purification, characterization and localization of linamarase in Cassava. Plant Physiol. 93: 176–181.
53. ITHO-NASHIDA, T., HIRAIWA, M., UDA, Y. 1987. Purification and properties of β-D-glucosidase (linamarase) from the butter bean, *Phaseolus lunatus*. J. Biochem. 101: 847–854.
54. MUELLER, E., NAHRSTEDT, A. 1990. Purification and characterization of a alpha-hydroxynitrile lyase from the hemolymph of the larvae of *Zygaena trifolii*. Planta Med. 56: 611–612.
55. SELMAR, D., LIEBEREI, R., BIEHL, B., CONN, E. E., 1989. alpha-Hydroxynitrile lyase in *Hevea brasiliensis* and its significance for rapid cyanogenesis. Physiol. Plant. 75: 97–101.
56. XU, L.-L., SINGH, B. K., CONN, E. E. 1988. Purification and characterization of acetone cyanohydrin lyase from *Linum usitatissimum*. Arch. Biochem. Biophys. 263: 256–263.
57. XU, L.-L., SINGH, B. K., CONN, E. E. 1986. Purification and characterization of mandelonitrile lyase from *Prunus lyonii*. Arch. Biochem. Biophys. 250: 322–328.
58. POULTON, J. E. 1988. Localization and catabolism of cyanogenic glycosides. In: (CIBA Foundation, ed.), Cyanide Compounds in Biology, Vol 140. J. Wiley, Chichester, pp. 67–91.
59. KAKES, P. 1985. Linamarase and other β-glucosidases are present in the cell walls of *Trifolium repens* L leaves. Planta 166: 156–160.
60. THAYER, S. S., CONN, E. E. 1981. Subcellular localization of dhurrin β-glucosidase and hydroxy nitrile lyase in the mesophyll cells of *Sorghum* leaf blades. Plant Physiol. 67: 617–622.
61. NAHRSTEDT, A., MUELLER, E. 1993. β-Glucosidase (linamarase) of the larvae of the moth *Zygaena trifolii* and its inhibition by some alkaline earth metal ions. In: (A. Esen, ed.),

β-Glucosidases. Biochemistry and Molecular Biology, Vol 533. American Chemical Society, Washington, pp. 132–44.
62. NISHIDA, R., ROTHSCHILD, M., MUMMERY, R. 1994. A cyanoglucoside, sarmentosin, from the magpie moth, *Abraxas grossulariata*, Geometridae : Lepidoptera. Phytochemistry 36: 37–38.
63. NISHIDA, R., ROTHSCHILD, M. 1995. A cyanoglucoside stored by a *Sedum*-feeding Apollo butterfly, *Parnassius phoebus*. Experientia 51: 267–269.
64. NAHRSTEDT, A., DAVIS, R. H. 1981. The occurrence of the cyanoglucosides linamarin and lotaustralin in *Acraea* and *Heliconius* butterflies. Comp. Biochem. Physiol. 68B: 575–577.

Chapter Nine

POLYPHENOL OXIDASE AS A COMPONENT OF THE INDUCIBLE DEFENSE RESPONSE IN TOMATO AGAINST HERBIVORES

C. Peter Constabel, Daniel R. Bergey, and Clarence A. Ryan

Institute of Biological Chemistry
Washington State University
Pullman, Washington 99164-6340

Introduction 231
Overview of Polyphenol Oxidase (PPO) 233
    PPO Reactions and Browning 233
    PPO Structure and Specificity 234
    Localization, Expression and Distribution of PPO in Plants 235
Proposed Functions of PPO 236
    Proposed Physiological Roles of PPO in Healthy Plants 236
    PPO and Plant Disease 237
    PPO as a Defense against Herbivores 239
Systemin, PPO, and the Octadecanoid Wound-Signaling Pathway 241
    Systemin and Systemic Wound Induction of PPO 241
    PPO and the Octadecanoid Wound-Signaling Pathway 244
Conclusions 246

INTRODUCTION

Polyphenol Oxidases (PPOs) are copper-containing enzymes that use molecular oxygen to catalyze the oxidation of monophenolic and *ortho*-diphenolic compounds (Fig. 1). Both PPOs and their phenolic substrates are widespread among higher plants and fungi. The PPO-generated oxidation products, *o*-quinones, spontaneously undergo further reactions to form compounds responsible for the commonly observed browning of injured or diseased plant tissues.[1] The physiological significance of PPO and these browning processes in plants has

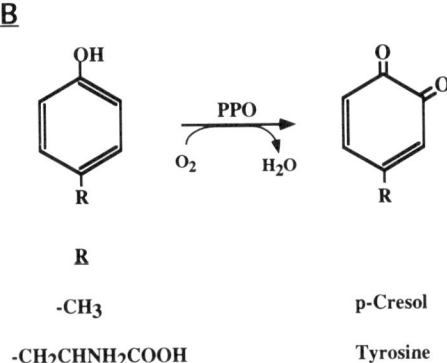

Figure 1. Reactions catalyzed by PPO and some common PPO substrates. A. Diphenolase (catechol oxidase) activity. B. Monophenolase (tyrosinase, cresolase) activity.

remained enigmatic despite intense research efforts, summarized in numerous reviews.[2–7] Although a wealth of studies has attempted to elucidate the role of PPO in pathogen defense,[2,8–10] to date the most convincing case has been made for a function of PPO in defense against insects.[7,11–14] Interestingly, PPO is important for both insect entrapment and antinutritive defensive mechanisms, suggesting that PPO is a versatile enzyme which may have other, yet to be discovered functions.

In this chapter we give an overview of the physiological and ecological functions which have been proposed for PPO, and then focus on the role of PPO as a systemic wound response protein with antinutritive effects against insect herbivores. Such a defensive function of PPO is supported by our recent findings[15] that PPO accumulation in tomato leaves is induced by systemic wound signals acting through the octadecanoid wound-signaling pathway. This signaling pathway was originally proposed based on its role in the induced accumulation of anti-herbivore proteinase inhibitors. Our results have thus revealed how one wound-signaling pathway induces distinct yet complementary antinutritive defense proteins. We conclude that, in tomato, PPO contributes to a broad spectrum of induced defenses which protect the plant from a diversity of herbivores and pests.

OVERVIEW OF POLYPHENOL OXIDASE (PPO)

PPO Reactions and Browning

PPO uses O_2 to oxidize *ortho*-diphenolic compounds to *o*-quinones (Fig. 1). In addition to this reaction, PPO from some species also catalyzes the hydroxylation of monophenols at the *ortho* position, producing the *o*-diphenol which is then rapidly oxidized to the *o*-quinone. The hydroxylase activity is variously referred to as monophenolase, cresolase or tyrosinase; whereas the diphenolase activity is known as catechol oxidase.[2] The complex of both activities has been called phenolase, or the phenolase complex.[3] The quinones produced by PPO are very reactive and in the absence of reducing agents will oxidize nucleophilic groups of many different cellular constituents and metabolites. Reactions of the quinones are complex and depend on the chemical environment and nature of the parent phenolic compound oxidized. Some well-characterized reactions involving the quinones include polymerization with other phenolics (Fig. 2A), and alkylation of thiol and amino groups of proteins and amino acids, which often leads to their covalent crosslinking (Fig. 2B).[16–18] These reactions are thought to be responsible for the browning of plant extracts and damaged plant tissues. Oxidative browning is especially common in fruit where it reduces quality and has been of concern to the food industry for many years.[1,20] In some cases, however, the enzyme-mediated browning may be a desirable trait for the food product, as in the case of raisins, dates, or figs.[21] Reactions of the *o*-quinones are also responsible for most of the proposed biological functions of PPO (see Section 3).

Plants also contain other enzymes capable of oxidizing phenolic compounds and causing tissue browning, in particular peroxidases and laccases.[1,3] Peroxidases are heme-containing proteins which use H_2O_2 as the oxidant, and are capable of oxidizing a wide range of substrates.[3,22] Laccases, like PPO,

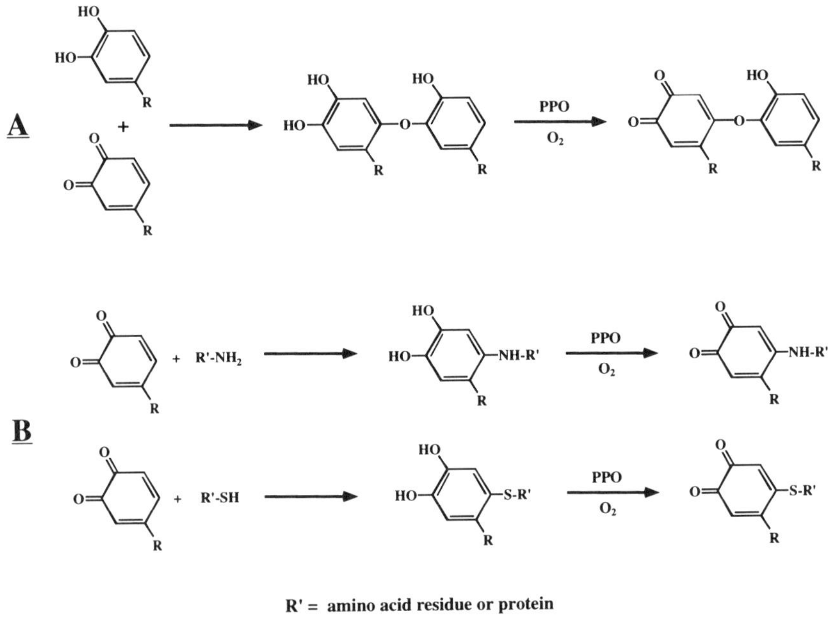

Figure 2. Condensation reactions of *o*-quinones and subsequent reoxidation by PPO. A. Condensation of *o*-quinones and *o*-diphenols. Reoxidation by PPO will lead to further reactions and polymerization. B. Reaction of *o*-quinone with amino and thiol side groups of amino acids and proteins. Further PPO-catalyzed oxidation of the phenolic adduct can lead to additional reactions with other proteins and crosslinking (modified from Reference 3).

contain copper and use O_2 to oxidize various phenolic compounds including mono- and *o*-diphenols. However, laccases are thought to belong to a different group of proteins from PPO,[23] and are distinguished from PPO by their ability to oxidize *p*-phenols, as well as their susceptibility to specific inhibitors.[24,25]

PPO Structure and Specificity

PPO has been purified and characterized from many plant species, especially from fruit, and the reported molecular weights for PPO protein have varied widely.[2,3,21] The recent molecular cloning of a number of plant PPO cDNAs, however, reveals mature polypeptide sequences with predicted molecular weights of between 54kD and 62kD.[26–34] The molecular weight variability of PPO has been suggested to be due to the enzyme's unusual susceptibility to limited proteolysis, or to modification by PPO-generated quinones.[7,35] An interesting case of proteolytic processing involves grape PPO where, following transit peptide cleavage and import into the chloroplast, the 60kD product is activated by further processing at the C-terminus to yield a 40kD mature form.[30,36] In broad bean, a similar C-terminal processing event occurs, but this appears to have no direct effect on PPO activity.[35]

PPO contains two copper atoms in the active site, and the protein domains containing the Cu-coordinating His residues are highly conserved among known plant PPOs[7,27,30–34] as well as animal and microbial tyrosinases and phenol oxidases.[37] These conserved regions were used to design oligonucleotide primers for the isolation of two PPO cDNAs from pokeweed using the polymerase chain reaction (PCR).[31] The overall sequence conservation among known plant PPO proteins is 45–60% amino acid identity over the entire length of the mature polypeptide.[33]

Plant PPOs generally have low substrate specificities and are thus able to utilize a wide range of *ortho*-diphenolic compounds, provided there are no substituent groups adjacent to the hydroxyl groups.[2] Caffeic acid and chlorogenic acid are common plant phenolics[38] and generally good PPO substrates (Fig. 1). Together with 3,4-dihydroxyphenylalanine (DOPA), catechol, and 4-methylcatechol, these compounds are widely used as assay substrates.[39,40] In general, PPO from a given plant will oxidize a range of phenolic compounds, but PPO from different sources may show different substrate specificities.[41] Flavonoids such as epi-catechin and rutin are oxidized by PPO from some species, for example tea leaves and pears.[42,43] As mentioned above, many plant PPOs demonstrate monophenolase activity and thus also utilize compounds such as tyrosine, *p*-cresol, and *p*-coumaric acid as substrates (Fig. 1). For example, potato tuber PPO readily oxidizes monophenols,[44] but PPO from trichomes of the wild potato *Solanum berthaulthii* shows absolute specificity for *ortho*-diphenolic compounds.[45] Likewise, monophenols are not recognized as substrates by PPO isolated from pear, sweet cherry, and tomato leaf.[11,43,46]

A perplexing feature of plant PPO has been the frequent observation that the enzyme from some sources is not immediately active following extraction, but present in a latent form. It can then be activated by aging, treatment with detergents, acids or bases, ammonium sulfate, cations, proteolytic enzymes, and denaturing agents.[2,3,7] The mechanism underlying the activation of PPO appears to be induced conformational changes in the enzyme.[47,48] Not all plant PPOs manifest latency, however, and PPO from bean, corn, and tomato is extracted in a fully active form.[49]

Localization, Expression and Distribution of PPO in Plants

PPO is widespread in higher plants and fungi,[2,5,50] as well as in the animal kingdom.[51] Sherman et al.[50] carried out an extensive survey of PPO activities in several members of all major green plant taxa using spectrophotometric, cytochemical, and activity-stained polyacrylamide gel assays. Almost all angiosperms, ferns and fern allies, most bryophytes, and some green algae were found to contain PPO activity. Surprisingly, no evidence of PPO activity was obtained for any of the eight gymnosperms assayed, although a later study suggested it was present in this group but not easily detectable due to different substrate preferences.[52]

There is general agreement in the literature that plant PPO is localized in plastids, a conclusion drawn from studies in a variety of species using cell fractionation, immunocytochemistry, and cytochemical methods based on PPO activity.[6,7] Furthermore, PPO is present in a variety of tissue-specific plastids including amyloplasts, leucoplasts, etioplasts and chromoplasts,[4] as well as in the plastids of epidermal cells and glandular trichomes.[53] All cloned plant PPOs encode proteins with N-terminal transit peptides typical of nuclear-encoded chloroplast proteins, further supporting the plastid localization of PPO.[26-34] Recently, a phenol oxidase with a substrate specificity consistent with PPO was reported from cell walls of mung bean hypocotyls. However, since this protein was recognized by anti-laccase antibodies, it appears that this phenol oxidase may be a PPO-like laccase.[54]

PPO activity has been found in almost all plant tissue types including leaves,[50] stems,[55,56] roots and hypocotyls,[56,57] bark,[58] and tubers.[44,59] High levels of PPO are often found in flowers and fruits.[21] Furthermore, PPO is developmentally regulated, with levels varying during leaf maturation[60,61] and fruit ripening.[21] Stress treatments, in particular pathogen infection and wounding, can induce PPO activity to high levels (see next section). Treatment of tomato roots with heat or chloroform also results in elevated PPO levels,[56] as does gamma irradiation of fruit.[21] Curing of tobacco leaves causes an initial rise in PPO activity followed by a slow decline.[60,62] The latency of PPO makes it difficult to distinguish whether stress induces PPO activity via enzyme activation or *de novo* protein synthesis. For example, during senescence of spinach leaves, an increase in PPO activity was shown to be due to enzyme activation,[63] but the wound-induction of PPO in sweet potato tubers was the result of *de novo* enzyme synthesis.[64] The recent availability of cloned PPO genes has allowed PPO induction to be assayed at the mRNA level, and those experiments indicate that wound-induced increases in PPO activity result from the transcriptional activation of PPO genes (this chapter).[7,33,65]

PROPOSED FUNCTIONS OF PPO

Proposed Physiological Roles of PPO in Healthy Plants

Animal tyrosinases have been shown to play a role in melanogenesis for protection against solar radiation, sclerotinization leading to hardening of arthropod exoskeletons, as well as defense of insects against microbial invasion.[66-69] For plant PPOs, however, such well-defined physiological roles have not been discovered. The co-occurence of phenolic compounds in tissues containing high levels of PPO, and the capacity of many PPOs to hydroxylate monophenols, led to suggestions that PPO is important in phenylpropanoid biosynthesis. As Mayer and Harel[2] have pointed out, however, it is unclear whether or not the diphenol is released from the enzyme *in vivo*, and to date there is no clear evidence supporting a role for PPO in phenylpropanoid biosynthesis. PPO does not convert tyrosine to dihydroxyphenylalanine (DOPA) in *Vicia faba*,[70] nor is it involved in the synthesis of caffeic acid and *ortho*-hydroxylated flavonoids in mung bean.[71] Suggestions that PPO may be required for suberization of tuber tissues or sclerotinization of seed coats remain unsubstantiated and, in fact, it appears rather that peroxidases may be involved.[72,73]

A problem in defining the role of PPO in healthy plants has been the observation that its phenolic substrates are localized in the vacuole,[74] whereas the enzyme is sequestered in the chloroplast. Also difficult to explain is the latency of PPO activity. These observations have led some researchers to suggest that *in vivo* PPO has a function unrelated to phenolic oxidation. One proposal suggests a role for PPO in photosynthesis as a regulator of the Mehler reaction during pseudocyclic photophosphorylation (i.e., ATP production with O_2 as the terminal electron acceptor).[4,49] Alternatively, PPO may function in removal of excess O_2 from the chloroplast.[2] The presence of PPO in the plastids of plants from many different taxa, including several algae and mosses,[50,52] lends some support to it playing a role in photosynthetic processes. To date, however, no suitable PPO substrate in chloroplasts has been found. A proposed role for PPO in photosynthesis also fails to account for the abundance of the enzyme in non-photosynthetic organs such as roots, tubers and flowers. Further evidence against a possible role for PPO in photosynthesis comes from genetically-engineered PPO-null plants described by Steffens et al.[7] These transgenic plants apparently grow and develop normally under greenhouse conditions, not requiring PPO for normal photosynthetic functions.[7]

PPO and Plant Disease

The conceptual problems concerning PPO's physical separation from its substrates, as well as its latency, are resolved if PPO remains inactive in healthy plants and becomes active only when the cellular compartments are destroyed, for example during wounding or infection. In fact, these stresses are often accompanied by visible browning of plant tissue. The association between tissue browning, plant disease, and PPO, has been a major force driving PPO research, and many attempts have been made to discover a direct role for PPO in plant defense.[3,8–10,75]

Several potential PPO-based defense mechanisms against pathogens have been discussed,[8,10,75] and all are based on the very reactive nature of the *o*-quinones. These have been shown *in vitro* to inactivate the lytic enzymes secreted by many pathogens as they invade plant tissues. Polygalacturonase, which hydrolyzes the pectin in plant cell walls, is inhibited *in vitro* by PPO plus the substrate DOPA.[76] Patil and Dimond[77] showed that potato PPO plus chlorogenic or caffeic acid reduces *in vitro* polygalacturonase activity from *Verticillium albo-atrum* by 50% within one hour. In order for enzyme inhibition to occur, the oxidation of the phenolics must occur in the presence of the polygalacturonase, and adding previously oxidized phenolics has no effect on the enzyme. The inhibition of polygalacturonase is prevented if free amino acids are added to the reaction mixture, suggesting that the mechanism of polygalacturonase inactivation involves oxidation of amino acid residues of the enzyme by the PPO-generated *o*-quinones.[77]

Friend[75] discusses several examples of PPO-oxidized phenolics as being directly toxic to pathogens via the oxidation of surface components. In particular, the modification of viral coat proteins by *o*-quinones has been extensively studied. Pierpoint et al.[17] found that Potato Virus X (PVX) is modified and partially inactivated when incubated with PPO and chlorogenic acid. The modified PVX coat protein contained covalently-bound chlorogenic acid, exhibited crosslinks not seen in the untreated protein, and was more resistant to digestion with trypsin. These data demonstrate the ability of PPO-generated quinones to covalently modify viral proteins, an effect that could have detrimental effects on other pathogens as well.[75]

The sealing of wounds and lesions by polymerizing phenolics to a melanin-like barrier which prevents further pathogen invasion, for example during the hypersensitive reaction, has also been discussed.[8] The localized tissue necrosis characterizing the hypersensitive reaction is triggered by pathogen invasion, but is under strict genetic control of the plant, as demonstrated by the isolation of mutant plants which spontaneously develop necrotic lesions in the absence of pathogen infection.[78,79] Lesion formation involves oxidative reactions, and PPO levels have been shown to increase in and around the necrotic zones.[80-83] Spraying leaves with the PPO substrate catechol prior to lesion development hastens lesion appearance and increases lesion size, while treatment with antioxidants such as ascorbic acid reduces lesion formation.[82] In some cases lesion reduction is also accompanied by reduced virus replication.[81]

Reviews of PPO and disease resistance cite many examples of the induction of PPO by fungal, bacterial, and viral pathogens.[8-10] In addition to its induction in leaves, PPO can also be induced in tubers,[84] roots,[56] and hypocotyls.[85] It is not always clear however, if this increase in PPO activity reflects the plant's attempt to ward off the pathogen, or is a secondary effect of tissue death and senescence. To address this question, many investigators have tried to correlate PPO levels with the degree of resistance, often using similar crop cultivars with different resistance characteristics. For example, Bashan et al.[86] compared peroxidase and PPO activities in tomato cultivars infected by *Pseudomonas syringae* pv tomato and found that PPO levels were higher in the resistant cultivar. Kalia and Sharma[87] used a genetic approach to demonstrate that resistance to powdery mildew correlates with higher PPO levels in different pea lines. Thukral et al.[88] found a similar correlation with downy mildew resistance of pearl millet. These cases, supporting a role of PPO in disease resistance, are countered by the many reports in which no correlation between resistance and elevated PPO levels was observed.[85,89–91] These contradictory results illustrate the common synopsis stated in PPO reviews, which concludes that there is no overall association of PPO with disease resistance.[2,6,10,92]

Nevertheless, these conflicting data do not preclude a significant function of PPO in pathogen defense.[93] Plant defense is complex and PPO induction is only one of many potential defense mechanisms, not all of which would be expected to be effective against all pathogens. Furthermore, the action of defensive proteins is likely to be synergistic, as has been shown *in vitro* for the antifungal activity of chitinase and β-glucanase.[94] A more definitive answer regarding the relationship between PPO and plant disease resistance should be attainable by testing transgenic plants modified in their ability to express PPO for resistance to pathogens. Such plants have been described[7,95] and should be powerful tools for clarifying the role of PPO in plant disease resistance.

PPO as a Defense against Herbivores

While the function of PPO in plant defense against pathogens remains unresolved, recent evidence indicates that PPO may be employed by plants as a defensive protein against insects in both entrapment and antinutritive strategies.[7,11,13,14] Entrapment of small-bodied insects occurs on leaf surfaces of certain *Solanum* and *Lycopersicon* species with high densities of glandular trichomes. When damaged by insects such as aphids, the trichomes release an exudate which undergoes a rapid darkening and hardening caused by oxidation and polymerization of the trichome contents. Insects coming into contact with this exudate as it is hardening are physically trapped and prevented from feeding via the occlusion of mouthparts.[14] As much as 50–70% of the total protein in the glandular trichome is PPO.[45] Peroxidase is also present in the trichomes and may contribute to the oxidation reactions, but PPO is thought to be the major protein facilitating this defense mechanism.[7]

The antinutritive effect of PPO on chewing insects has been proposed and documented by Duffey and coworkers.[11–13] This defensive mechanism is based on the propensity of PPO-generated *o*-quinones to covalently modify and crosslink dietary proteins during feeding. The resulting negative effects on food quality and amino acid availability for mammals have been described.[19,20] It is now becoming apparent that during plant-insect interactions, and likely in other plant-herbivore interactions as well, these processes can serve as antinutritive defense mechanisms.[11–13]

The defensive capacity of PPO has been most extensively studied in the tomato, which contains significant amounts of both PPO protein and *ortho*-diphenolic compounds.[11,96,97] Using the tomato fruit worm *Heliothis zea* and the beet army worm *Spodoptera exigua* as test organisms, Felton et al.[11] found that larval growth rate varied inversely with the amount of PPO found in the leaves and fruit on which the insects were feeding. Adding PPO inhibitors to the insect food abolishes this effect on growth. In artificial diets containing both PPO and chlorogenic acid, a significant reduction in growth rate is observed only if the phenolic oxidation occurs in the presence of the dietary protein. Allowing the enzyme and substrate to react prior to their addition to the diet has no significant impact on the insects. Therefore, the inhibitory effect is not derived from a chlorogenic acid polymer, but from the interaction of chlorogenoquinone with dietary protein. If insects are fed radiolabelled chlorogenic acid together with PPO in artificial diets, almost 50% of the radiolabel recovered is covalently bound to protein in the insect feces.[11] Together these experiments suggest that in the insect gut, PPO produces *o*-quinones by oxidizing chlorogenic acid and other phenolic compounds, which then covalently bind free amino acids and dietary protein. The covalently modified amino acid residues in the protein are less easily assimilated by the insect's digestive system,[12] and they may interfere with digestive enzymes or reduce protein solubility.[98] In all cases, the net effect for the insect is a loss of available dietary protein. The effectiveness of PPO as an antinutritive defense protein is further enhanced because those amino acids (lysine, histidine, cysteine, and methionine) that are generally limiting in folivore insect diets are also among the most susceptible to attack by *o*-quinones.[12,19]

The antinutritive functions of PPO are diverse and depend on conditions such as pH and detergency of the insect gut. Tomato PPO is active within the pH range 6–10,[11] but its reactions with proteins are thought to be optimal at a slightly basic pH. Insects with acidic digestive systems, for example coleopterans such as the Colorado potato beetle, might thus be able to circumvent PPO defense.[96] Higher pH and detergency improve the solubility of protein in the gut, which may partially alleviate the antinutritive effects of ingested PPO.[98] Additional considerations influencing the effects of PPO on herbivores include the levels of oxidizing and reducing agents in the gut. Antioxidants such as ascorbic acid or glutathione reduce the quinones before they can alkylate dietary protein.[13] Finally, the effectiveness of other defensive proteins such as the proteinase inhibitors must be considered in light of their potential crosslinking and modification by PPO-generated quinones.[99,100] Despite these limitations, however, PPO can clearly be an effective defense against insect herbivores. Preliminary results described by Steffens et al.[7] indicate that Colorado potato beetle larvae grow more slowly and have higher mortality rates when feeding on transgenic tomato plants overexpressing PPO, as compared to control larvae feeding on wild-type plants. In contrast, when feeding on transgenic PPO-null plants, the larvae grow faster and have lower mortality rates than the controls.

SYSTEMIN, PPO, AND THE OCTADECANOID WOUND-SIGNALING PATHWAY

Systemin and Systemic Wound Induction of PPO

The idea of PPO functioning as an anti-herbivore defense protein in tomato is supported by observations that tomato leaf PPO is inducible by wounding and insect damage.[86,96,101] We have recently established that this wound induction of PPO is systemic (occuring throughout the entire plant) even if the damage is localized.[15] Systemic wound signaling in tomato leaves has been studied intensively and is beginning to be understood in some detail.[102–104]

Research on defense signaling in tomato began with the observation that in response to insect feeding or mechanical wounding, tomato leaves accumulate proteinase inhibitors.[105] These small proteins inhibit trypsin and chymotrypsin, digestive enzymes commonly found in animals including insects, and have been shown to be effective antinutritive defenses.[106,107] The accumulation of the proteinase inhibitors is systemic and occurs rapidly in both the wounded and unwounded leaves on a wounded plant,[105] implying the existence of a mobile wound signal. The search for this proteinase-inhibitor inducing factor produced a number of proposed candidate signals including oligomers of galacturonic acid,[108] linolenic acid,[103] jasmonic acid,[109] abscisic acid,[110] electrical signals,[111] and an 18-amino acid peptide called systemin.[102] To date, the available evidence indicates that the systemic signal inducing the proteinase inhibitors in tomato is systemin.[102] Systemin is a potent inducer of the proteinase inhibitors (half-maximal induction at 25 fmol/plant) and is translocated rapidly within the plant, thus fulfilling the requirements of a systemic signal.[102]

The gene encoding the systemin precursor protein, prosystemin, has been cloned and characterized,[112] which has permitted the manipulation of prosystemin and systemin expression using transgenic plants. Transgenic tomato plants expressing the prosystemin cDNA in the antisense orientation are severely inhibited in systemic wound induction of the proteinase inhibitors, and are consumed much more rapidly by leaf-eating tobacco hornworm larvae than are wild-type control plants.[113] By contrast, transgenic tomato plants engineered to overexpress prosystemin in the correct ("sense") orientation show the opposite phenotype. These plants accumulate extraordinarily high levels of proteinase inhibitors in the absence of wounding, presumably due to the unregulated release of systemin from its overexpressed precursor prosystemin.[114] Overall, the above results demonstrate that prosystemin and systemin are essential components of the tomato defense signaling mechanism.

During the analysis of the prosystemin-overexpressing plants, it was observed that leaf extracts of these plants browned very rapidly and became viscous. The browning could be inhibited by reducing agents, suggesting that

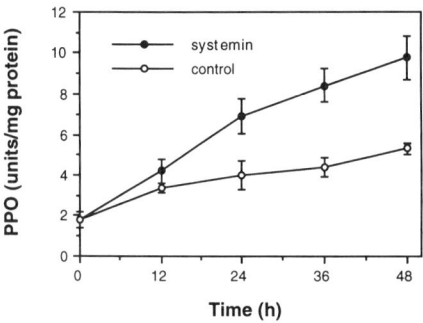

Figure 3. Induction of PPO activity in tomato leaves by systemin. Two-week-old plants were supplied 2.5 pmol systemin (upper curve), or buffer (lower curve) over a 2 h period, incubated under continuous light, and extracted and assayed for PPO activity at the times indicated. Each point represents the mean of five plants. Bars represent S.E.M.

oxidizing reactions underlie this effect. Enzyme assays for two common oxidative enzymes, peroxidase and PPO, were carried out on both control and transgenic leaf extracts. Up to 70-fold higher levels of PPO were measured in extracts of the transgenic leaves compared to control leaves.[15] Two-dimensional gel electrophoresis of total leaf proteins identified a protein of 66kD which was much more prevalent in transgenic leaves than in control leaves. Purification and N-terminal sequencing identified this induced protein as a tomato PPO, which suggested that the higher PPO activity in the transgenic plants was a consequence of the elevated levels of PPO protein, and not activation of a latent enzyme.[15]

In genetic crosses with the prosystemin-overexpressing transgenic tomato line, elevated levels of PPO activity always co-segregated with high proteinase inhibitor levels (Gregg Howe, personal communication), suggesting that the high PPO levels were also controlled by overexpression of the prosystemin transgene. To test if exogenously supplied systemin could induce PPO in wild-type tomato plants, two-week-old plants were excised, supplied with a solution of systemin or buffer, and assayed for PPO activity at various times. Treatment with systemin stimulated a five-fold increase in PPO activity within 48 h (Fig. 3). The pattern of PPO induction was similar to that observed for the proteinase inhibitors, with the difference that unwounded tomato plants already contain significant constitutive levels of PPO but not proteinase inhibitors. This result suggests that the increased PPO levels in the prosystemin-overexpressing plants were due to the elevated levels of systemin, rather than a secondary effect of prosystemin-overexpression. The buffer treatment also induced PPO approximately two-fold. As described below, wounding also induces PPO activity, so this increase could be attributed to the excision process.

The capacity of systemin to induce PPO prompted us to examine potential wound induction of this enzyme. When two-week-old tomato plants were wounded on one leaf by crushing with a hemostat, a treatment that strongly induces proteinase inhibitors in the entire plant, PPO activity increased in both the wounded and unwounded leaves, demonstrating that wound-induction of PPO is systemic (Fig. 4). Exposure of plants to methyl jasmonate, a powerful

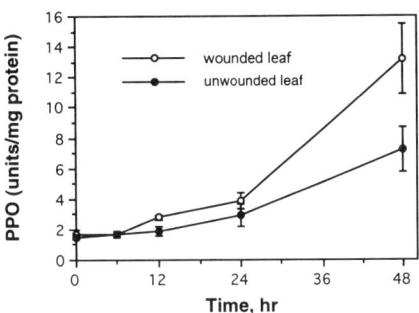

Figure 4. Induction of PPO activity in tomato leaves by wounding. Two-week-old tomato plants were wounded on one leaf, incubated under continuous light for various times, and assayed for PPO activity in both the wounded and unwounded leaves (upper and lower curves respectively). Each point represents the mean of five plants. Bars represent S.E.M.

inducer of the proteinase inhibitors in tomato, also caused a dramatic increase in PPO activity, reaching levels 20-fold higher than controls.[15] This potent inductive effect of methyl jasmonate, exceeding the wound-inducible levels, can also be observed for the proteinase inhibitors.[109]

Western blot analysis of protein extracts from methyl jasmonate-treated and prosystemin-overexpressing plants confirmed that increases in PPO activity reflected an increased abundance of PPO protein. A 62kD protein corresponding to constitutively expressed PPO was present in all extracts analyzed, but in methyl jasmonate-treated and prosystemin-overexpressing transgenic plants, the PPO antibody recognized a second induced protein migrating at 65kD, the wound-inducible PPO (Fig.5, lanes TG and MJ). Since there are at least seven PPO genes in tomato,[29] this induced PPO band most likely results from differential PPO gene expression.

The induction of PPO synthesis by wounding, by supplying excised plants with systemin, or by treatment with methyl jasmonate appears to be the result of transcriptional activation of one or more PPO genes, as shown by RNA hybridi-

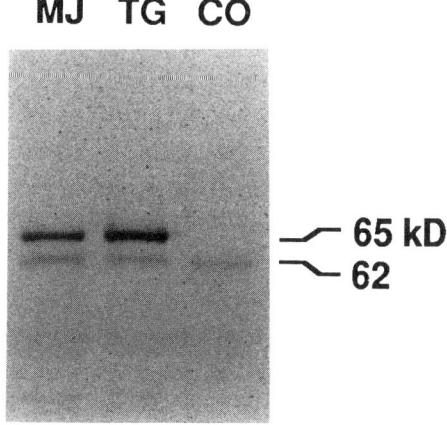

Figure 5. Western Blot analysis of induced PPO accumulation in leaves of tomato plants treated with methyl jasmonate (MJ), prosystemin-overexpressing transgenic (TG), or control (CO) plants. Total leaf proteins were separated by SDS-PAGE, blotted to nitrocellulose and incubated with an anti-PPO antibody. The bands were visualized using anti-rabbit IgG antibodies conjugated to alkaline phosphatase, and developed with nitroblue tetrazolium and 5-bromo-4-chloro-3-indolyl phosphate.

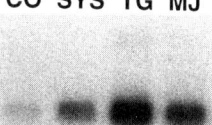

CO SYS TG MJ

Figure 6. Accumulation of PPO mRNA in wild-type tomato plants supplied with buffer (CO) or systemin (SYS), treated with methyl jasmonate (MJ), and in plants overexpressing a prosystemin transgene (TG). RNA was extracted 24 h after treatment, separated on a 1.4% agarose gel, blotted, and hybridized with a PPO cDNA.

zation analysis (Fig. 6). In control leaves, a small amount of PPO mRNA is detectable, but the signal is much stronger in systemin-treated, methyl jasmonate-treated and prosystemin-overexpressing tomato plants. A similar response is seen with the proteinase inhibitors, which accumulate in wounded plants due to their transcriptional activation.[115] Overall, PPO and proteinase inhibitor induction are coordinately regulated, suggesting that both are controlled by the same signaling mechanism.

PPO and the Octadecanoid Wound-Signaling Pathway

Systemin and methyl jasmonate are components of the octadecanoid signal transduction pathway.[103] In this pathway, wounding triggers the release of systemin, which is transported throughout the plant and interacts with specific receptors in target cells. This interaction is thought to activate a lipase which releases linolenic acid from membranes. Through a series of enzymatic steps linolenic acid is converted to jasmonic acid, which then interacts with other cellular factors, possibly transcriptional activators, to induce transcription of the proteinase inhibitor genes. This signaling pathway has been supported by demonstrations that several of the biosynthetic precursors of jasmonic acid activate proteinase inhibitor accumulation.[103] Studies using the octadecanoid pathway inhibitors diethyldithiocarbamic acid (DIECA)[116] and salicylic acid,[117] as well as a mutant tomato line deficient in an octadecanoid signaling step, have demonstrated that blockage of this pathway prevents proteinase inhibitor induction by wounding and systemin. The proteinase inhibitor-inducing activity of carbohydrate elicitors that are known to be involved in defense gene activation near sites of pathogen attacks (i.e., plant cell wall-derived oligogalacturonides and fungal cell wall-derived chitosan), are also blocked by salicylic acid and DIECA, suggesting that these carbohydrate inducers also act via the octadecanoid signaling pathway.[116,118]

We conducted experiments to see if PPO induction by systemin could be prevented by treating the excised plants with 1 mM salicylic acid. In these plants, PPO induction by systemin was significantly reduced, while salicylic acid alone did not reduce basal levels of PPO activity (Table 1). The JL-5 tomato mutant,[119]

Table 1. Inhibition of systemin-induced PPO activity in young tomato plants by salicylic acid (SA)[*]

Treatment	PPO activity (Units/mg protein)
Control	0.70 (0.15)
Systemin	5.42 (0.83)
Systemin + SA	1.26 (0.35)
SA	0.83 (0.10)

[*]Two-week-old tomato plants were excised and placed in a 1 mM salicylic acid solution (lower two samples) or phosphate buffer solution (upper two samples) for 60 min, and then supplied with test solutions for 120 min. The plants were transferred to water or 1 mM salicylic acid, incubated under continuous light for 40 h, and the leaf extracts assayed for PPO activity. Each value represents the mean of five samples (S.E.M).

which does not accumulate significant levels of the proteinase inhibitors after wounding (Howe et al., submitted), also does not accumulate PPO following wounding (Table 2). These experiments provide direct evidence that PPO is controlled by the octadecanoid wound signaling pathway.

The effectiveness of proteinase inhibitors in plant defense against insect herbivores has been documented.[106,107,120] Our results demonstrating the regulation of both the proteinase inhibitors and PPO via the same signaling pathway imply that PPO may also be an important element of tomato defense against insect herbivores, and complement previous data demonstrating the antinutritive effects of PPO on insects.[11,13]

Table 2. Induction of PPO in wild-type and JL-5 mutant tomato plants[*]

	PPO activity (Units/mg protein)	
	Control Plants	Wounded Plants
Wild-type		
lower leaf	1.1	6.8
upper leaf	1.2	2.5
JL-5		
lower leaf	1.0	1.1
upper leaf	0.9	1.1

[*]Two-week-old plants were wounded twice on the lower leaf, and leaf extracts assayed for PPO activity in both the lower and the upper (unwounded) leaf after 24 h. Each value represents the PPO activity of five pooled plant samples.

The discovery that PPO, along with the proteinase inhibitors, is induced in tomato by wounding and systemin, demonstrates that systemin and the octadecanoid pathway are key control elements in the regulation of a broad spectrum of defensive proteins. The importance of multiple lines of defense has been highlighted by recent observations describing insect herbivores that can change the proteinase composition of their digestive systems to overcome the effects of proteinase inhibitors.[121] This adaptive capability puts selective pressure on plant species to evolve diverse defensive strategies.

Plant phenolics and other natural products have been investigated intensively as chemical mediators of ecological interactions, and a major thrust within the chemical ecology field has been to search for anti-herbivore roles of phytochemicals.[38,122] The prevalence of oxidative enzymes such as PPO in wounded tissues necessitates caution in studies and bioassays examining secondary metabolites alone, since the net biological effects of these compounds may vary with their interaction with such enzymes. Therefore, studies aiming to reveal potential defensive functions of phytochemicals should include experiments which can assay these in their *in planta* biochemical context.

5. CONCLUSIONS

The induction of PPO in tomato by wounding and systemin via the octadecanoid pathway supports previous suggestions that PPO activity functions as an enzymatic defense in tomato against insect herbivores. The extent of this function in other plants is still an open question. Preliminary data indicate that PPO is not strongly wound- or methyl jasmonate-inducible in many species outside the Solanaceae (unpublished data). The wide distribution of PPO in the plant kingdom, its potential for a role in pathogen defense, and the general reactivity of the PPO-generated quinones hint at other possible roles. Thus, PPO may have been adopted for different functional roles by diverse plant species in distinct ecological situations.

ACKNOWLEDGMENTS

The authors thank John Steffens, Cornell University, for his gift of PPO antibody. This work was supported in part by the Washington State University College of Agriculture and Home Economics, grants from the National Science Foundation, and a fellowship from the Natural Science and Engineering Research Council of Canada (CPC).

REFERENCES

1. JOSLYN, M.A., PONTIG, J.D. 1951. Enzyme-catalyzed browning of fruit products. Adv. Food Res. 3: 1–37.

2. MAYER, A.M., HAREL, E. 1979. Polyphenol oxidases in plants. Phytochemistry 18: 193–215.
3. BUTT, V.S. 1980. Direct oxidases and related enzymes. In: Biochemistry of Plants, Vol 2 (P.K. Stumpf and E.E. Conn, eds.), Academic Press, New York, N.Y., pp. 81–123.
4. VAUGHN, K.C., DUKE, S.O. 1984. Function of polyphenol oxidase in higher plants. Physiol. Plant. 60: 106–112.
5. MAYER, A.M. 1987. Polyphenol oxidases in plants-recent progress. Phytochemistry 26: 11–20.
6. VAUGHN, K.C., LAX, A.R., DUKE, S.O. 1988. Polyphenol oxidase: the chloroplast enzyme with no established function. Physiol. Plant. 72: 659–665.
7. STEFFENS, J.C., HAREL, E., HUNT, M. 1994. Polyphenol oxidase. In: Genetic Engineering of Plant Secondary Metabolism. Rec. Adv. Phytochemistry, Vol 28 (B.E. Ellis et al., eds.), pp. 276–304.
8. FARKAS, G.L., KIRÁLY, Z. 1962. Role of phenolic compounds in the physiology of plant disease and disease resistance. Phytopathol. Z. 44: 105–150.
9. RUBIN, B.A., ARTSIKHOVSKAYA, E.V. 1964. Biochemistry of pathological darkening of plant tissues. Annu. Rev. Phytopath. 2: 157–178.
10. KOSUGE, T. 1969. The role of phenolics in host response to infection. Annu. Rev. Phytopath. 7: 195–222.
11. FELTON, G.W., DONATO, K., DEL VECCHIO, R.J., DUFFEY, S.S. 1989. Activation of plant foliar oxidases by insect feeding reduces nutritive quality of foliage for noctuid herbivores. J. Chem. Ecol. 15: 2667–2693.
12. FELTON G.W., DONATO, K.K., BROADWAY R.M., DUFFEY, S.S. 1992. Impact of oxidized plant phenolics on the nutritional quality of dietary protein to a noctuid herbivore, *Spodoptera exigua*. J. Insect Physiol. 38: 277–285.
13. DUFFEY, S.S., FELTON, G.W. 1991. Enzymatic antinutritive defenses of the tomato plant against insects. In: Naturally Occurring Pest Bioregulators (P.A. Hedin, ed.), ACS, Washington DC, pp. 167–197.
14. TINGEY, W. 1991. Potato glandular trichomes. In: Naturally Occurring Pest Bioregulators (P.A. Hedin, ed.), ACS, Washington DC, pp. 128–135.
15. CONSTABEL, C.P., BERGEY, D.R., RYAN, C.A. 1995. Systemin activates synthesis of wound-inducible tomato leaf polyphenol oxidase via the octadecanoid defense signaling pathway. Proc. Natl. Acad. Sci. USA 92: 407–411.
16. PIERPOINT, W.S. 1966. The enzymatic oxidation of chlorogenic acid and some reactions of the quinone produced. Biochem. J. 98: 56–580.
17. PIERPOINT, W.S., IRELAND, R.J., CARPENTER, J.M. 1977. Modification of proteins during the oxidation of leaf phenols: Reaction of potato virus X with chlorogenoquinone. Phytochemistry 16: 29–34.
18. LEATHAM, G.F., KING, V., STAHMANN, M.A. 1980. In vitro polymerization by quinones or free radicals generated by plant or fungal oxidative enzymes. Phytopathology 70: 1134–1140.
19. PIERPOINT, W.S. 1983. Reactions of phenolic compounds with proteins, and their relevance to the production of leaf protein. In: Leaf Protein Concentrates (L. Telek and H.D. Graham, eds.), AVI, Westport, Conn. pp. 235–267.
20. HURRELL, R.F., FINOT, P.A. 1984. Nutritional consequences of the reactions between proteins and oxidized phenolic acids. In: Nutritional and Toxicological Aspects of Food Safety (M. Friedman, ed.), Plenum Press, New York, pp. 423–435.
21. MACHEIX, J-J., FLEURIET, A., BILLOT, J. 1990. Fruit Phenolics, CRC Press, Boca Raton, Fl, pp. 295–378.
22. FRIC, F. 1976. Oxidative enzymes. In: Encyclopedia of Plant Physiology, New Series, Vol 4 (R. Heitefuss and H.P. Williams, eds.), Springer-Verlag, N.Y., New York, pp. 616–631.
23. O'MALLEY, D.M., WHETTEN, R., BAO, W., CHEN, C.-C., SEDEROFF, R.R. 1993. The role of laccase in lignification. Plant J. 4: 751–757.

24. SARONUO, R., KATO, F., IKENO, T. 1979. Kojic acid, a tyrosinase inhibitor from *Aspergillus albus*. Agric. Biol. Chem. 43: 1337–1338.
25. MURAO, S., HINODE, Y., MATSUMARA, E., NUMATA, A., KAWAI, K. OHISHI, H., OYAMA, H., SHIN, T. 1992. A novel laccase inhibitor, N-hydroxyglycine, produced by *Penicillium citrinum* YH-31. Biosci. Biotech. Biochem. 56: 987–988.
26. SHAHAR, T., HENNIG, N., GUTFINGER, T., HAREVEN, D., LIFSCHITZ, E. 1992. The tomato 66.3-kD polyphenol oxidase gene: molecular identification and developmental expression. Plant Cell 4: 135–147.
27. CARY, J.W., LAX, A.R., FLURKEY, W.H. 1992. Cloning and characterization of cDNAs coding for *Vicia faba* polyphenol oxidase. Plant Mol. Biol. 20: 245–253.
28. HUNT, M.D., EANNETTA, N.T., YU, H., NEWMAN, S.M., STEFFENS, J.C. 1993. cDNA cloning and expression of potato polyphenol oxidase. Plant Mol. Biol. 21: 59–68.
29. NEWMAN, S.M., EANNETTA, T., YU, H., PRINCE, J.P., DE VICENTE, M.C., TANKSLEY, S.D., STEFFENS, J.C. 1993. Organization of the tomato polyphenol oxidase gene family. Plant Mol. Biol. 21: 1035–1051.
30. DRY, I.B., ROBINSON, S.P. 1994. Molecular cloning and characterization of grape berry polyphenol oxidase. Plant Mol. Biol. 26: 495–502.
31. JOY IV, R.W., SUGIYAMA, M., FUKUDA, H., KOMAMINE, A. 1995. Cloning and characterization of polyphenol oxidase cDNAs of *Phytolacca americana*. Plant Physiol. 107: 1083–1089.
32. HIND, G., MARSHAK, D.R., COUGHLAN, S.J. 1995. Spinach thylakoid polyphenol oxidase: cloning, characterization, and relation to a putative protein kinase. Biochemistry 34: 8157–8164.
33. BOSS, P.K., GARDNER, R.C., JANSSEN, B.-J., ROSS, G.S. 1995. An apple polyphenol oxidase cDNA is up-regulated in wounded tissues. Plant Mol. Biol. 27: 429–433.
34. THYGESEN, P.W., DRY, I.B., ROBINSON, S.P. 1995. Polyphenol oxidase in potato. Plant Physiol. 109: 525–531.
35. ROBINSON, S.P., DRY, I.B. 1992. Broad bean leaf polyphenol oxidase is a 60-kilodalton protein susceptible to proteolytic cleavage. Plant Physiol. 99: 317–323.
36. RATHJEN, A.H., ROBINSON, S.P. 1992. Aberrant processing of polyphenol oxidase in a variegated grapevine mutant. Plant Physiol. 99: 1619–1625.
37. FUJIMOTO, K., OKINO, N., KAWABATA, S-I., IWANAGA, S., OHNISHI, E. 1995. Nucleotide sequence of the cDNA encoding the proenzyme of phenol oxidase A1 of *Drosophila melanogaster*. Proc. Natl. Acad. Sci. USA 92: 7769–7773.
38. HARBORNE, J.B. 1993. Introduction to Ecological Biochemistry, 4th Edition. Academic Press, London, p. 318
39. ESTERBAUER, H., SCHWARZL, E., HAYN, M. 1977. A rapid assay for catechol oxidase and laccase using 2-nitro-5-thiobenzoic acid. Anal. Biochem. 77: 486–494.
40. GAUILLARD, F., RICHARD-FORGET, F., NICOLAS, J. 1993. New spectrophotometric assay for polyphenol oxidase activity. Anal. Biochem. 215: 59–65.
41. YASUNOBU, K.T. 1959. Mode of action of tyrosinase. In: Pigment Cell Biology (M. Gordon, ed.), Academic Press, New York, pp. 583–608.
42. BERKOWITZ, J.E., COGGON, P., SANDERSON, G.W. 1971. Formation of epitheaflavic acid and its transformation to thearubigins during tea fermentation. Phytochemistry 42: 2271–2278.
43. DE JESUS RIVAS, N., WHITAKER, J.R. 1973. Purification and some properties of two polyphenol oxidases from bartlett pears. Plant Physiol. 52: 501–507.
44. PATIL, S.S., ZUCKER, M. 1965. Potato phenolases. J. Biol. Chem. 240: 3938–3943.
45. KOWALSKI, S.P., EANNETTA, N.T., HIRZEL, A.T., STEFFENS, J.C. 1992. Purification and characterization of polyphenol oxidase from glandular trichomes of *Solanum berthaultii*. Plant Physiol. 100: 677–684.

46. LANZARINI, G., PIFFERI, P.G., ZAMORANI, A. 1972. Specificity of an o-diphenol oxidase from *Prunus avium* fruits. Phytochemistry 11: 89–94.
47. SWAIN, T., MAPSON, L.W., ROBB, D.A. 1966. Activation of *Vicia faba* (L.) tyrosinase as effected by denaturing agents. Phytochemistry 5: 469–482.
48. MOORE, B.M., FLURKEY, W.H. 1990. Sodium dodecyl sulfate activation of a plant polyphenol oxidase. J. Biol. Chem. 265: 4982–4988.
49. TOLBERT, N.E. 1973. Activation of polyphenol oxidase of chloroplasts. Plant Physiol. 51: 234–244.
50. SHERMAN, T.D., VAUGHN, K.C., DUKE, S.O. 1991. A limited survey of the phylogenetic distribution of polyphenol oxidase. Phytochemistry 30: 2499–2506.
51. GORDON, M. 1959. Pigment Cell Biology. Academic Press, New York, p. 647
52. SHERMAN, T.D., LE GARDEUR, T., LAX, A.R. (1995) Implications of the phylogenetic distribution of polyphenol oxidase in plants. In: Enzymatic Browning and Its Prevention (C.Y. Lee, J.R. Whitaker, eds.), ACS, Washington, DC, pp. 103–119.
53. YU, H., KOWALSKI, S.P., STEFFENS, J.C. 1992. Comparison of polyphenol oxidase expression in glandular trichomes of *Solanum* and *Lycopersicon* species. Plant Physiol. 100: 1885–189.
54. CHABANET, A., GOLDBERG, R., CATESSON, A.-M., QUINET-SZELY, M., DELAUNAY, A.M., FAYE, L. 1994. Characterization and localization of a phenoloxidase in mung bean hypocotyl cell walls. Plant Physiol. 106: 1095–1102.
55. STAFFORD, H.A., GALSTON, A.W. 1970. Ontogeny and hormonal control of polyphenol oxidase isozymes in tobacco pith. Plant Physiol. 46: 763–767.
56. GENTILE, I.A., FERRARIS, L., MATTA, A. 1988. Variations of phenol oxidase as a consequence of stresses that induce resistance to *Fusarium* wilt of tomato. J. Phytopath. 122: 45–53.
57. MUELLER, W.C., BECKMAN, C.H. 1978. Ultrastructural localization of polyphenol oxidase and peroxidase in roots and hypocotyls of cotton seedlings. Can. J. Bot. 56: 1579–1587.
58. TAKAI, S., HUBBES, M. 1973. Polyphenol-oxidase activity and growth inhibition of *Hypoxylon pruinatum* (Klotzsche) Cke. by aspen bark meal. Phytopath. Z. 78: 97–108.
59. IKEDIOBI, C.O., CHELVARAJAN, R.L., UKOHA, A.I. 1989. Biochemical aspects of wound healing in yams (*Dioscorea* spp). J. Sci. Food Agric. 48: 131–139.
60. ANDERSON, R.A., LOWE, R., VAUGHN, T.A. 1969. Plant phenol and polyphenol oxidase in *Nicotiana tabacum* during greenhouse growth, field growth and air-curing. Phytochemistry 8: 2139–2147.
61. TAKEO, T., BAKER, J.E. 1973. Changes in multiple forms of polyphenol oxidase during maturation of tea leaves. Phytochemistry 12: 21–24.
62. SHEEN, S.J., CALVERT, J. 1969. Studies on polyphenol content, activities and isozymes of polyphenol oxidase and peroxidase during air-curing in three tobacco types. Plant Physiol. 44: 199–204.
63. MEYER, H-U., BIEHL, B. 1981. Activation of latent phenolase during spinach leaf senescence. Phytochemistry 20: 955–959.
64. HYODO, H., URITANI, I. 1966. A study on increase in o-diphenol oxidase activity during incubation of sliced sweet potato tissue. Plant Cell Physiol. 7: 137–144.
65. THIPYAPONG, P., HUNT, M.D., STEFFENS, J.C. (1995) Systemic wound induction of potato (*Solanum tuberosum*) polyphenol oxidase. Phytochemistry 40: 673–676.
66. SUGUMARAN, M. 1991. Molecular mechanisms for mammalian melanogenesis; comparison with insect cuticular sclerotization. FEBS Lett. 293: 4–10.
67. HEARING, V.J., TSUKAMOTO, K. 1991. Enzymatic control of pigmentation in mammals. FASEB J. 5: 2902–2909.
68. KRAMER, K.J., MORGAN, T.D., HOPKINS, T.L., CHRISTENSEN, A., SCHAEFER, J. 1991. Insect cuticle tanning. In: Naturally Occurring Pest Bioregulators (P.A. Hedin, ed.), ACS, Washington DC, pp. 87–105.

69. SÖDERHALL, K., CERENIUS,L., JOHANSSON, M.W. 1995. The prophenoloxidase activating systemin and its role in invertebrate defense. Ann. N.Y. Acad. Sci. 712: 166–161.
70. GRIFFITH, T., CONN, E.E. 1973. Biosynthesis of 3,4-dihydroxyphenylalanine in *Vicia faba*. Phytochemistry 12: 1651–1656.
71. DUKE, S.O., VAUGHN, K.C. 1982. Lack of involvement of polyphenol oxidase in *ortho*-hydroxylation of phenolic compounds in mung bean seedlings. Physiol. Plant 54: 381–385.
72. ESPELIE, K.E., FRANCESCHI, V.R., KOLATTUKUDY P.E., 1986. Immunocytochemical localization and time course of appearance of an anionic peroxidase associated with suberization in wound-healing potato tuber tissue. Plant Physiol. 81: 487–492.
73. EGLEY, G.H., PAUL Jr., R.N., VAUGHN, K.C., DUKE, S.O. 1983. Role of peroxidase in the development of water-impermeable seed coats in *Sida spinosa* L. Planta 157: 224–232.
74. WIERMANN, R. 1981. Secondary products in cell and tissue culture. In: The Biochemistry of Plants, Vol 7 (P.K. Stumpf and E.E. Conn, eds.), Academic Press, New York, N.Y., pp. 85–116.
75. FRIEND, J. 1981. Plant phenolics and lignification in plant disease. Prog. Phytochem. 7: 197–261.
76. DEVERALL, B.J. 1961. Phenolase and pectic enzyme activity in the chocolate spot disease of beans. Nature 189: 311.
77. PATIL, S.S., DIMOND, A.E. 1967. Inhibition of *Verticillium* polygalacturonase by oxidation products of polyphenols. Phytopathology 57: 492–496.
78. GREENBERG, J.T., GUO, A., KLESSIG, D.F., AUSUBEL, F.M. 1994. Programmed cell death in plants: a pathogen-triggered response activated coordinately with multiple defense functions. Cell 77: 551–563.
79. DIETRICH, R.A., DELANEY, T.P., UKNES, S.J., WARD, E.R., RYALS, J.A., DANGLE, J.L. 1994. Arabidopsis mutants simulating disease resistance response. Cell 77: 565–577.
80. FARKAS, G.L., KIRALY, Z., SOLYMOSY, F. 1960. Role of oxidative metabolism in the localization of plant viruses. Virology 12: 408–421.
81. HAMPTON, R.E., FULTON. R.W. 1961. The relation of polyphenol oxidase to instability *in vitro* of prune dwarf and sour cherry necrotic ringspot viruses. Virology 13: 44–52.
82. PARISH, C.L., ZAITLIN, M., SIEGEL, A. 1965. A study of necrotic lesion formation by tobacco mosaic virus. Virology 26: 413–418.
83. AVDIUSHKO, S.A., YE, X.S., KUC, J. 1993. Detection of several enzymatic activities in leaf prints of cucumber plants. Physiol. Mol. Plant Path. 42: 441–454.
84. FRIEND, J., THORNTON, J.D. 1974. Caffeic acid-o-methyl transferase, phenolase and peroxidase in potato tuber tissue inoculated with *Phytophthora infestans*. Phytopath. Z. 81: 56–64.
85. ARORA, Y.K., BAJAJ, K.L. 1985. Peroxidase and polyphenol oxidase associated with induced resistance of mung bean to *Rhizoctonia solani* Kuhn. Phytopath. Z. 114: 325–331.
86. BASHAN, Y., OKON, Y., HENIS, Y. 1985. Peroxidase, polyphenol oxidase, and phenols in relation to resistance against *Pseudomonas syringae* pv. *tomato* in tomato plants. Can. J. Bot. 65: 366–372.
87. KALIA, P., SHARMA, S.K. 1988. Biochemical genetics of powdery mildew resistance in pea. Theor. Appl. Genet. 76: 795–799.
88. THUKRAL, S.K., SATIJA, D.R., GUPTA, V.P. 1986. Biochemical genetic basis of downy mildew resistance in pearl millet. Theor. Appl. Genet. 71: 648–651.
89. AHL-GOY, P., FELIX, G., MÉTRAUX, J.P., MEINS, Jr., F. 1992. Resistance to disease in the hybrid *Nicotiana glutinosa* x *Nicotiana debneyi* is associated with high constitutive levels of β-1,3-glucanase, chitinase, peroxidase and polyphenol oxidase. Physiol. Mol. Plant Path. 41: 11–21.
90. PRYOR, A. 1976. Polyphenol oxidase and hypersensitive resistance: catechol oxidase is not involved in rust resistance in *Zea mays* L. Physiol. Plant Path. 8: 307–311.
91. FACCIOLI, G. 1979. Relation of peroxidase, catalase, and polyphenol oxidase to acquired resistance in plants of *Chenopodium amaranticolor* locally infected by tobacco necrosis virus. Phytopath. Z. 95: 237–249.

92. MANIBHUSHANRAO, K., ZUBER, M., MATSUYAMA, N. 1988. Phenol metabolism and plant disease resistance. Acta Physiol. Entom. Hung. 23: 103–114.
93. BELL, A.A. 1981. Biochemical mechanisms of disease resistance. Annu. Rev. Plant Physiol. 32: 21–81.
94. MAUCH, F., MAUCH-MANI, B., BOLLER, T. 1988. Antifungal hydrolases in pea tissue. II. Inhibition of fungal growth by combinations of chitinase and β-1,3-glucanase. Plant Physiol. 88: 936–942.
95. BACHEM, C.W.B, SPECKMANN, G.-J., VAN DER LINDE, P.C.G., VERHEGGEN, F.T.M., HUNT, M.D., STEFFENS, J.C., ZABEAU, M. 1994. Antisense expression of polyphenol oxidase inhibits enzymatic browning of potato tubers. Bio/technology 12: 1101–1105.
96. FELTON, G.W., WORKMAN, J., DUFFEY, S.S. 1992. Avoidance of antinutritive plant defense: role of midgut pH in Colorado Potato beetle. J. Chem. Ecol. 18: 571–583.
97. ELLIGER, C.A., WONG, Y., CHAN, B.G., WAISS Jr., A.C. 1981. Growth inhibitors in tomato (*Lycopersicon*) to tomato fruit worm (*Heliothis zea*). J. Chem. Ecol. 7: 753–758.
98. FELTON, G.W., DUFFEY, S.S. 1991. Reassessment of the role of gut alkalinity and detergency in insect herbivory. J. Chem. Ecol. 17: 1821–1836.
99. LUDLUM, C.T., FELTON, G.W., DUFFEY, S.S. 1991. Plant Defenses: chlorogenic acid and polyphenol oxidase enhance toxicity of *Bacillus thruringiensis* subsp. *kurstaki* to *Heliothis zea*. J. Chem. Ecol. 17: 217–237.
100. FELTON, G.W., BROADWAY, R.M., DUFFEY, S.S. 1989. Inactivation of protease inhibitors by plant-derived quinones: complications for host-plant resistance against noctuid herbivores. J. Insect Physiol. 35: 981–990.
101. FELTON, G.W., SUMMERS, C.B., MUELLER, A.J. 1994. Oxidative responses in soybean foliage to herbivory by bean leaf beetle and three-cornered alfalfa hopper. J. Chem. Ecol. 20: 639–650.
102. PEARCE, G., STRYDOM, D., JOHNSON, S., RYAN, C.A. 1991. A polypeptide from tomato leaves induces wound-inducible proteinase inhibitor proteins. Science 253: 895–898.
103. FARMER, E.E., RYAN, C.A. 1992. Octadecanoid precursors of jasmonic acid activate the synthesis of wound-inducible proteinase inhibitors. Plant Cell 4: 129–134.
104. RYAN, C.A. 1992. The search for the proteinase-inhibitor inducing factor, PIIF. Plant Mol. Biol. 19: 123–133.
105. GREEN, T.R. RYAN, C.A. 1972. Wound-induced proteinase inhibitor in plant leaves: a possible defense against insects. Science 175: 776–777.
106. JOHNSON, R., NARVAEZ, J., AN, G., RYAN, C.A. 1989. Expression of proteinase inhibitors I and II in transgenic tobacco plants: Effects on natural defense against *Manduca sexta* larvae. Proc. Natl. Acad. Sci. USA 86: 9871–9875.
107 RYAN, C.A. 1990. Proteinase inhibitors in plants: genes for improving defenses against insects and pathogens. Annu. Rev. Phytopath. 28: 425–429.
108. BISHOP, P.D., MAKUS, D.J., PEARCE, G., RYAN, C.A. 1981. Proteinase inhibitor-inducing activity in tomato leaves resides in oligosaccharides enzymatically released from cell walls. Proc. Natl. Acad. Sci. USA 78: 3536–3540.
109. FARMER E.E., RYAN, C.A. 1990. Interplant communication: airborne methyl jasmonate induces synthesis of proteinase inhibitors in plant leaves. Proc. Natl. Acad. Sci. USA 87: 7713–7716.
110. PEÑA-CORTES, H., SANCHEZ-SERRANO, J.J., MERTENS, R., WILLMITZER, L. 1989. Abscisic acid is involved in the wound-induced expression of the proteinase inhibitor II gene in potato and tomato. Proc. Natl. Acad. Sci. USA 86: 9851–9855.
111. WILDON, D.F., THAIN, J.F., MINCHIN, P.E.H., GUBB, I.R., REILLY, A.J., SKIPPER, Y.D., DOHERTY, H.M., O'DONNELL, P.J., BOWLES, D.J. 1992. Electrical signaling and systemic proteinase inhibitor induction in the wounded plant. Nature 360: 62–65.

112. MCGURL, B., PEARCE, G., OROZCO-CARDENAS, M., RYAN, C.A. 1992. Structure, expression, and antisense inhibition of the systemin precursor gene. Science 255: 1570–1573.
113. OROZCO-CARDENAS, M., MCGURL, B., RYAN, C.A. 1993. Expression of an antisense prosystemin gene in tomato plants reduces resistance toward *Manduca sexta* larvae. Proc. Natl. Acad. Sci. USA 90: 8273–8276.
114. MCGURL, B., OROZCO-CARDENAS, M., PEARCE, G., RYAN, C.A. 1994. Overexpression of the prosystemin gene in transgenic tomato plants generates a systemic signal that constitutively induces proteinase inhibitor synthesis. Proc. Natl. Acad. Sci. USA 91: 9799–9802.
115. GRAHAM J.S., HALL, G., PEARCE, G., RYAN, C.A. 1986. Regulation of synthesis of proteinase inhibitor I and II mRNAs in leaves of wounded tomato plants. Planta 169: 399–405.
116. FARMER, E.E., CALDELARI, D., PEARCE, G., WALKER-SIMMONS, M.K., RYAN, C.A. 1994. Diethyldithiocarbamic acid inhibits the octadecanoid signaling pathway for the wound induction of proteinase inhibitors in tomato leaves. Plant Physiol. 106: 337–342.
117. DOARES, S.H., NARVAEZ-VASQUEZ, J., CONCONI, A., RYAN, C.A. 1995. Salicylic acid inhibits synthesis of proteinase inhibitors in tomato leaves induced by systemin and jasmonic acid. Plant Physiol. 108: 1741–1746.
118. DOARES, S.H., SYROVETS, T., WEILER, E.W., RYAN, C.A. 1995. Oligogalacturonides and chitosan activate plant defensive genes through the octadecanoid pathway. Proc. Natl. Acad. Sci. USA 92: 4095–4098.
119. LIGHTNER, J., PEARCE, G., RYAN, C.A., BROWSE, J. 1993. Isolation of signaling mutants of tomato (*Lycopersicon esculentum*). Mol. Gen. Genet. 241: 595–601.
120. HILDER, V.A., GATEHOUSE, A.M.R., SHEERMAN, S.E., BARKER, R.F., BOULTER D. 1987. A novel mechanism of insect resistance engineered into tobacco. Nature 330: 160–163.
121. JONGSMA, M.A., BAKKER, P.L., PETERS, J., BOSCH, D., STIEKEMA, W.J. 1995. Adaptation of *Spodoptera exigua* larvae to plant proteinase inhibitor by induction of gut proteinase activity insensitive to inhibition. Proc. Natl. Acad. Sci. USA 92: 8041–8045.
122. SWAIN, T. 1977. Secondary compounds as protective agents. Annu. Rev. Plant Physiol. 28: 479–501.

Chapter Ten

THE ROLE OF BENZOIC ACID DERIVATIVES IN SYSTEMIC ACQUIRED RESISTANCE

Scott Uknes, Shericca Morris, Bernard Vernooij, and John Ryals

Agricultural Biotechnology Research Unit
Ciba Geigy Corporation
P. O. Box 12257, Research Triangle Park, North Carolina 27709

Introduction 253
What Is Systemic Aquired Resistance (SAR)? 253
Altered Gene Expression Causes the Resistant State 254
Salicylic Acid Is Required for SAR but Is Not the Long Distance Signal . .256
Biosynthesis of SA 257
SA Is Important in Gene-for-Gene Resistance 257
Dissecting the Signal Transduction Cascade with *Arabidopsis* Mutants . . .259
Summary: The Future Role of SAR in Agriculture. 260

INTRODUCTION

Diseases of major crops limit crop quality, yield and grower choice. To address this issue, we have taken an approach that capitalizes on the natural phenomenon of systemic acquired resistance (SAR) to develop disease resistant crops. We found chemicals which induce SAR, examined the activity of SAR genes/anti-fungal peptides expressed in plants responding to pathogens, and isolated varieties with constitutive SAR.

WHAT IS SYSTEMIC ACQUIRED RESISTANCE (SAR)?

SAR is a natural resistance phenomenon first noted by botanists in the early 1900s (for review of the early literature see Ref. 1). Ross brought these early botanical observations into the laboratory by performing experiments with

tobacco mosaic virus (TMV) on a local lesion host, tobacco. He showed that after inoculation of several lower leaves of a tobacco plant with TMV, the remainder of the plant body developed resistance. He termed this resistance, systemic acquired resistance.[2] Significantly, SAR was determined to be broad spectrum, working against
viral, bacterial and fungal pathogens, independent of the type of inducing pathogen.[3,4] SAR appears to be a ubiquitous higher plant defense response. SAR has been demonstrated in plants as varied as soybean, potato, tomato, pearl millet, alfalfa, cucurbits, green bean, *Arabidopsis thaliana*, rice and barley (for review see Ref. 4).

Treatments such as wounding or application of most phytohormones have been found to have no significant effect on the development of SAR[5]. However, certain synthetic chemicals, such as 2,6-dichlorisonicotinic acid (INA), induce the same spectrum of resistance in plants while having no direct anti-microbial activity.[4,6] These chemicals appear to induce the plant's natural defense system, SAR, and soon will be available for growers to "immunize" their crops against disease.

ALTERED GENE EXPRESSION CAUSES THE RESISTANT STATE

Many of the best characterized defense responses induced by pathogen infection (e.g. local gene expression, phytoalexin accumulation, cell wall cross-linking and oxygen free-radical formation) are only detected adjacent to the pathogen.[7,8] Therefore, such responses are not associated with the resistant state found in the non-inoculated portion of the infected plant. Van Loon and Antoniw[9] demonstrated that the acidic-extracellular subset of the pathogenesis-related proteins (PR proteins) in tobacco accumulate during the onset of resistance. More recently, mRNA accumulation of at least nine families of genes was shown to be coordinately induced in the non-inoculated, resistant leaves of pathogen-infected plants; these gene families are known as "SAR genes".[10] Several of the SAR genes encode proteins with direct anti-microbial activity. These include β-1,3-glucanases, chitinases and cysteine-rich proteins related to thaumatin (see Ref. 4 for review). We have assessed the effect of many of the SAR genes when expressed constitutively in transgenic tobacco, alone and in combination (Table 1). Individual genes appear to confer resistance to a distinct spectrum of pathogens. For example, constitutive high-level expression of tobacco PR-1a in transgenic tobacco (*Nicotiana tabacum* cv. *Xanthi* nc) results in increased tolerance to the oomycete pathogens *Peronospora tabacina* (causal agent of blue mold disease) and *Phytophthora parasitica* (causal agent of black shank disease).[11] Similarly, transgenic tobacco and *Brassica* seedlings that express a chitinase from bean (SAR gene homolog) are significantly protected against

Table 1. Activities of SAR genes

Gene Tested \ Pathogen:	TMV	PVY	P. syringae	C. nicotiana	P. parasitica	P. tabacina	B. cinerea	R. solani	H. virescens
PR-1a (acidic)					R	R			
Basic PR-1			r						
PR-2a (acidic glucanase)					r				
PR-2c (acidic glucanase)								r	
PR-Q' (acidic glucanase)				r					
Basic glucanase-vtp		r			r		r		
Class I chitinase (basic)								r	
PR-3 (Class II acidic chitinase)				r	r			r	
Cuc acidic Class III chitinase								r	
Tob basic Class III chitinase			r						R
PR-5 (acidic)			r	r			r		
SAR 8.2					R	r			
NahG	s		s	s	s				
CGA 245,704 (chemical treatment)	R		R	R	R	R			

The gene indicated on the left of the table was driven by a strong constitutive promoter. At least three independent transgenic plant lines (5 plants per line) that have high expression (usually determined by Western) were tested in a 'blind' disease assay that included a positive and a negative control. Pathogens are indicated at the top of the table, grouped by type. The last row are results from 1 mg/mL CGA 245,704 (a chemical inducer of SAR) treatment 7 days prior to inoculation of non-transgenic tobacco with the indicated pathogen.
A blank in the table indicates that a valid test has not been performed or data not shown.
r indicates statistically significant resistance was observed in at least two out of three of the lines in several independent tests.
R indicates statistically significant resistance in many experiments with more than three independent lines of transgenic plants (more than 100 total individual plants) as well as a general correlation of resistance with antifungal protein abundance.
s indicates statistically significant susceptibility to the indicated pathogen.

damping-off caused by *Rhizoctonia*12. We have observed similar resistance to *Rhizoctonia* in tobacco transformed with acidic and basic chitinases from tobacco and cucumber (Table 1).

More recently, a PR-5 gene from tobacco was shown to delay symptoms to *Phytophthora* in transgenic potatoes.[13] Combinations of genes often demonstrated more than additive activity, indicating synergy[14] (and unpublished results). Therefore, the broad spectrum resistance observed during SAR appears to be the result of the independent and synergistic anti-fungal activities of many SAR genes acting in concert. Several SAR genes are common among plants, however, differences in expression of specific genes are observed in each species. In tobacco, PR-1 is the most abundant SAR gene while in cucumber an acidic class III chitinase is most abundant.[6,10] Such differences may reflect evolutionary or breeding constraints that have selected for the most effective SAR response. On the whole, plants have developed an inducible pyramid of defenses encompassing multiple modes of activity. Since both inducibility and pyramiding of activities are mechanisms that favor stability over evolutionary time, SAR is likely to be a stable form of resistance in a natural setting.

SALICYLIC ACID IS REQUIRED FOR SAR BUT IS NOT THE LONG DISTANCE SIGNAL

In 1979, White showed that the exogenous application of salicylic acid (SA) and other benzoic acid derivatives could induce resistance to TMV and the accumulation of PR-proteins.[15] In 1990, SA levels were shown to increase dramatically following pathogen attack.[16,17] Recently, the involvement of SA in signaling of SAR has been clearly demonstrated. A gene encoding salicylate hydroxylase (nahG) from *Pseudomonas putida* was employed to catalyze the conversion of SA to catechol, a compound with no SAR inducing activity.[18] Transgenic tobacco that express nahG do not accumulate SA in response to pathogen infection, do not have SAR gene expression and show no SAR.[18] In contrast, wild type tobacco accumulate SA during the induction of SAR in response to TMV infection. Therefore, in tobacco it appears that SA is required for the development of normal SAR, but whether SA is transported to distal portions of the plant to induce gene expression and resistance is unknown. To address this question, Vernooij et al. performed grafting experiments using wild type and salicylate-hydroxylase-expressing plants.[19] When inoculated with TMV, NahG transgenic rootstocks induced SAR in wild type scion tissue equally as well as wild type rootstocks did. Since the NahG root-stocks do not accumulate detectable SA, a different molecule must be responsible for the resistance that develops in the scion. Additionally, scions that expressed salicylate hydroxylase were unable to become resistant following TMV inoculation of the rootstock, regardless of the rootstock to which they were grafted. These results are consistent with those obtained in leaf removal experiments in cucumber.[20] Taken together, the simplest interpretation is that SA is not the long-distance signal for SAR induction, but SA accumulation is required for the long-distance signal to be transduced into gene expression and resistance.

Recently, Chen and Klessig have shown that a SA binding protein identified from tobacco is a catalase.[21] Other SA-related compounds that induce SAR genes were also shown to inhibit this catalase. Therefore, Chen and Klessig hypothesized that SA may induce SAR by catalase inhibition, which would result in the generation of reactive oxygen species.[22] Supporting this hypothesis, hydrogen peroxide and herbicides, that may result in the formation of reactive oxygen species (ROS), were shown to induce an SAR gene.

To test the idea that ROS is involved in SA signal transduction, we measured hydrogen peroxide levels in tobacco plants exhibiting SAR. We also measured SAR gene expression and resistance in control and NahG plants after treatment with hydrogen peroxide or the herbicides used by Chen and Klessig. We found that hydrogen peroxide levels were not elevated in systemic tissues that exhibited SAR gene expression and resistance.[8] In addition, only one of the herbicides, paraquat, was able to induce significant resistance. Even in this case, the presence of the NahG gene eliminated the induction of SAR gene expression and resistance.[8] In summary, our results showed that only paraquat could induce SAR, and that SA accumulation was required for this induction. Importantly, hydrogen peroxide was unable to induce gene expression or resistance in Xanthi or NahG plants. Therefore, catalase inhibition by SA may have a role in oxygen free-radical formation at the site of pathogen infection but it does not appear to be involved in SA induction of SAR.

BIOSYNTHESIS OF SA

Salicylic acid is a simple molecule derived from phenylalanine. At least two possible biosynthetic pathways via trans-cinnamic acid have been postulated. Recently, Raskin and co-workers have obtained evidence that SA is produced from trans-cinnamic acid via oxidation to benzoic acid (BA) followed by hydroxylation to SA23. First, since no o-coumaric acid accumulated following feeding of plant tissue with trans-cinnamic acid, and since applied o-coumaric acid did not induce SAR, biosynthesis of SA via an o-coumaric intermediate is unlikely. In contrast, plant cells fed labeled trans-cinnamic acid accumulated both labeled BA and a small amount of labeled SA23. Second, BA but not o-coumaric acid application to plants resulted in the induction of SAR genes and resistance. Third, the 2-hydroxylase enzyme activity that converts BA to SA is induced several fold by pathogen infection, further supporting this pathway for SA production in vivo.[24]

Importantly, both SA and BA have conjugate forms, mostly glucose for SA, that may represent caches for active SA.[25,26] Supporting this idea, BA-glucose conjugate levels are very high (100µg/g fresh weight) in non-infected tobacco, and decline following pathogen-induced necrosis, with concomitant accumulation of free BA and SA.[23] Further experimentation is required to clarify the role of BA and conjugate forms of BA and SA in SAR signaling.

SA IS IMPORTANT IN GENE-FOR-GENE RESISTANCE

Plants exhibiting SAR show the same physiological responses to pathogens that are observed in cases where plants express single, dominant resistance genes.

These responses include increased deposition of papilla-like material and rapid plant cell death at the attempted site of penetration.[27–30] In order to address whether SA may play a role in resistance conferred by dominant resistance genes, we assayed bacterial and fungal disease progression in NahG plants.[31] Disease symptoms and pathogen growth occurred on NahG *Arabidopsis* even when inoculated with avirulent bacteria or races of *Peronospora parasitica* to which *Arabidopsis* is normally resistant. When avirulent *Peronospora parasitica* was used, the pathogen was able to grow and sporulate before cell collapse occurred (hypersensitive-response). In addition, NahG tobacco plants showed more severe disease symptoms than controls when inoculated with TMV, *Pseudomonas syringae* pv. *tabaci*, and *Phytophthora parasitica* (Table 1). NahG *Arabidopsis* also showed greater disease symptoms compared to the control when inoculated with virulent *Pseudomonas syringae* pv. *tomato* (DC 3000) and *Peronospora parasitica*.

Taken together, these results underscore the importance that SA and SA-dependent responses play in general plant health. The picture that emerges is that in normal plants, if SA or SAR gene expression levels are low after the plant has come into contact with a pathogen, then the likely result is disease (Fig. 1). If the SA and SAR gene expression levels are high when pathogen contact is initiated, then the likely result is resistance. In nature, the full spectrum of responses from complete resistance to systemic disease (death) is observed.[32] Virulent pathogens may have evolved activities that actively repress the plant's SAR response, perhaps by interfering with the systemic signal or SA. In contrast, the role of some dominant resistance genes may be to rapidly induce SAR genes via SA. Alternatively, SAR may act to stimulate and prime the signal transduction pathway leading from recognition (determined by a resistance gene) to the development of the resistant state.

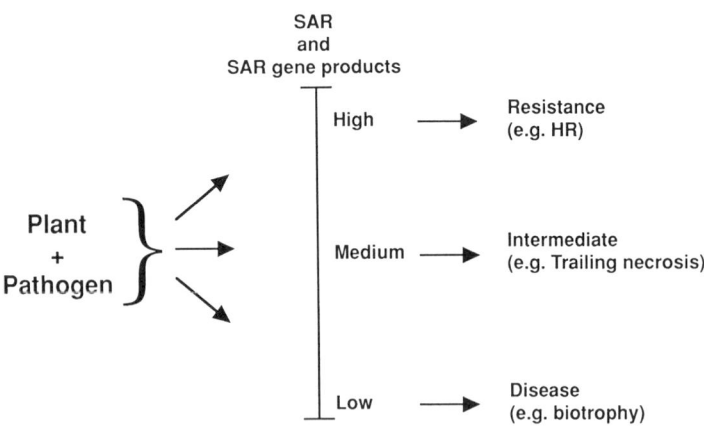

Figure 1. Simplified effects of SAR on plant health.

DISSECTING THE SIGNAL TRANSDUCTION CASCADE WITH *ARABIDOPSIS* MUTANTS

Arabidopsis mutants with altered SAR gene expression were isolated in order to genetically dissect the signal transduction mechanism leading to SAR.[33–37] Mutants were isolated (*nims* for no immunity) that can not induce SAR gene expression and resistance in response to pathogens or chemical treatment.[37] Plants that have high constitutive SAR gene expression and resistance are called *cims* (constitutive immunity). *cim Arabidopsis* mutants fall into two categories: plants with lesions that simulate disease (lsd)[38] and plants with no obvious macroscopic phenotype. All *cim* plants are distinguished by the resistance demonstrated after pathogen inoculation. *cim Arabidopsis* mutants demonstrate the feasibility of constitutive SAR for breeding broad-spectrum disease resistance into crop species. *lsd* Mutants appear similar to other so called lesion mimics of important crop species such as maize and wheat.[39] In addition, *lsd* mutants display all the known characteristics of pathogen infected tissues including, auto-florescence, callose deposition and SAR gene expression, while lesion mimics without constitutive SAR only exhibit macroscopic lesions.[38] More recently, we have analyzed lesion mimic mutants of maize (*les* mutants). Several maize *les* mutants developed lesions that resemble the disease symptoms of specific fungal infections. For example, the *les* 4 phenotype resembles infection of non-mutant plants by *Bipolaris maydis*, causal agent of Southern Corn Leaf Blight (Fig. 2, for review see Ref 39).

*les*4 Plants also have very high expression of SAR genes when the lesion phenotype is expressed (Figure 3). Identification of the genes affected in *les* mutants may help us understand the processes involved in disease and resistance. Toward this aim, Johal has identified transposon tagged alleles of at least two maize lesion mimic mutants.[40]

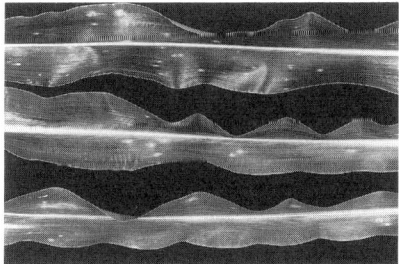 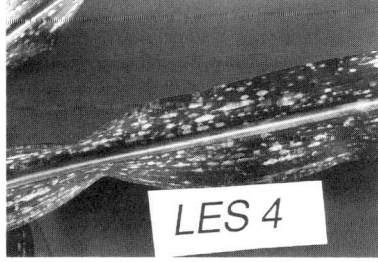

Figure 2. Similarity of lesions present in les 4 and Southern Corn Leaf Blight. Left-Maize leaf infected with *Bipolaris maydis*. Right-*les*4 mutant of maize.

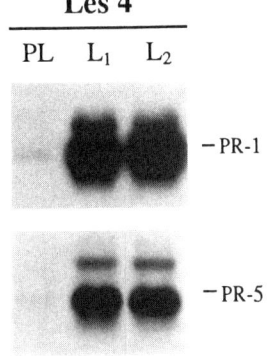

Figure 3. SAR gene expression in *les*4 maize. RNA blot analysis of PR-1 and PR-5 maize gene expression. PL-RNA from a *les*4 plant before lesions become visible. L1-RNA from a *les*4 plant approximately seven days after the initiation of lesions. L2-RNA from a *les*4 plant approximately 3 weeks after initiation of lesions. RNA blot analysis carried out as described previously.[10]

SUMMARY: THE FUTURE ROLE OF SAR IN AGRICULTURE

SAR appears to be a ubiquitous higher plant defense response. SAR is tightly correlated with the expression of specific genes that, when expressed constitutively from strong promoters or viagenetic alteration (*cim*), confer resistance to disease. Furthermore, when SAR is eliminated, through the enzymatic conversion of SA, generalized severe disease susceptibility occurs. Therefore, SAR and SA appear to be essential components of the defensive repertoire that ensures plant health in nature. Chemicals that activate SAR in the field as well as crop varieties with constitutive SAR gene expression will provide new solutions to disease problems for growers in the near future.

ACKNOWLEDGMENTS

We would like to dedicate this manuscript to the memory of Dr. Joseph Varner, a most exquisite teacher and student of biology. We especially thank Danielle Chandler, Henry-York Steiner, Nicole Specker, Michelle Hunt, Urs Neuenschwander, Kay Lawton, Mark Starrett, Leslie Friedrich, Kris Weymann, Sandy Dincher, David Negrotto, Michael Lanahan, John Salmeron and Eric Ward. We also thank M. G. Neuffer and E. Coe for the maize lesion mimic mutants.

REFERENCES

1. CHESTER, K.S. 1933. The problem of acquired physiological immunity in plants. Quart. Rev. Biol. 8: 275–324
2. ROSS, A.F. 1961. Systemic acquired resistance induced by localized virus infections in plants. Virology. 14: 340–358
3. KUC, J. 1982. Induced immunity to plant disease. BioScience. 32: 854–860

4. KESSMANN, H., STAUB, T., HOFMANN, C., MAETZKE, T., HERZOG, J., WARD, E., UKNES, S., RYALS, J. 1994. Induction of systemic acquired resistance in plants by chemicals. Ann. Rev. Phytopathol. 32: 439–59
5. UKNES, S., DINCHER, S., FRIEDRICH, L., NEGROTTO, D., WILLIAMS, S., TAYLOR, H., POTTER, S., WARD, E., RYALS, J. 1993. Regulation of pathogenesis-related protein-1a gene expression in tobacco. Plant Cell. 5: 159–169
6. MÉTRAUX, J.P., AHL GOY, P., STAUB, T., SPEICH, J., STEINEMANN, A., RYALS, J., WARD, E. 1991. Induced resistance in cucumber in response to 2,6-dichloroisonicotinic acid and pathogens. In: Induced resistance in cucumber in response to 2,6-dichloroisonicotinic acid and pathogens, Vol I. (H. Hennecke, D.P.S. Verma, eds.), Dordrecht, Kluwer, pp. 432–439.
7. LAMB, C.J., LAWTON, M.A., DRON, M., DIXON, R.A. 1989. Signals and transduction mechanisms for activation of plant defenses against microbial attack. Cell. 56: 215–224.
8. NEUENSCHWANDER, U., VERNOOIJ, B., FRIEDRICH, L., UKNES, S., KESSMANN, H., RYALS, J. 1995. Is hydrogen peroxide a second messenger of salicylic acid in systemic acquired resistance? Plant Journal 8: 227–233.
9. VAN LOON, L.C., ANTONIW, J.F. 1982. Comparison of the effects of salicylic acid and ethephon with virus-induced hypersensitivity and acquired resistance in tobacco. Neth. J. Pl. Path. 88: 237–256.
10. WARD, E.R., UKNES, S.J., WILLIAMS, S.C., DINCHER, S.S., WIEDERHOLD, D.L., ALEXANDER, D.C., AHL-GOY, P.,MÉTRAUX, J.-P., RYALS, J.A. 1991. Coordinate gene activity in response to agents that induce systemic acquired resistance. Plant Cell. 3: 1085–1094.
11. ALEXANDER, D., GOODMAN, R.M., GUT-RELLA, M., GLASCOCK, C., WEYMANN, K., FRIEDRICH, L., MADDOX, D., AHL-GOY, P., LUNTZ, T., WARD, E., RYALS, J. 1993. Increased tolerance to two Oomycete pathogens in transgenic tobacco expressing pathogenesis-related protein 1a. Proc. Natl. Acad. Sci, USA. 90: 7327–7331.
12. BROGLIE, K., CHET, I., HOLLIDAY, M., CRESSMAN, R., BIDDLE, P., KNOWLTON, C., MAUVAIS, C.J., BROGLIE, R. 1991. Transgenic plants with enhanced resistance to the fungal pathogen *Rhizoctonia solani*. Science. 254: 1194–1197.
13. LIU, D., RAGHOTHAMA, K.G., HASEGAWA, P.M., BRESSAN, R.A. 1994. Osmotin overexpression in potato delays development of disease symptoms. Proc. Natl. Acad. Sci. 91: 1888–1892.
14. ZHU, Q., MAHER, E., MASOUD, S., DIXON, R., LAMB, C. 1994. Enhanced protection against fungal attack by constitutive co-expression of chitinase and glucanase genes in transgenic tobacco. Bio/Technology. 12: 807–812.
15. WHITE, R.F. 1979. Acetylsalicylic acid (aspirin) induces resistance to tobacco mosaic virus in tobacco. Virology. 99: 410–412.
16. MALAMY, J., CARR, J.P., KLESSIG, D.F., RASKIN, I. 1990. Salicylic acid: a likely endogenous signal in the resistance response of tobacco to viral infection. Science. 250: 1002–1004.
17. MÉTRAUX, J.-P, SIGNER, H., RYALS, J., WARD, E., WYSS-BENZ, M., GAUDIN, J., RASCHDORF, K., SCHMID, E., BLUM, W., INVERARDI, B. 1990. Increase in salicylic acid at the onset of systemic acquired resistance in cucumber. Science. 250: 1004–1006.
18. GAFFNEY, T., FRIEDRICH, L., VERNOOIJ, B., NEGROTTO, D., NYE, G., UKNES, S., WARD, E., KESSMANN, H., RYALS, J. 1993. Requirement of salicylic acid for the induction of systemic acquired resistance. Science. 261: 754–756.
19. VERNOOIJ, B., FRIEDRICH, L., MORSE, A., REIST, R., KOLDITZ-JOWHAR, R., WARD, E., UKNES, S., KESSMANN, H., RYALS, J. 1994. Salicylic acid is not the translocated signal responsible for inducing systemic acquired resistance but is required in signal transduction. Plant Cell. 6: 959–965.
20. RASMUSSEN, J.B., HAMMERSCHMIDT, R., ZOOK, M.N. 1991. Systemic induction of salicylic acid accumulation in cucumber after inoculation with *Pseudomonas syringae* pv *syringae*. Plant Physiol. 97: 1342–1347.

21. CHEN, Z., RICIGLIANO, J., KLESSIG, D.F. 1993. Purification and Characterization of a soluable salicylic acid binding-protein from tobacco. Proc. Natl. Acad. Sci. USA. 90: 9533–9537.
22. CHEN, Z., SILVA, H., KLESSIG, D. 1993. Involvement of reactive oxygen species in the induction of systemic acquired resistance by salicylic acid in plants. Science. 242: 883–886.
23. YALPANI, N., LEON, J., LAWTON, M., RASKIN, I. 1993. Pathway of salicylic acid biosynthesis in healthy and virus-inoculated tobacco. Plant Physiol. 103: 315–321.
24. LEON, J., YALPANI, N., RASKIN, I., LAWTON, M. 1993. Induction of benzoic acid 2-hydroxylase in virus-inoculated tobacco. Plant Physiol. 103: 323–328
25. ENYEDI, A.J., YALPANI, N., SILVERMAN, P., RASKIN, I. 1992. Localization, conjugation and function of salicylic acid in tobacco during the hypersensitive reaction to tobacco mosaic virus. Proc. Natl. Acad. Sci. 89: 2480–2484.
26. MALAMY, J., HENNIG, J., KLESSIG, D.F. 1992. Temperature-dependent induction of salicylic acid and its conjugates during the resistance response to tobacco mosaic virus infection. The Plant Cell. 4: 359–366.
27. DEAN, R.A., KUC, J. 1988. Rapid lignification in response to wounding and infection as a mechanism for induced systemic protection in cucumber. Physiol. Mol. Plant Pathol. 31: 69–81.
28. STEIN, B.D., KLOMPARENS, K.L., HAMMERSCHMIDT, R. 1993. Histochemistry and ultrastructure of the induced resistance response of cucumber plants to Colletotrichum lagenarium. J. Phytopathol. 137: 177–188
29. UKNES, S., MAUCH-MANI, B., MOYER, M., WILLIAMS, S., DINCHER, S., CHANDLER, D., POTTER, S., SLUSARENKO, A., WARD, E., RYALS, J. 1992. Acquired resistance in *Arabidopsis*. The Plant Cell. 4: 645–656
30. UKNES, S., WINTER, A., DELANEY, T., VERNOOIJ, B., MORSE, A., FRIEDRICH, L., POTTER, S., WARD, E., RYALS, J. 1993. Biological induction of systemic acquired resistance in *Arabidopsis*. Mol. Plant Microbe Interact. 6: 680–685.
31. DELANEY, T., UKNES, S., VERNOOIJ, B., FRIEDRICH, L., WEYMANN, K., NEGROTTO, D., GAFFNEY, T., GUT-RELLA, M., KESSMANN, H., WARD, E., RYALS, J. 1994. A central role of salicylic acid in plant disease resistance. Science. 266: 1247–1250.
32. CRUTE, I. 1985. The genetic bases of relationships between microbial parasites and their hosts. In: The genetic bases of relationships between microbial parasites and their hosts., (R. Fraser, ed.) Boston, Kluwer Academic Publishers Group, pp 80–142.
33. LAWTON, K., UKNES, S., FRIEDRICH, L., GAFFNEY, T., ALEXANDER, D., GOODMAN, R., MÉTRAUX, J.-P., KESSMANN, H., AHL-GOY, P., GUT RELLA, M., WARD, E., RYALS, J. 1993. The Molecular Biology Of Systemic Acquired Resistance. In: The Molecular Biology Of Systemic acquired resistance., (B. Fritig, M. Legrand eds.), Dordrecht, Kluwer Academic Publishers, pp.410–420.
34. UKNES, S., LAWTON, K., WARD, E., GAFFNEY, T., FRIEDRICH, L., ALEXANDER, D., GOODMAN, R., MÉTRAUX, J.-P., KESSMANN, H., AHL GOY, P., GUT RELLA, M., RYALS, J. 1993. The molecular biology of systemic acquired resistance. In: The molecular biology of systemic acquired resistance., Vol. 2, (Gresshoff, P. ed.) Boca Raton, CRC Press, pp. 1–10.
35. BOWLING, S.A., GUO, A., CAO, H., GORDON, A.S., KLESSIG, D. F., DONG, X. 1994. A mutation in arabidopsis that leads to constitutive expression of systemic acquired resistance. Plant Cell. 6: 1845–1857.
36. CAO, H., BOWLING, S.A., GORDON, S., DONG, X. 1994. Characterization of an *Arabidopsis* mutant that is nonresponsive to inducers of systemic acquired resistance. Plant Cell. 6: 1583–1592.
37. DELANEY, T., FRIEDRICH, L., RYALS, J. 1995. *Arabidopsis* signal transduction mutant defective in chemically and biologically induce disease resistance. Proc. Natl. Acad. Sci. 92: 6602–6606.

38. DIETRICH, R.A., DELANEY, T.P., UKNES, S.J., WARD, E.R., RYALS, J.A., DANGL, J.L. 1994. *Arabidposis* mutants simulating disease resistance response. Cell. 77: 565–577.
39. WALBOT, V., HOISINGTON, D., NEUFFER, M. 1983. Disease lesion mimic mutations. In: Disease lesion mimic mutations., (T. Kosuge, C. Meredith, A. Hollaender, eds.), New York, Plenum Publishing Co., pp. 431–441.
40. JOHAL, G., LEE, E., CLOSE, P., COE, E., NEUFFER, M., BRIGGS, S. 1994. A tale of two mimics; transposon mutagenesis and characterization of two disease lesion mimic mutations of maize. Maydica. 39: 69–76

Chapter Eleven

NATURAL PRODUCTS, COMPLEXITY, AND EVOLUTION

Bruce B. Jarvis[1] and J. David Miller[2]

[1] Department of Chemistry and Biochemistry
University of Maryland, College Park, Maryland 20742
[2] Plant Research Centre
Agriculture Canada
Ottawa, Ontario, Canada K1A OC6

> *The more we know, the more fantastic the world becomes and the profounder the surrounding darkness.*
> Aldous Huxley (1925)

Introduction 265
Origins 268
Examples of Natural Product-Mediated Interactions 273
Natural Product Metabolism in Plants vs. Fungi 273
Limitation of Herbivory: Fungi and Insects 277
Limitation of Herbivory: Fungi and Plants 279
Limitation of Herbivory: Plants and Insects 282
Discussion 284

INTRODUCTION

As we approach the end of the 20th Century, it is useful to look back and reflect on the important fundamental properties of nature that have been elucidated by the science of our times. Although it is always problematic to make predictions, we nonetheless today can see important studies emerging that center on the questions surrounding the origins of self-organizing systems, the so-called science of complexity.[1,2]

Communities of molecules self-assemble into structures like membranes and micelles, and at some stage, such communities can support true complexity when a series of interconnected chemical events occur (or "emerge" in the terms of complexity). These are characterized by participation in complex interlocked cycles involving feedback mechanisms controlled by an elaborate chemical signalling system, a unicellular organism -life. But it does not stop there -a series of hierarchical systems evolve. Multicellular organisms, whose cells are organized into a series of interconnected local units, give rise to organisms that themselves are organized into communities. These in turn yield societies and other associated complexities. Currently in biology, interest in understanding how complex systems develop looms above all the others: the mechanics of embryology, development of neural networks, and evolution.

In this context we wish to approach the long standing question of the function served by natural products to those organisms which produce them. Natural products have long fascinated chemists and biologists alike, and, although their interests typically differ, they all share an interest in the question of why these chemicals are biosynthesized. A number of suggestions have been proposed, and the question has been reviewed recently by Williams et al.[3] and by Maplestone et al.[4] The conclusion reached is that these compounds are selected during the course of evolution because they increase the survival fitness of the producing organisms. The evidence supports this conclusion, but a number of interesting and fundamental questions remain concerning the roles of natural products as viewed from the perspective of complexity.

Often the most interesting questions are related directly to human activities. Although we often focus on the lethal, the vast majority of interactions between organisms are of a positive nature and of mutual benefit. One need only note that there are no axenic complex multicellular organisms on earth. All have associative microorganisms (principally bacteria) that are important for the well being of these higher organisms. Many are mediated by natural products, a case in point being the symbiotic *Rhizobium* bacteria (a diazotrophic microorganism) which colonizes the roots of legumes. The bacteria are attracted to the roots by specific plant flavonoids (e.g. cyanidin and luteolin), and in turn species-specific lipopolysaccharides found on the cell surface of the bacteria bind to species-specific carbohydrate-binding glycoproteins (lectins) of the plant roots, assuring a specific bacterium-plant association (Fig. 1).[5,6]

Mammals, particularly humans, appear to lack what might be thought of as classical natural products [low molecular weight compounds (< 3000 Daltons) which seem "to have no explicit role in the internal economy of the organism that produces it"].[3] This condition may exist because mammals have evolved an immune system providing protection from pathogens. Similarly, mammals such as humans are able to monitor and control their environment through activities of the brain. However, by adopting a more conservative notion of what constitutes a natural product, one can point out that we also employ such weapons in

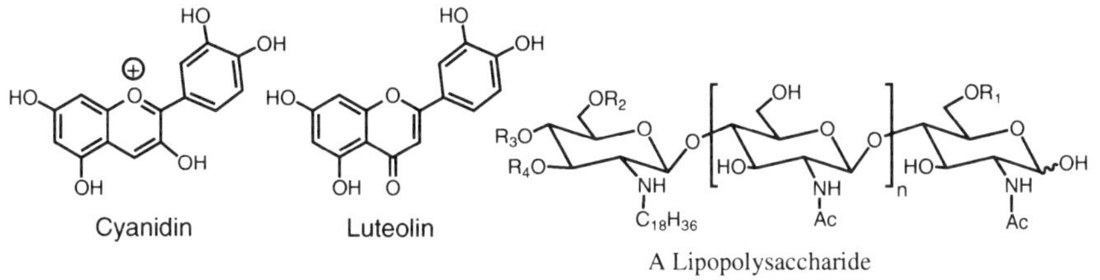

NATURAL PRODUCTS, COMPLEXITY, AND EVOLUTION 267

Figure 1. *Rhizobium* signaling molecules.

our arsenal. The recent finding that nitric oxide is released by macrophages as cytotoxins in response to bacterial infections appears to be an example -though nature has found additional uses for this highly reactive chemical.[7]

There are several inherent problems with drawing general conclusions based on isolated accounts of the observed biological and biochemical properties of natural products. 1) We are often ignorant of the details of the interactions (both chemical and biological) of compounds produced by one organism with other organisms in the environment. 2) Little information is available on the genetics of natural products production, although significant advances have been made in the last decade.[8] 3) Although convenient to think in terms of individual organisms, it is increasingly evident that many organisms have highly coupled interactions with other organisms,[9] in which it is impossible to sort out all the interrelated chemistry.[10] Often, it is not just two or more individual organisms interacting exogenously in a complex fashion, but also internally as well, i.e. organisms have intimate symbiotic arrangements. Even eukaryotic cells themselves appear to be the descendants of an ancient protistal-bacterial symbiotic collaboration that has led to the present day eukaryotic cells containing organelles (e.g. mitochondria and chloroplasts) whose bacterial DNA origins are evident.[9,11]

This paper is based on a number of hypotheses: 1. Precursors of extant natural products arose from mutations of enzymes involved in the synthesis of primary metabolites. 2. The chemicals that emerged were modified by variants of other enzymes of primary metabolism in a gene-for-gene fashion with other biological components of the ecosystem. 3. Under the pressure of natural selection, natural products have evolved as messages, defined in broad terms, i.e. come hither, don't eat me, germinate, divide, attach, disengage, etc. 4. Horizontal transfer of genes (which code for biosynthetic pathways) between species has occurred and conferred selective advantage to organisms. 5. The actions of natural products in human physiology are often not relevant to their function(s) in the producing organisms.

Although we make a distinction between external communication (pheromones, antifeedants, mediators of host-parasite interactions, etc.) and internal communication (hormones, immune modulators, neural transmitters, etc.), they are in fact part and parcel of the same thing. Natural products are chemicals released within a *system* by one component which convey *information* and *instructions* to another component(s) within the system. They are a natural outgrowth and consequence of an increase in complexity, and they are a critical part of the chemical "glue" that holds systems together.[12] That natural products usually tend to be small organic molecules is a consequence of the functions they serve: messengers that must survive long enough to shuttle between the various components of the system.

ORIGINS

Of the four major classes of biochemicals (carbohydrates, proteins, nucleic acids, and lipids), experimentation has shown that the first three classes could have readily arisen through prebiotic chemistry.[13] Impressive results are reported in carbohydrate chemistry where Müller et al.[14] have shown that simple changes in reaction conditions can lead to surprising specificity in the synthesis of ribose derivatives. Furthermore, Eschenmoser and his colleagues have shown that it is no accident has 2-deoxyribose has been chosen (i.e. through natural selection) to be the sugar in the backbone of DNA. For only this sugar (of all the possible pentoses and hexoses) has the inherent conformational and steric properties to function properly in a complex molecular information system like DNA.[15]

It is now evident that any theory of the prebiotic origin of present day biomolecules must address three fundamental criteria: 1) were favorable conditions present to lead naturally to the formation of the compounds? 2) were there pressures to naturally select these compounds over a long period of time? and 3) did the compounds possess the ***inherent*** property to self-assemble? In addition, some must also have had the ability to interact specifically (recognition factors) with other molecules and, importantly, to function as catalysts.[16] The present day consensus is that RNA (e.g. ribozymes)[17,18] was the first biopolymer to serve the latter function and to survive into modern times.

As with all molecules of life, the exact origin of secondary metabolites is unknown. Whether these compounds are of strictly biogenic origin or whether their origins can actually be traced to prebiotic chemistry is a question of fundamental interest. Although it is generally accepted that current natural products are of biogenic origin, it may not have been so historically. Davies[19] has speculated that natural products can be traced to prebiotic times where they served a number of roles including effectors or co-factors which stabilized active conformations of biopolymers and perhaps even functioned as primitive catalysts. Although this is an interesting and provocative suggestion, it seems unlikely.[20] First, proteins and nucleic acids in general have little need for exogenous molecules (other than water) to effect their conformations -this is an inherent property of the molecules themselves. Second, the nature of natural products is such that unlike carbohydrates, proteins, and nucleic acids, they defy any basic descriptions of their molecular architecture, i.e. not constructed of basic units and made complex by variation in the number and/or arrangements of these basic units, as are the proteins, carbohydrates, and nucleic acids. In fact, the quintessential property of natural products is their near infinite variety. Although the biosynthesis of many can be traced back to acetate (e.g. fatty acid, terpene, and polyketide biosynthesis) or amino acids (e.g. alkaloid biosynthesis), there are many whose biosynthetic origins are either obscure or result from a complex combination of pathways. Furthermore, there seem to have been no general driving forces in prebiotic condensation reactions leading **selectively** to natural products since few if any of these compounds exhibit a tendency to self-assemble.

A fundamental problem in analyzing the role of natural products in evolution from the view of complexity is that we have no effective way of establishing the order of evolutionary events. We are faced with a system composed of numerous interlocked cycles, and we have no means to trace its history. Although the accepted rationale for how such systems arise is through natural selection acting over long periods of time, from the perspective of complexity, self-organization also plays a important role. At a critical stage, a large number of components that are interacting with one another in a non-systematic way "crystallize' (or another term used, "undergo a phase transition"), and form a stable system of highly interconnected cycles.[1] Once formed, these cycles are robust but difficult to analyze from a traditional reductionist point of view. While, this does not preclude our studying the role of individual components, it does suggest that we must take care not to overinterpret our data.

Stories connected with natural products hold our interest. We are amazed by how ingenious nature is in solving problems and constructing systems that survive and evolve. Organic chemists in particular are struck by the ingenuity of nature's unexpected chemistry. Even when chemists "invent" new chemistry, we often discover later that nature had long ago evolved the chemistry we thought unique to the laboratory.

For example, in the early 1970's, chemists at Cal Tech reported an unusual cycloaromatization of an enediyne into a reactive 1,4-benzenoid diradical which readily abstracts hydrogen atoms to yield aromatic compounds.[21] At the time, this reaction was something of a curiosity and languished in the literature (although it continued to be of interest to the physical organic chemists)[22] until the discovery of the complex of novel antibiotic dynemicins, calicheamicins, esperamicins, kedarcidins, and the neocarzinostatins.[23] These antibiotics all share the enediyne substructure, and their mode of action relies on the cycloaromatization reaction (Fig. 2). The enediyne system is fairly stable in these antibiotics until activation by a metabolic transformation (e.g. reduction of the anthraquinone ring in dynemicin A) which brings about a conformational change

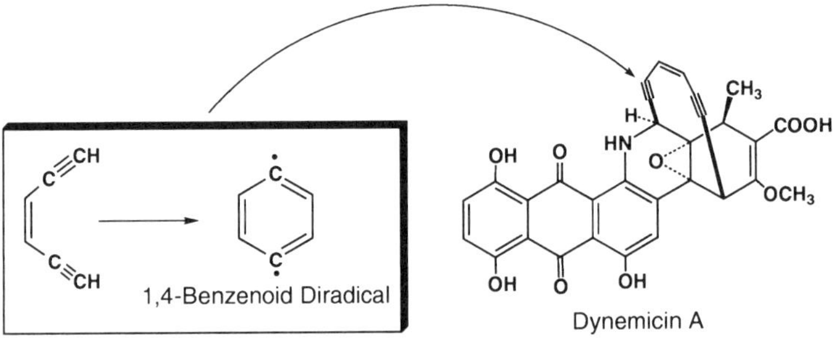

Figure 2. Cyclization of enediynes to benzenoid diradicals: structural relationship to the anitbiotic dynemicin A.

(via opening of the epoxide ring in dynemicin) that induces the cycloaromatization. If these events occur when the antibiotic is intercalated in the nuclear DNA, hydrogen abstraction by the 1,4-diradical initiates double-stranded cleavage.

The properties of the enediyne antibiotics are commonly shared by many of the diverse DNA-binding antibiotics. Typically, these antibiotics selectively bind to the DNA (e.g. intercalation and/or binding to the minor grove of the DNA double helix). The antibiotic then undergoes metabolic activation to yield reactive intermediates which disrupt DNA structure and inhibit replication and/or transcription. These sophisticated processes argue strongly for their having evolved to damage the DNA of the target organism.[3,4] A point to which we will return is that the producing organism itself must have a mechanism(s) to protect itself from these potent antibiotics.

Many natural products have been investigated because of their physiological action in humans. Do the effects in humans have any relevance to the roles of the metabolites in the life of the producing organisms? Because all organisms on earth share (substantially) the same biochemistry, it is not unreasonable to suggest that indeed such relevance exists. For example, from an evolutionary perspective, it seems reasonable that the binding of many of our own "secondary metabolites" (e.g. the steroidal hormones) has antecedents in our plant and fungal cousins. However, in the vast and complex array of chemical interactions between species, there will be many "accidents" which convey no insight into ecological roles, e.g. physiological effects in non-target organisms that have no relevance to the biochemistry of these natural products in the target organisms. On the other hand, some microbial and plant natural products that exhibit specific binding to our own receptors may do so as a consequence of a "distant memory" of binding to similar targets in other species. There seem to be two explanations for plant or fungal natural products exhibiting high specific binding to human protein receptors. Either they are an accident of nature or some distant evolutionary connection has been

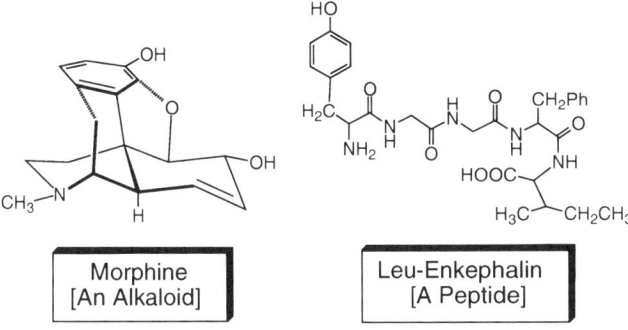

Figure 3. Ligands for the opiate receptors in brain tissue.

conserved. Two examples of the former are the opium alkaloids and tetrahydrocannabinol (THC), the active principle in marijuana.

When the opiate receptors were first discovered in vertebrate brains,[24] it was realized that the presence of such receptors surely demanded a natural ligand. Indeed, within a few years, such natural ligands were found and their structures determined.[25] Surprisingly, they turned out to be polypeptides (the endorphins and enkephalins) whose structures appeared to have little in common with those of the opiate alkaloids (e.g. morphine) (Fig. 3). In a similar fashion, the natural receptor for THC was eventually discovered in brain tissue,[26] and recently, the natural ligand for this receptor has been shown to be anandamide, a simple amide derivative of the important C-20 fatty acid, arachidonic acid (Fig. 4).[27] Furthermore, the mammalian THC receptor is closely related to a receptor found in sea urchins.[28] It would be difficult to argue any evolutionary significance or connection in these examples. They are simply examples of natural products from "lower" organisms that have marked physiological effects in humans as a result of an accidental fit with our receptors, which normally exhibit binding only to very specific ligands.

Although this may be the usual explanation when exogenous natural products show high binding to human receptors, one should not dismiss all such

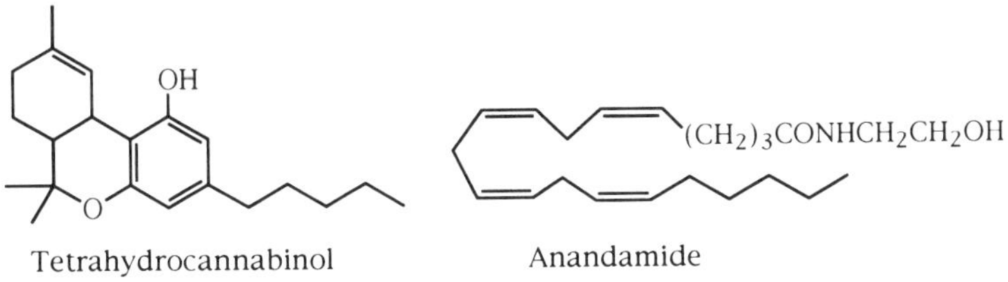

Figure 4. Ligands for the tetrahydrocannabinol receptors in brain tissue.

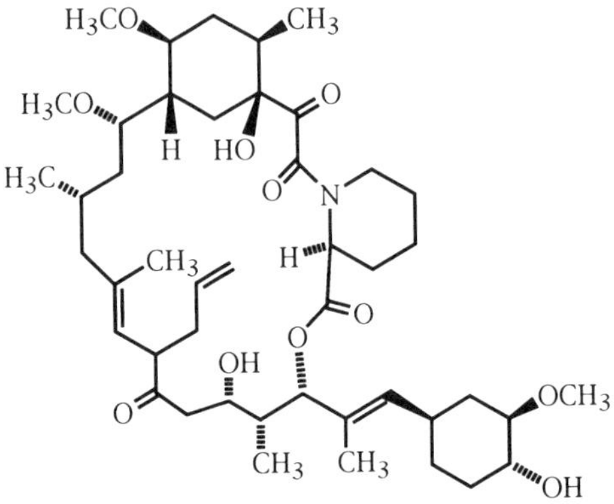

Figure 5. FK506, an immunosuppressant fungal metabolite.

cases as mere accidents of nature. Protein receptors are found throughout the cells, and many natural products exhibit high and specific binding to these proteins. A basic problem in cell biology has been to understand how a signal induced by a ligand-receptor binding on the outer cell membrane surface can convey information to the inner cellular material and bring about chemical transformations -a process referred to as signal transduction. There are a series of cytoplasmic proteins that are coupled to the membrane receptors that convey a signal to an intracellular target, often a gene. Attempts to understand this transduction system have uncovered a number of interesting pharmacological properties of natural products which result from their binding to specific intracellular protein receptors.[29,30]

One such transduction system involves the immune system which has evolved from a basic property of cells - the absolute requirement to distinguish between what is the same and what is different.[31,32] Cells must be able to determine and properly respond to other cells with respect to whether they "belong together." The success of organ transplantation depends upon turning off the immune system to a degree sufficient to prevent rejection of transplanted tissue. This is possible today only because of the use of immunosupressive agents such as cyclosporin and FK506, fungal natural products (Fig. 5). These immunosuppressants act in T cells by binding to intracellular receptors known as immunophilins and inhibiting intracellular signal transduction pathways.[33] Immunophilins are evolutionarily ancient proteins found in both eukaryotes and prokaryotes.[34] It is likely that there are low molecular weight cellular chemicals which are the natural ligands for these proteins.[29,30] There are a number of natural products which interfere with specific cellular functions, some of whose targets are known, but for many (so-called orphan ligands),[29,30] the protein receptors are unknown. From the perspective of evolution, production of these natural products may increase the fitness of the producing organism through their interactions with the cellular receptors of other organisms.

EXAMPLES OF NATURAL PRODUCT-MEDIATED INTERACTIONS

It will be useful to detail some examples drawn from chemical ecology which illustrate a common trait of complexity -that of unexpected complicated patterns arising from seemingly simple rules.[35,36] Although this discussion will center on a given organism and the apparent function to that organism served by natural products, it should be kept in mind that all organisms are embedded in a local ecology (which in turn is embedded in a broader ecology, etc.). One needs to be cautious in drawing conclusions without considering the system as a whole. This is a departure from the standard reductionist view that has served science so well over the years, whereby one develops a basic understanding of a complicated system by dismantling the structure and examining in great detail all the components of the system. Since eventually we will need to understand natural products at the genetic level, an additional point needs to be emphasized. The genome is not a blueprint for an organism but rather more a recipe -context is critical, not only in the development of an organism, but also in the expression of those genes coding for natural products.

A word of caution needs to be interjected. There are likely to be few allelochemic relationships that depend only on the interaction of a single chemical or class of chemicals. Rather, in the typical situation, such as the chemical defense of plants against insect attack, the process is likely mediated by a complex array of chemicals which jointly exert selection pressures. Thus in the absence of the knowledge as to how other relevant natural products collectively also affect the system, it is difficult to analyze the evolutionary development of a phenotype from the perspective of a single compound.

NATURAL PRODUCT METABOLISM IN PLANTS VS. FUNGI

The evolutionary pressures that led to the development of natural products in fungi and plants are similar in some respects and different in others. Plants cannot move away from a threat, and thus poisons and armour are essential in dealing with herbivory. Volatile metabolites released by conifers and shrubs are known to help determine population structure within ecosystems.[37] Other volatile metabolites released by insect-damaged plants may recruit predators of those insects.[38] In plants, natural products generally occur in specific organs or tissues. In cereals, phenolic acids, flavonoids, and conjugated amines are located in the pericarp and aleurone layers.[39] Compounds that appear to be important in resisting stored-product insects are mainly located in the pericarp.[39] The production and regulation of natural products in plants are associated with genetic controls of the *differentiation* of specific tissues. Sequences that code for metabolites are expressed while the genes for particular tissues are being expressed.[40,41]

There is an interesting connection between the size of the genome of an organism and the number of tissue types found in the organism (number of tissue types ~ square root of the number of genes).[1] This has been attributed to the inherent nature of self-organized systems. Since the production of natural products is often tissue-specific, the regulation of secondary metabolism is coupled to the expression of cellular phenotype and must be a part of the self-organized system.

Another important feature of plant secondary metabolism is that highly reactive compounds are located in organs that remain physiologically active. To accommodate this potential problem, the biosynthesis and storage of active compounds is highly compartmentalized. Compounds are synthesized in one part of the cell, stored in vacuoles and transported to other parts of the plant. During transport, modifications can occur. Several cell types can be involved.[42] Many toxic compounds are stored as non-toxic glycosides that are hydrolyzed by plant enzymes should, for example, an insect bite into the plant.[43]

Compartmentalization of toxin production, especially via symbiotes is common in marine organisms as well.[44] The dinoflagellate *Prorocentrum lima*, which produces okadaic acid, a potent inhibitor of eukaryote phosphatases PP2A and PP1, appears to protect itself by a novel mechanism. Okadaic acid is actually biosynthesized in the chloroplasts of *P. lima* where the prokaryote phosphatases are unaffected by this inhibitor (J. L. C. Wright, private communication). Before release to the cytoplasm, okadaic acid is transformed to the diol ester, which is inactive as a eukaryote phosphatase inhibitor,[45] and the diol ester is further transformed to the water-soluble DTX-4 toxin,[46] which is 10–50 fold less active than okadaic acid (Fig. 6). This compound is then transported through the cytoplasm to the outer membrane where it is cleaved to okadaic acid before release out to the aqueous environment.

Most of our knowledge of the metabolites of fungi comes from studies of filamentous Ascomycetes and Deuteromycetes (molds) as opposed to species that produce large reproductive structures such as mushrooms, bracket fungi, and puff-balls. This may be because the former are easier to grow in the laboratory, or it may be that they are truly more prolific producers of secondary metabolites. All of the agriculturally-important mycotoxins are produced by molds.[47,48] Although the term natural product is applied to plants, mushrooms and filamentous fungi, many aspects of the biology of the accumulation of the compounds involved are different between the first two and the latter organisms. This primarily is due to the growth habit of molds -the mycelium. Higher plants are

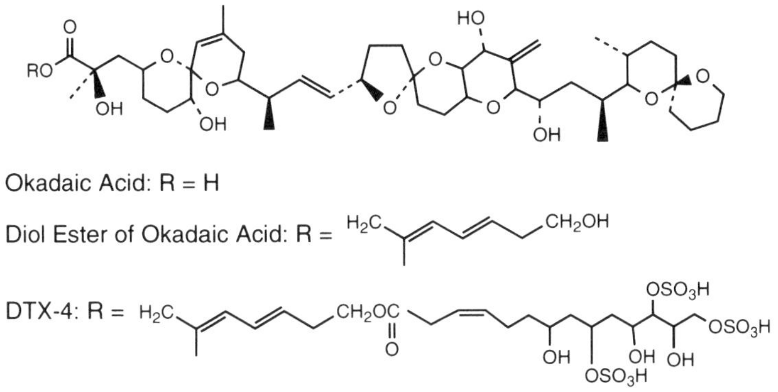

Figure 6. Okadaic acid derivatives, marine toxins from the dinoflagellate *Prorocentrum lima*.

complicated structures with many component organs. Although the macroscopic fungal colony *seems* analogous, it is not. Every facet of the biology of filamentous fungi revolves around the fact that only the terminal few cells in the mycelium are active. The production of natural products is the only appreciable non-reproductive *differentiation* in molds.[49]

Fungal natural products are induced by a nutrient limitation. As the hypha grows through or across the substrate, the terminal cell grows, stops, flattens out and grows again in short cycles. This has been explained in terms of the mechanism by which mycelia are produced[50] and in terms of the physical constraint of the mycelium as it moves through e.g. plant tissue.[51] The few cells immediately behind the terminal cell excrete enzymes and otherwise import nutrients into the active area of the mycelium. The nutrients available to these cells become limiting and natural products are released into these microenvironments. The spaces between the nutrient-deprived cells and the substrate contain high concentrations of metabolites.[52] Which fungi grow in a given substrate depends primarily on the water activity (relative humidity) and quality of nutrients present.

In fermentors, semisynchronous populations of cells are grown under controlled conditions of appropriate pH, oxygen and osmotic tensions to produce sufficient biomass of cells coincident with the exhaustion of the required nutrient limitation. This pattern of growth, nutrient exhaustion and metabolite production is the same *in vitro* and in a fermentor, only the scale is different. Osmotic tension (adjusted by sugar alcohols and salt), temperature and usually pH must mimic the optimal conditions for growth of the fungi. Which nutrient limitation is important depends on the biosynthetic origin of the fungus.[53] Amino acid-derived compounds are produced under carbon limitation; compounds derived from acetate are produced under nitrogen limitation. It is the specific nutrient

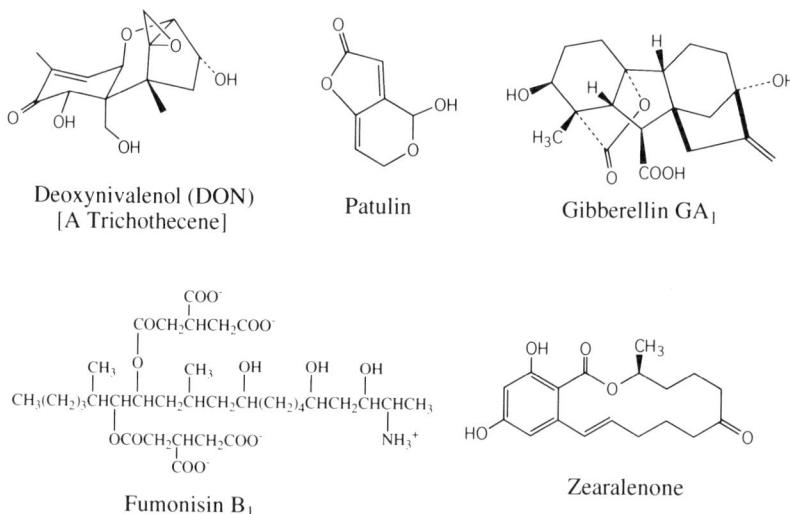

Figure 7. Structures of selected fungal metabolites, all of which are mycotoxins except the plant growth hormone gibberellin GA_1.

limitation rather than growth limitation per se that modulates metabolite production.[54]

Polyketides and sesquiterpenes (e.g. aflatoxin, fumonisin, gibberellins, patulin, trichothecenes, and zearalenone, Fig. 7) are produced under nitrogen limitation.[55–60] In the cases of patulin and gibberellins, nitrogen addition to a producing culture results in cessation of metabolite production.[54,61] This is due to repression of the expression of polyketide synthetases.[62,63] The amino acid derived penicillin is produced under carbon limitation. Some of the enzymes involved are modulated by carbon catabolite repression.[63]

Oxygen tension is often very specific to the metabolite to be produced. The *Fusarium graminearum* metabolites zearalenone and deoxynivalenol are both derived from acetate and produced under nitrogen limiting conditions. Zearalenone is produced under high oxygen tensions[58] and deoxynivalenol (and other trichothecenes) under low oxygen tensions.[59] Generally, pH is not a factor except that pH normally is maintained within the optimum range for growth (pH has an important effect on nutrient uptake). An exception to this rule is fumonisin which appears to require acid pH's. This may reflect the fact that fumonisin is prone to chelation.[60]

Such generalities are plagued by numerous inconsistencies in experimental data. *In vitro* experiments demonstrate considerable variability in response of different strains to the same culture conditions. This reflects the genetic variability of natural strains in terms of primary metabolism, the large numbers of genes involved, and the peculiar behavior of the regulatory genes for secondary pathways.[64] For example, mutants of *Aspergillus flavus* exist that apparently produce aflatoxin more or less concomitant with growth (under non-homogenous still culture conditions).[65] Nonetheless, careful analysis of each of these apparent exceptions shows that they are just that.

There is little known about the enzymology of fungal natural products. Traditional views have been that enzymes arose from modified enzymes of primary metabolism. Another hypothesis is that modifier proteins were produced that changed the active sites of the enzymes. Recent data on the polyketide synthetase involved in patulin production shows that this enzyme has appreciable sequence differences compared to fungal lipid synthetases.[66]

As in plants, the biosynthesis of toxic metabolites by fungi is done in compartments.[67] A difference is that fungal natural products are excreted directly into the environment. In trichothecenes, for example, the final reactions are carried out by membrane-bound oxidases.[68] Thus the step that confers the greatest increase in toxicity occurs as the compound is leaving the (dying) cell.

LIMITATION OF HERBIVORY: FUNGI AND INSECTS

Aflatoxin is one of the most important mycotoxins and a major health hazard in some developing countries.[47] Discovered more than 30 years ago, aflatoxin is one of the most studied natural products.[69] Much is known about its toxicity, biochemistry, and molecular biology as well as the natural history of the fungi that produce it. This research effort revolves around the extremely potent carcinogenicity of aflatoxin to humans and domestic animals.[70] The genotoxic aflatoxins exert their effect at the DNA level where, after intercalation and metabolization (cytochrome P-450) to their C15,C16-epoxides, they act selectively to alkylate the N-7 of the guanine residues (Fig. 8).[71,72] Although this adduct may be excised and the DNA repaired, enough unrepaired damage can occur to result in mutations and/or carcinogenesis.[73] Although arguments have been made to explain the evolution of aflatoxin on the basis of its mutagenicity (which illustrates human preoccupation with the carcinogenicity of aflatoxin),[74,75] we propose to take a broader view.

Aflatoxin is produced by three species of *Aspergillus*, *A. flavus*, *A. parasiticus* and *A. nomius*.[70] *A. flavus* and *A. parasiticus* are associated with corn and groundnuts in the Americas and Africa with the former more associated with corn.[47] In Asia, A. *parasiticus* is rare and thought to have been introduced from the U.S.A.[76] *A. nomius* was described in 1987 from strains primarily isolated from insects.[77] *A. flavus* is capable of growth at 37 °C and can be pathogenic to insects and mammals.[78] *A. flavus*, *A. parasiticus* and *A. nomius* are very closely related and are distinguished by differences in temperature tolerance, aflatoxin production, habitat preference, and animal pathogenicity.[77] *A. flavus* is thought

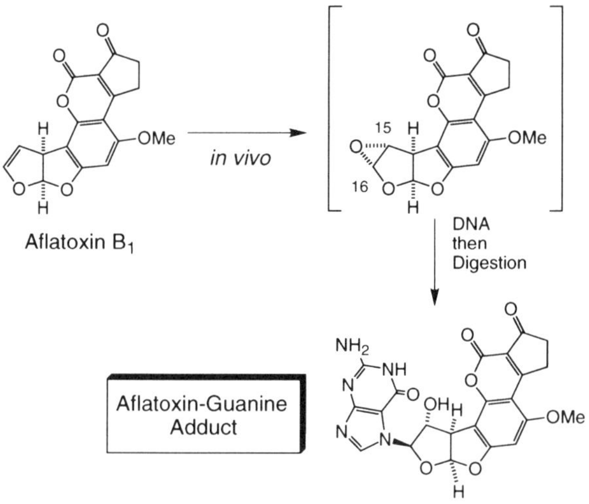

Figure 8. Metabolic activation of the potent carcinogen aflatoxin B_1 followed by conjugation with DNA.

to be the ancestor of all three based on DNA relatedness and the order of aflatoxin synthesis genes.[77,79]

It appears that *A. flavus* and related species evolved as insect pathogens and later to use other high oil/fat substrates such as groundnuts and corn. Pitt (personal communication) suggests that crop plant saprophytes are evolving very quickly as new substrates become widespread. Few animals store seeds presumably because of mycotoxin problems and those that do have sophisticated storage practices.[80] Aflatoxin is toxic to most insects.[81,82] Insect species that use *A. flavus*-infested food sources have gut enzymes or gut-associated bacteria capable of degrading aflatoxin.[81] It has recently been discovered *A. flavus* produces a variety of antifeedant chemicals which are accumulated in the conidia and sclerotia.[80,83]

As noted, an evolutionary rationale for the selection of the potent mutagenicity of aflatoxin was proposed by Lillehoj.[75,84] He argues that the production of mutagens confers a selective advantage, but does not explain how this would affect fungal population structure.[80] Aflatoxin is genotoxic primarily because of the DNA adducts that are formed.[70] The aflatoxin precursors are less genotoxic.[75]

In contrast to the gene-for-gene interaction in the evolution of new pathogenic races of fungal diseases,[85] the evolution of insect pathogenicity factors or antiherbivory factors could be associated with aflatoxin precursors regardless of their low toxicity to insects today.[82] As noted above, insects that feed on aflatoxin-contaminated corn have evolved detoxification systems illustrating a gene-for-gene response. Aflatoxins are not phytotoxic[86] and have low antimicrobial activity,[87] and their carcinogenicity may be accidental.

A number of genera produce the aflatoxin precursor sterigmatacystin.[88] The enzymatic genes involved are clustered although the order and direction of transcription are not identical (Keller, personal communication). Regulatory genes also appear to be conserved at least in the aspergillii.[88] In *A. flavus*, an additional two conversions are needed to produce aflatoxin B1 from sterigmatocystin,[89] and the genes involved are organized in a similar way as in the sterigmatocystin-producing species.[79] Sequence data for the polyketide synthetases, aryl alcohol dehydrogenases and fatty acid synthetases involved in both *A. nidulans*[88] and *A. flavus/parasiticus* are similar to the corresponding enzymes in primary metabolism.[79]

LIMITATION OF HERBIVORY: FUNGI AND PLANTS

With respect to natural products, historically plants are known as producers of alkaloids, although these nitrogenous metabolites are also produced by animals and fungi. The alkaloids have a long and rich history as natural products of interest in human history and pharmacology. Although many have interesting and potent bioactivities and could be discussed in terms of their roles in evolution, we confine our attention to a few examples, which are illustrative of some unexpected chemical-based interrelationships.

The ergot alkaloids are well known mycotoxins produced by the fungus *Claviceps purpurea* and are the toxins responsible for ergotism, a serious human disease widespread in the Middle Ages (also known as Saint Anthony's Fire) but whose history can be traced back to biblical times. Infection of cereals (mainly rye) by *C. purpurea* can be readily detected as the sclerotia (ergot) forms on the seed heads. A variety of maladies (e.g. "fescue foot," gangrene of the extremities)[90] in animals grazing on tall fescue grass have been traced to ergot alkaloids even though there is no apparent fungal activity on the plant material. The source of these alkaloids is traced to an endophytic fungus (*Acremonium coenophialum*) actually living inside the grass. A similar endophyte, *Lolium perenne* L., is responsible for "ryegrass staggers," a neurological disorder in grazing animals whose cause appears to be fungal-produced tremorgens such as lolitrem B (Fig. 9).[91]

This plant-fungus relationship carries benefits for both. The fungi depend upon the plant for nutrition and dissemination,[92] while the endophyte-infected plants are protected from insect predation by the toxins. Different toxins produced by the fungus are active against different species of insects attacking.[93] Infected plants exhibit better drought tolerance and have increased resistance to fungal disease.[92] Vegetative reproduction (the primary mode in most grasses) is enhanced.[93] In the absence of pathogen/insect pressure, under greenhouse con-

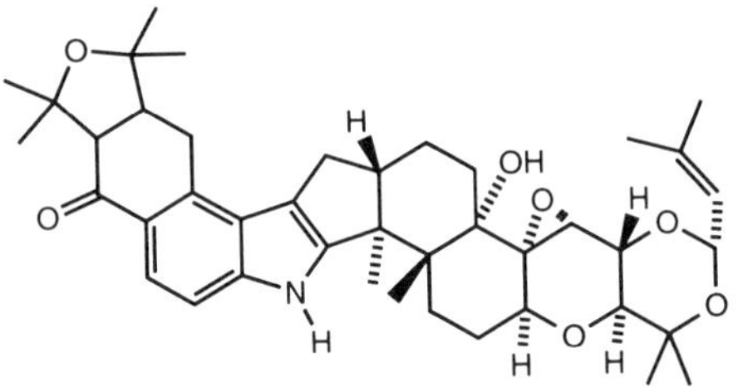

Figure 9. Lolitrem B, a tremorgenic alkaloid produced by the endophyte *Lolium perenne* and the cause of ryegrass staggers.

ditions with high nitrogen fertility, endophyte-removed plants grow more vigorously.[93] Overall, it is believed that there is a competitive advantage to endophyte-infected grasses, although the studies in this area are few and sometimes conflicting.[94] This grass-fungal mutualism is not unique. Leaf endophytes of other plant species produce toxins that appear to limit herbivory.[95] Like grasses, conifer trees produce few insecticidal compounds.[95] Many antifeedant compounds have been isolated from conifer needle endophytes.[96,97] The presence of endophyte species that produce such toxins has been shown to increase the fitness of the tree population.[98] Some seaweeds also have fungal endophytes (termed mycophycobioses). When *Turgidosculum ulvae* is present in the thalli of the green alga *Blidingia minima* var vexata, intertidal animals will not eat the alga.[99]

Similar relationships also occur with vascular endophytes. Taxol (Fig. 10), an alkaloid found in the bark of the yew tree is of current interest because of its effectiveness in the treatment of certain less tractable cancers such as ovarian and breast cancers.[100,101] Taxol and related taxanes are present in relatively high amounts in various plant parts in the genus *Taxus* [102] and can be produced in tissue cultures of *T. brevifolia*.[103] In an extraordinary turn of events, Stierle et al. report that a fungus, *Taxomyces andreanae*, isolated from the inner bark of *T. brevifolia*., produces taxol and related baccatin III in culture, though the level of

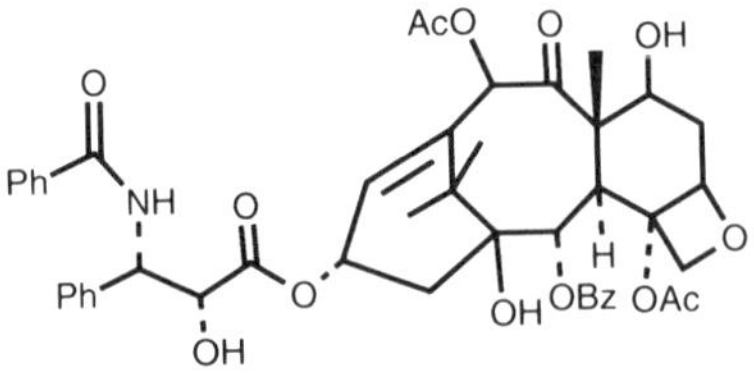

Figure 10. Taxol, an anticancer drug from the bark of the American Yew tree (*Taxus brevifolia*).

production was only *ca.* 50 ng/L.[104] Nonetheless, this raises the interesting question of the biosynthetic origin(s) of the taxanes. This discovery is reminiscent of the finding of gibberellins in plants, which at the time were thought to be produced exclusively by fungi.[105] Later work established that the biosynthetic pathway to the gibberellins up through GA_{12} was the same for plants and fungi.[106]

An example of the ultimate resolution of a fungal-plant mutualism -the elimination of the fungus- may be the plant-derived macrocyclic trichothecenes found in two species of the Brazilian shrub *Baccharis megapotamica* and *B. coridifolia* (Fig. 11). Trichothecenes are well known mycotoxins produced by a number of plant pathogenic and saprophytic fungi such as *Fusarium, Myrothecium,* and *Stachybotrys*, which present serious health problems for animals and humans.[107] Trichothecenes are among the most potent protein synthesis inhibitors, acting at the ribosomal level of transcription and inhibiting peptidyl transferase in eukaryotic systems.[108] Among their most notable biological activities is their potent phytotoxicity.[109]

Plant defensive strategies against trichothecenes has been illustrated in studies of the pathology of *Fusarium graminearum* diseases of wheat and corn. The trichothecene deoxynivalenol (Fig. 7) is produced by this mold and is a significant virulence factor.[110] Genotypes of wheat and corn that are highly resistant to the disease are highly resistant to the membrane damaging effects of the trichothecene possibly because of a mutant protein antibiotic pump in the cell membrane.[111,112] In wheat, the peptidyl transferase has a mutation rendering it less sensitive to the trichothecene.[113] In both species, the toxin is degraded.[114,115] Corn also produces compounds that inhibit toxin production without inhibiting mycelial growth.[116]

The original hypothesis accounting for the presence of trichothecene fungal toxins in the *Baccharis* plants was that a fungus (likely *Myrothecium*) was associated with the plant.[117] Later studies,[118,119] however, strongly pointed to the *Baccharis* plants themselves as the biogenic source. Perhaps the most unusual biological properties are found in *B. coridifolia* where in the female plants pollination triggers a major increase in macrocyclic trichothecene (e.g. roridins

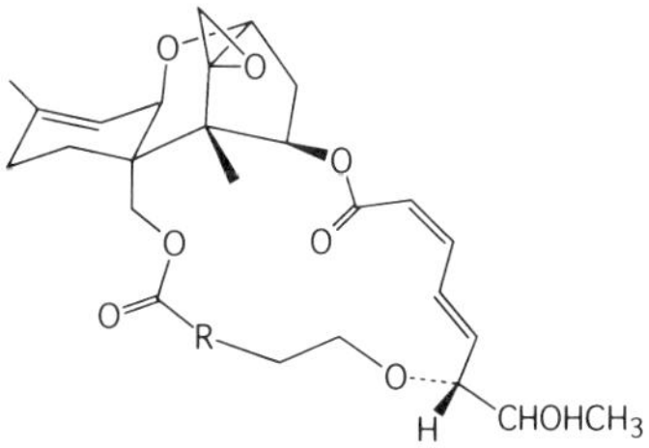

Figure 11. General structure of the macrocyclic trichothecenes, potent cytotoxins isolated from both fungal culture of *Myrothecium* and from extracts of Brazilian shrubs of the genus *Baccharis*.

Roridin A: R = $CHOHCHCH_3$
Roridin E: R = $CH=CCH_3$

A and E) production (Fig. 11). Cross pollination with other species of *Baccharis* does not elicit this response.[120] Furthermore, in both *B. coridifolia* and *B. megapotamica*, the macrocyclic trichothecenes were shown to function as plant hormones, in that they were required for seed germination.[120] The conclusion reached was that these two species of *Baccharis* have acquired the genes, via horizontal gene transfer,[121] from a *Myrothecium* fungus. Over twenty other *Baccharis* species have been assayed for trichothecenes, and none has been found to contain trichothecenes.[122] *Baccharis* species, as well as all other plants studied to date, with the exceptions of *B. coridifolia* and *B. megapotamica*, have been shown to be especially sensitive to roridin mycotoxins.[109,119,122]

The latter cases appear to involve horizontal gene transfer between plants and fungi,[121] for which there are no documented examples in the literature. Chapman and Ragan[123] suggested that plants may have acquired fungal genes involved in gibberellin biosynthesis via horizontal transfer from a fungus. Pirozynski[124] pointed out the correspondence between woodiness and the presence of endomycobionts in plants. In trees, mycotrophism is obligate but tends to be facultative to occasional, and often absent in herbaceous and aquatic plants. A key to this relationship may be that boron was required in the evolution of xylem and the concomitant lignification. Plants appear be able to assimilate boron only through the mediation of their fungal symbionts. Laboratory transfer of fungal natural product genes into plants has been described,[125] and natural horizontal gene transfers into plants from viruses and bacteria are well known.[126,127] When one discovers compounds present in one type of organism that normally are associated with a different type of organism, several different general mechanisms need to be considered: chemical transfers from one organism to another; biosynthetic generation through convergent evolution; and biosynthetic generation though horizontal gene transfer.[128] The latter would appear to be common in the microbial world[129,130] but is apparently less common in more complex multicellular eukaryotic organisms. Interest in this phenomenon as a source of evolutionary diversity in plants remains high.[131]

LIMITATION OF HERBIVORY: PLANTS AND INSECTS

Insect-plant relationships are strongly modulated by natural products.[132,133] An example is found in the allelochemical interactions of certain moths and butterflies with pyrrolizidine alkaloid-producing plants. Pyrrolizidine alkaloids (PAs) exhibit a wide range of toxicity in livestock and humans including being carcinogenic, teratogenic, and genotoxic. A number of insects that feed on PA-containing plants are able to tolerate these toxins and sequester them for their own defensive purposes.[134] Arctiid moths have utilized the plant-derived pyrrolizidine alkaloids and their derivatives as male sex pheromones, and fe-

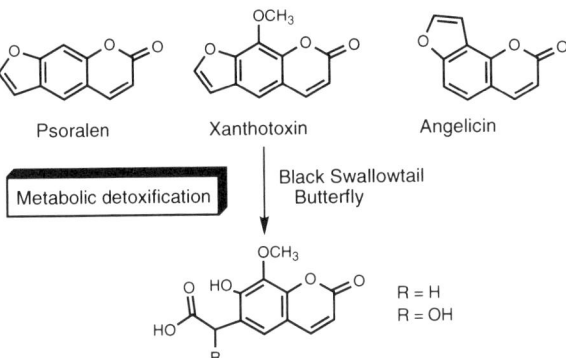

Figure 12. Furanocoumarins from the family Umbelliferae and their metabolism by the black swallowtail butterfly.

males use them as a measure of the fitness of males for mating.[135] Much can be made of the evolutionary significance of these findings.[132–135]

In 1964, Ehrlich and Raven[136] suggested a direct evolutionary relationship between insect herbivory and the chemical diversity found in the angiosperms. Herbivory by insects should exert pressures on plants to evolve increasingly potent chemical deterrents, and in turn, some insects will evolve toward being specialists which have mechanisms to handle plant toxins which normal insect generalists can not tolerate. Larvae of the black swallowtail butterfly, for example, have become specialists on plants of the family Umbelliferae, which produce high levels of the furanocoumarins. The butterflies have evolved a capacity to metabolically detoxify the toxins.[137,138] However, evolutionarily younger members of this plant family have evolved a structural variation in angelicin, an angular instead of the normal linear furanocoumarin (Fig. 12). Angelicin is significantly less tolerated by the larvae than the linear furanocoumarins such as xanthotoxin,[139] even though angular furanocoumarins are considerably less toxic than linear ones. These data point to a coevolutionary relationship, though one must be careful to analyze such systems in terms of interacting populations rather than families.[140]

The multipurpose nature of natural products also is illustrated by the coumarins and furanocoumarins (e.g. xanthotoxin).[137] In addition to being insect toxins, some function as phytoalexins (inducible plant-produced fungitoxic and fungistatic agents) that have been shown *in vitro* to inhibit the biosynthesis of fungal-produced phytotoxic trichothecenes.[141] Such inhibition offers some protection to the plant from pathogenic *Fusarium* which produce trichothecenes.[142] On the other hand, many plant-produced furanocoumarins are metabolically deactivated by fungal-produced enzymes which are elicited in response to these plant metabolites.[142] The coumarins act by binding to DNA and, in the presence

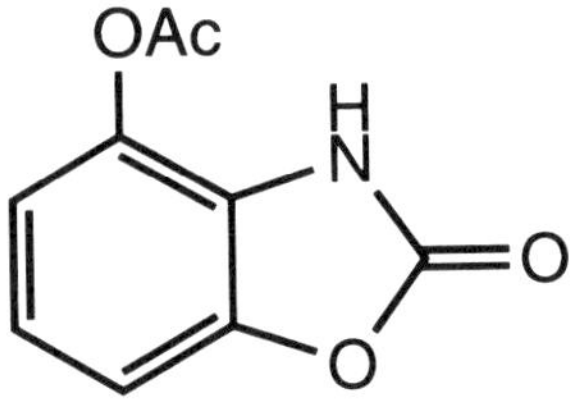

Figure 13. 4-Acetylbenzoxazolin-2-one (4-ABOA), an allelochemical found in corn.

of UV light, they undergo photocyclization reactions principally with the pyrimidine components. Thus, their toxicities are greatly enhanced in the presence of strong sunlight.[137] Other plant allelochemicals including 4-acetylbenzoxazolin-2-one (4-ABOA) (Fig. 13) also illustrate multipurpose roles.[116] This chemical (found in corn) is a potent inhibitor of deoxynivalenol synthesis in fungi and is active against some insects that attack corn (Miller, Dowd, Noroton, Fielder, Collins, unpublished data).

DISCUSSION

We are locked into the present. Although we may make educated guesses about past biochemical systems and their evolution to our present day systems,[16] there are severe restrictions on our ability to provide compelling analyses. Nonetheless, several general features about secondary metabolites emerge from the above data. The production of natural products has all the hallmarks of a system under developmental control. The genes required for biosynthesis are many and clustered. They are highly regulated to respond to exogenous signals some of which are specific but many of which are general. Many systems are highly plastic. At all levels, there is evidence of self-organizing driving forces operating: natural products exhibit binding to receptors; various cellular structures appear where these chemicals are produced, stored, and transported; the chemicals are critical components in larger systems (e.g. cell-cell, tissue-cell (tissue), organism-organism(s) interactions). The components appear to work together and likely have evolved together. Nature has linked various components through chemical communication, and the coin of the realm is often secondary metabolism.

Often the more biologically potent antibiotics are found toward the end of the biosynthesis trail.[143] Proceeding down the biosynthetic pathway (corresponding to increasingly more recent evolutionary events?), the metabolites become more biologically active.[82] This correlation with bioactivity is broad, both with respect to specific activities (e.g. mutagenicity, cytotoxicity, carcinogenicity, $LD_{50's}$, etc.) and with respect to the bioactivities observed in the animal models and cell lines.[84] That natural products have a tendency to be more biologically active as they appear later in the course of biosynthesis appears to be a general phenomenon and is consistent with evolutionary principles.

There are problems, however, with this hypothesis that become evident from genetic analysis. There is a fundamental difference between the arrangement of genes involved in primary metabolism and those involved in secondary metabolism. In primary metabolism, genes encoding enzymes of metabolic pathways are not usually organized in clusters but are typically scattered about the genome, often appearing on different chromosomes.[144,145] Genes involved in natural product biosynthetic pathways usually appear to be clustered.[146] In filamentous fungi, the latter appears to be the case for trichothecenes,[147,148] aflatoxins,[6,79] and β-lactam antibiotics.[146]

The β-lactam antibiotics are particularly interesting because they are produced both by bacteria (*Streptomyces*) and by eukaryotes (fungi). There is evidence to suggest, though this has not been established unequivocally, that the eukaryotes acquired the β-lactam genes via horizontal transfer from a prokaryote.[149] Interestingly, the penam coding genes and the cepham coding genes in *Streptomyces* occur together in the same cluster, but in the fungus, *Acremonium chrysogenum*, the genes are located on separate chromosomes.[149]

At least nine of the genes involved in the aflatoxin biosynthetic pathway also are clustered.[79] Trail et al.[89] have suggested that these genes are arranged sequentially in parallel with the enzyme reactions catalyzed by their gene products. Recent data, however (Keller, personal communication), show that there are at least 25 clustered genes in the pathway, but that they are not arranged strictly in order of the enzyme reactions in the pathway.

The origins of gene clustering presents a dilemma. If mutations are truly random, how can one account for gene clustering? How can genes self-organize into clusters? One can argue that the genes were acquired through horizontal transfer from another organism, but that still leaves unanswered the question as to how the cluster arose in the first place.

The diversity of genes in a cluster may arise by a combination of 1) gene duplication, followed by divergent mutation-selection processes leading to related (and possibly clustered) genes with different functions, and 2) recruiting of new (and unrelated) genes into the cluster.[150] Both of these processes, especially the latter, suggest that there are self-organizing processes operating to ensure the success of the phenomenon. This issue of gene organization-phenotypic outcome is important in the question of what drives evolution. In addition to the random mutation/natural selection paradigm of Darwinian evolution, strong arguments can be made to include the drive for self-organization (complexity) as well.[151]

Many of the natural products discussed in this paper are acetate-derived. The most common are formed *via* polyketide biosynthesis. A central gene in the polyketide biosynthesis pathway is polyketide synthase (PKS), which shows sequence similarity to conserved regions of the fatty acid synthase genes[152] and which is organized in a modular fashion allowing for an enormous variety of products.[153] However, it is peculiar that the PKS which is involved in the biosynthesis of 6-methylsalicylic acid in *Penicillium urticae* is evolutionarily more closely related to the fatty acid synthase genes of animals than it is to its own fungal fatty acid synthase gene.[66] This appears to reflect a common situation - natural product genes show weaker sequence homology with their own primary metabolite genes than with primary metabolite genes found in other organisms.[154] In other words, these genes appeared to have been imported from other organisms rather than evolving from primary metabolite genes of the organism itself.

That the *raison d'être* for the production of secondary metabolites by plants and animals is for defensive and competitive purposes is well established. There is also sufficient evidence to support this same thesis for the microbial world.[155] Although the mechanisms for evolving diversity of natural products in nature are built on Darwinian principles of mutation-selection, it appears that self-organizing properties may also play an important role. The sum of the data ranging from molecular structure-function relationships, gene structure and organization, to phenotypic and environmental properties of the producing organisms suggests that complexification increases at all levels.[156] Each trophic level, from molecules to man has its own inherent tendency to self-organize. The properties of natural products allow them to play integral parts in these self-organizing systems and thus make them indispensable to life on this planet.

REFERENCES

1. KAUFFMAN, S. 1995. At Home in the Universe: The Search for the Laws of Self-Organization and Complexity. Oxford University Press, New York, p. 321.
2. COVENEY, P., HIGHFIELD, R. 1995. Frontiers of Complexity: The Search for Order in a Chaotic World. Fawcett Columbine, New York, p. 462.
3. WILLIAMS, D. H., STONE, M. J., HAUCK, P. R., RAHMAN, S. K. 1989. Why are secondary metabolites (natural products) synthesized? J. Nat. Prod. 52: 1189–1208.
4. MAPLESTONE, R. A., STONE, M. J., WILLIAMS, D. H. 1992. The evolutionary role of secondary metabolites - a review. Gene 115: 151–157.
5. FISHER, R. F., LONG, S. P. 1992. *Rhizobium*-plant signal exchange. Nature 357: 655–660.
6. VAN RHIJN, P., VANDERLEYDEN, J. 1995. The *Rhizobium*-plant symbiosis. Microbial. Rev. 59: 124–142.
7. KOPROWSKI, H., MAEDA, H. (eds.) 1995. The Role of Nitric Oxide in Physiology and Pathophysiology, Springer-Verlag, Berlin, p. 90.
8. SKORY, C. D., CHANG, P.-K., CARY, J., LINZ, J. E. 1992. Isolation and characterization of a gene from *Aspergillus parasiticus* associated with the conversion of versicolorin A to sterigmatocystin in aflatoxin biosynthesis. Appl. Environ. Microbiol. 58: 3527–3537.
9. MARGULIS, L., FESTER R. (eds.) 1991. Symbiosis as a Source of Evolutionary Innovation 1991, The MIT Press, Cambridge, MA, p. 435.

10. BU'LOCK, J. D. 1985. Genetic aspects of mycotoxin formation. In: Regulation of Secondary Metabolite Function., (H. Kleinkauf, H. V. Döhren, H. Dornauer, G. Nesemann, eds.), VCH Publishers, Weinheim, Germany, pp. 1–12.
11. MARGULIS, L. (Ed.) 1993. Symbiosis in Cell Evolution, 2nd ed., 1993 (L. Margulis), W. H. Freeman, New York, p. 452.
12. PIEPERSBERG, W. 1992. Metabolism and cell individualization. In: Secondary Metabolites: Their Function and Evolution, (J. Davies, ed.), John Wiley & Sons, Chichester, England, pp. 294–299.
13. BENGSTON, S. (ed.) 1994. Early Life (S.Bengston, ed.), Columbia University Press, New York, p. 630.
14. MÜLLER, D., PITSCH, S., KITTAKA, A., WAGNER, E., WINTNER, C. E., ESCHENMOSER, A. 1990. Chemie von a-aminonitrilen. aldomerisierung von glycoaldehyd-phosphat zu racemischen hexose-2,4,6-triphosphaten und (in gegenwart von formaldehyd) racemischen pentose-2,4-diphosphaten: *rac*-allose-2,4,6-triphosphat und *rac*-ribose-2,4-diphosphat sind die reaktionshauptprodukte. Helv. Chem. Acta. 73: 1410–1468.
15. ESCHENMOSER, A., LOEWENTHAL, E. 1992. Chemistry of potentially prebiotic natural products. Chem. Soc. Rev. 92: 1–16.
16. DE DUVE, C. 1995. Vital Dust: Life as a Cosmic Imperative, Basic Books, New York, p. 362.
17. ALTMAN, S. 1990. Enzymatic cleavage of RNA by RNA (Nobel Lecture). Angew. Chem. Int. Ed. Engl. 29: 749–758.
18. CECH, T. R. 1990. Self-splicing and enzymatic activity of an intervening sequence RNA from *Tetrahymena* (Nobel Lecture). Angew. Chem. Int. Ed. Engl. 29: 759–768.
19. DAVIES, J. 1990. What are antibiotics? Archaic functions for modern activities. Mol. Microbiol. 4: 1227–1232.
20. CAVALIER-SMITH, T. 1992. Origins of secondary metabolism. In: Secondary Metabolites: Their Function and Evolution, (J. Davies, ed.), John Wiley & Sons, Chichester, England, pp. 64–87.
21. BERGMAN, R. G. 1973. Reactive 1,4-dehydroaromatics. Acc. Chem. Res. 6: 25–31.
22. LOCKHART, T. P., BERGMAN, R. G. 1981. Evidence for the reactive spin state of 1,4-dehydrobenzene. J. Am. Chem. Soc. 103: 4091–4096.
23. Doyle T.W., Kadow, J.F. (eds.) 1994. Recent progress in the chemistry of enediyne antibiotics. In: *Tetrahedron* Symposium-in-Print Number 53, pp. 1311–1538.
24. PERT, C. B., SNYDER, S. H. 1973. Opiate receptor: demonstration in nervous tissue. Science 179: 1011–1014.
25. HUGHES, J., SMITH, T. W., KOSTERLITZ, H. W., FOTHERGILL, L. A., MORGAN, B. A., MORRIS, H. R. 1975. Identification of two related pentapeptides from the brain with potent opiate agonist activity. Nature 258: 577–579.
26. DEVANE, W. A., DYSARZ, F. A., III, JOHNSON, M. R., MELVIN, L. S., HOWLETT, A. C. 1988. Determination and characterization of a cannabinoid receptor in rat brain. Mol. Pharmacol. 34: 605–613.
27. DEVANE, W. A., HANUS, L., BREUER, A., PERTWEE, R. G., STEVENSON, L. A., GRIFFIN, G., GIBSON, D., MANDELBAUM, A., ETINGER, A., MECHOULAM, R. 1992. Isolation and structure of a brain constituent that binds to the cannabinoid receptor. Science 258: 1946–1949.
28. CHANG, M. C., BERKERY, D., SCHUEL, R., LAYCHOCK, S. G., ZIMMERMAN, A. M., ZIMMERMAN, S., SCHUEL, H. 1993. The sea urchin *Strongylocentrotus purpuratus* has cannabinoid receptor remarkably similar to that in mammals. Mol. Reprod. Dev. 36: 507–516.
29. SCHREIBER, S. L. 1992. Using the principles of organic chemistry to explore biology. Chemical & Engineering News, October 26, 1992, pp. 22–32.
30. SPENCER, D. M., WANDLESS, T. J., SCHREIBER, S. L., CRABTREE, G. R. 1993. Controlling signal transduction with synthetic ligands. Science 262: 1019–1024.

31. ROTH, J., LEROITH, D., COLLIER, E. S., WATKINSON, A., LESNIAK, M. A. 1986. The evolutionary origins of of intercellular communication and the Maginot lines of the mind. Ann. NY Acad. Sci. 463: 1–11.
32. TRAVIS, J. 1993. Tracing the immune system's evolutionary history. Science 261: 164–165.
33. SCHREIBER, S. L. 1991. Chemistry and biology of the immunophilins and their immunosuppressive ligands. Science 251: 283–287.
34. SIGAL, N. H., DUMONT, S. J. 1992. Cyclosporin A, FK-506, and Rapamycin: pharmacologic probes of lymphocyte signal transduction. Ann. Rev. Immunol. 10: 519–560.
35. WALLDROP, M. M. 1992. Complexity: the Emerging Science at the Edge of Order and Chaos, Simon & Shuster, New York, NY, p. 380.
36. LEWIN, R. 1992. Complexity: Life at the Edge of Chaos, Macmillen, New York, NY, p. 208.
37. BELL, E.A. 1980. The possible significance of secondary compounds in plants. In: Encyclopedia of Plant Physiology, Vol. 8, (E. A. Bell, B. V. Charlwood, eds.), Springer-Verlag, New York, NY, pp. 11–21.
38. TURLINGS, T. C. J., LOUGHRIN, J. H., MCCALL, P. J., ROSE, U. S. R., LEWIS, W. J., TUMLINSON, J. H. 1995. How caterpillar-damaged plants protect themselves by attracting parasitic wasps. Proc. Natl. Acad. Sci. 92: 4169–4174.
39. SEN, A., BERGVINSON, D., MILLER, S. S., ATKINSON, J., FULCHER, R.G., ARNASON, J. T. 1994. Distribution and microchemical detection of phenolic acids, flavanoids, and phenolic acid amides in maize kernels. J. Agric. Food Chem. 42: 1879–1883.
40. CONE, K. C., COCCIOLONE, S. M., BURR, F. A., BURR, B. 1993. Maize anthocyanin regulatory gene pl is a duplicate of c1 that functions in the plant. The Plant Cell 5: 1795–1805.
41. DOONER, H. K., ROBBINS, T. P., JORGENSEN, R. A. 1991. Genetic and developmental control of anthyocyanin biosynthesis. Ann. Rev. Genet. 25: 173–199.
42. LUCKNER, M. 1980. Expression and control of secondary metabolism. In: Encyclopedia of Plant Physiology, Vol. 8, (E. A. Bell and B. V. Charlwood, eds.), Springer-Verlag, New York, NY, pp. 23–63.
43. LYONS, P. C., HIPSKIND, J. D., WOOD, K. V., NICHOLSON, R. L. 1988. Separation and quantification of cyclic hydroxamic acids and related compounds by high-pressure liquid chromatography. J. Agric. Food Chem. 36: 57–60.
44. KOBAYASHI, J., ISHIBASHI, M. 1993. Bioactive metabolites of symbiotic marine microorganisms. Chem. Rev. 93: 1753–1769.
45. HU, J., MARR, J., DEFREITAS, A. S. W.,QUILLIAM, M. A., WALTER, J. A., WRIGHT, J. L. C. 1992. New diol esters isolated from cultures of the dinoflagellates *Prorocentrum lima* and *Prorocentrum lconcavum*. J. Nat. Prod. 55: 1631–1637.
46. HU, T., CURTIS, J. M., WALTER, J. A., WRIGHT, J. L. C. 1995. Identification of DTX-4, a new water-soluble phosphatase inhibitor from the toxic dinoflagellate *Prorocentrum lima*. J. Chem. Soc., Chem. Commun. 597–599.
47. MILLER, J. D. 1995. Fungi and mycotoxins in grain: implications for stored product research. J. Stored Prod. Res. 31: 1–16.
48. TURNER, W.B., ALDRIDGE, D.C. 1983. Fungal Metabolites. Academic Press, New York, NY, p. 631.
49. ZAHNER., H, ANKE, H., ANKE, T. 1983. Evolution and secondary pathways. In: Secondary Metabolism and Differentiation in Fungi, Vol.5 (J. W. Bennett and A. Ciegler, eds.), Marcel Dekker, Inc, New York, NY, pp. 153–171.
50. LOPEZ-FRANCO, R., BARTNICKI-GARCIA, S., BRACKER, C. E. 1994. Pulsed growth of fungal hyphal tips. Proc. Natl. Acad. Sci. 91: 12228–12232.
51. HALE, M. D., EATON, R. A. 1985. Growth of soft rot fungi in wood. Trans. Brit. Mycol. Soc. 84: 277–281.

52. MILLER, J. D., GREENHALGH, R. 1988. Metabolites of fungal pathogens and plant resistance. In: Biotechnology for Crop Protection, Vol. 379 (P. A. Hedin, J. J. Menn, and R. M. Hollingworth, eds.), American Chemical Society, Washington, DC, pp. 117–129.
53. BU'LOCK, J.D. 1975. Secondary metabolism in fungi and its relationship to growth and development. In: The Filamentous Fungi, Industrial Mycology, Vol. 1 (J. E. Smith and D. A. Berry, eds.), Academic Press, Inc., New York, NY, pp. 33–58.
54. CANDAU, R., AVALOS, J., CERDA-OLMEDO, E. 1992. Regulation of gibberellin biosynthase in *Gibberella fujikuroi*. Plant Physiol. 100: 1184–1188.
55. BORROW, A., JEFFREYS, E. G., KESSEL, R., LLOYD, E. C., LLOYD, P. B., NIXON, I. S. 1961. The metabolism of *Gibberella fujikuroi* in stirred culture. Can. J. Microbiol. 7: 227–276.
56. BORROW, A., BROWN, S., JEFFREYS, E. G., KESSEL, R., LLOYD, E. C., LLOYD, P. B., ROTHWELL, A., ROTHWELL, B., SWAIT, J. C. 1964. The kinetics of metabolism of *Gibberella fujikuroi* in stirred culture. Can. J. Microbiol. 10: 407–444.
57. BUCHANAN, R. L., BENNETT, J. W. 1988. Nitrogen regulation of polyketide mycotoxin production. In: Nitrogen Source Control of Microbial Processes, (S. Sanchez-Esquival, ed.), CRC Press, Boca Raton, FL, pp. 137–149.
58. HIDY, P. H., BALDWIN, R. S., GREASHAM, R. L., KIETH, C. L., MCMULLEN, J. R. 1977. Zearalenone and derivatives: production and biological activities. Adv. Appl. Microbiol. 22: 54–82.
59. MILLER, J. D., BLACKWELL, B. A. 1986. Biosynthesis of 3-acetyldeoxynivalenol and other metabolites by *Fusarium culmorum* HLX 1503 in a stirred jar fermenter. Can. J. Bot. 641: 1–5.
60. MILLER, J. D., SAVARD, M. E., RAPIOR, S. 1994. Production and purification of fumonisins from a stirred jar fermenter. Natural Toxins, 2: 354–359.
61. ROLLINS, M. J., GAUCHER, G. M. 1994. Ammonium repression of antibiotic and intracellular proteinase production in *Penicillium urticae*. Appl. Microbiol. Biotechnol. 41: 447–455.
62. BRUCKNER, B., BLECHSCHMIDT, D. 1991. Nitrogen regulation of gibberelin biosynthesis in *Gibberella fujikuroi*. Appl. Microbiol. Biotechnol. 35: 646–650.
63. ESPESO, E. A., PENALVA, M. A. 1992. Carbon catabolite repression can account for the temporal pattern of expression of a penicillin biosynthetic gene *Aspergillus nidulans*. Molecular Microbiol. 6: 1457–1465.
64. MILLER, J. D., GREENHALGH, R. 1985. Nutrient effects on the biosynthesis of trichothecenes and other metabolites by *Fusarium graminearum*. Mycologia 771: 130–136.
65. GENDLOFF, E. H., CHU, F. S., LEONARD, T. J. 1992. Variation in regulation of aflatoxin biosynthesis among isolates of *Aspergillus flavus*. Experientia 48: 84–87.
66. WANG, I.-K., REEVES, C., GAUCHER, G. M. 1991. Isolation and sequencing of a genomic DNA clone containing the 3' terminus of the 6-methylsalicyclic acid polyketide synthase gene of *Penicillium urticae*. Can. J. Microbiol. 37: 86–95.
67. LENDENFELD, T., GHALI, D., WOLSCHEK, M., KUBICEK-PRANZ, E., KUBICEK, C. P. 1993. Subcellular compartmentation of penicillin biosynthesis in *Penicillium chrysogenum*: The amino acid precursors are derived from the vacuole. J. Biol. Chem. 2681: 665–671.
68. BEREMAND, M.N., MCCORMICK, S. P. 1992. Biosynthesis and regulation of trichothecene production by *Fusarium* species. In: Handbook of Applied Mycology, Vol. 5, (D. Bhatnagar, E. B.Lillehoj, and D. K. Arora, eds.), Marcel Dekker, Inc., New York, NY, pp. 359–384.
69. COULOMBE, R. A., JR. 1991. Aflatoxins. In: Mycotoxins and Phytoalexins, (R. P. Sharma and D. K. Salunkhe, eds.), CRC Press, Boca Raton, FL, pp. 103–143.
70. Some naturally occurring substances: food items and constituents, heterocyclic aromatic amines and mycotoxins. 1993. Vol. 56, *IARC*, Lyon, France. p. 599.
71. MISRA, R. P., MUENCH, K. F., HUMAYUN, M. Z. 1983. Covalent and noncovalent interactions of aflatoxins with defined deoxyribonucleic acid sequences. Biochemistry 22: 3351–3359.
72. EATON, D. L., GALLAGHER, E. P. 1994. Mechanisms of aflatoxin carcinogenesis. Ann. Rev. Pharmacol. Toxicol. 34: 135–172.

73. TASHIRO, F., MORIMURA, S., HAYASHI, K., MAKINO, R., KAWAMURA, H., HORIK-OSHI, N., NEMOTO, K., OHTSUBO, K., SUGIMURA, T., UENO, Y. 1986. Expression of the c-Ra-ras and c-myc genes in aflatoxin B_1-induced heptocellular carcinomas. Biochem. Biophys. Res. Commun. 138: 858–864.
74. LILLEHOJ, E. B. 1980. Secondary metabolites as chemical signals between species in ecological niches. In: Proc. 6th Int. Fermentation Symp. Advances in Biotechnology, Vol. 3, (M. Moo-Young, C. Vezina, and K. Singh, eds.), Pergamon Press, Elmsford, NY, pp. 397–423.
75. LILLEHOJ, E. B. 1991. Aflatoxin: an ecologically elicited genetic activation signal. In: Mycotoxins and Animal Foods, (J. E. Smith and R. S. Henderson, eds.), CRC Press, Boca Raton, FL, pp. 1–35.
76. PITT, J. L., HOCKING, A. D., BHUDHASAMAI, K., MISCAMBE, B. F., WHEELER, K. A., TANBOON-EK, P. 1993. The normal mycoflora of commoditied from Thailand. 1. Nuts and oilseeds. Int. J. Food Microbiol. 20: 211–226.
77. KURTZMAN, C. P., HORN, B. W., HESSELTINE, C. W. 1987. *Aspergillus nominus*, a new aflatoxin-producing species related to *Aspergillus flavus* and *Aspergillus tamarii*. Antoine van Leeuwenhoek 53: 147–158.
78. KLICH, M. A., PITT, J. I. 1988. A Laboratory Guide to Common *Aspergillus* species and their telomorphs. Commonwealth Scientific and Industrial Research Organisation, Australia. p. 116.
79. YU, J., CHANG, P.-K., CARY, J. W., WRIGHT, M. BHATNAGAR, D., CLEVELAND, T. E., PAYNE, G. A., LINZ, J. E. 1995. Comparative mapping of aflatoxin pathway gene clusters in *Aspergillus parasiticus* and *Aspergillus flavus*. Appl. Environ. Microbiol. 61: 365–2371.
80. WICKLOW, D. T. 1994. The mycology of stored grain: an ecological perspective. In: Stored Grain Ecosystems, (D. S. Jayas, N. D. G. White, and W. E. Muir, eds.), Marcel Dekker, New York, NY, pp. 197–249.
81. DOWD, P. F. 1992. Insect interactions with mycotoxin-producing fungi and their hosts. In: Handbook of Applied Mycology, Vol. 5, (D. Bhatnagar, E. B. Lillehoj, and D. K. Arora, eds.), Marcel Dekker, Inc. pp. 137–155.
82. JARVIS, J. L., GUTHRIE, W. D., LILLEHOJ, E. B. 1984. Aflatoxin and selected biosynthetic precursors: effects on the European corn borer in the laboratory. J. Agric. Entomol. 1: 354–359.
83. WICKLOW, D. T., SHOTWELL, O. L. 1982. Intrafungal distribution of aflatoxins among conidia and sclerotia of *Aspergillus flavus* and *Aspergillus parasiticus*. Can. J. Microbiol. 29: 1–5.
84. WICKLOW, D. T. 1988. Metabolites in the coevolution of fungal chemical defence systems. In: Coevolution of Fungi with Plants and Animals, (K. A. Pirizynski and D. Hawksworth, eds.), Academic Press, London. pp. 173–201.
85. BRONSON, C. R. 1991. The genetics of phytotoxin production by plant pathogenic fungi. Experientia 47: 771–776.
86. MCLEAN, M. 1994. The phytotoxic effects of aflatoxin B1: a review (1984–1994). South African J. Sci. 90: 385–390.
87. BOUTIBONNES, P. 1980. Antibacterial activity of some mycotoxins. IRCS Pharmacology 8: 850–851.
88. KELLER, N. P., BROWN, D., BUTCHKO, R. A. E., FERNANDES, M., KELKAR, H., NESBITT, C., SEGNER, S., BHATNAGAR, D., CLEVELAND, T. E., ADAMS, T, H. 1995. A conserved polyketide mycotoxin gene cluster in *Aspergillus nidulans*. In: Molecular Approaches to Food Safety Issues Involving Toxic Microorganisms, (J. L .Richard, ed.), Alaken Inc., Fort Collins, CO, pp. 263–277.
89. TRAIL, F., MAHANTI, N., LINZ, J. 1995. Molecular biology of aflatoxin biosynthesis. Microbiol. 141: 755–765.
90. POWELL, R. G., PETROSKI, R. J. 1992. Alkaloid toxins in endophyte-infected grasses. Nat. Toxins 1: 163–170.

91. GALLAGHER, R. T., HAWKES, A. D., STEYN, P. S., VLEGGAAR, R. 1984. Tremorgenic neuraltoxins from perennial ryegrass causing ryegrass staggers disorder of livestock: structure elucidation of lolitrem B. J. Chem. Soc. Chem. Commun. 614–616.
92. BACON, C. W., DE BATTISTA, J. 1991. Endophytic fungi of grasses. In: Handbook of Applied Mycology, Vol. I, (D. K. Arora, B. Rai, K. G. Mukerji, and G. R. Knudsen, eds.), Marcel Dekker, Inc., New York, NY, pp. 231–256.
93. CLAY, K. 1990. Fungal endophytes of grasses. Ann. Rev. Ecol. Syst., 21: 275–295.
94. HAMMON, K. E., FAETH, S. H. 1992. Ecology of plant-herbivore communities: a fungal component? Nat. Toxins 1: 197–208.
95. BENTLEY, M. D., LEONARD, D. E., LEACH, S., REYNOLDS, E., STODDARD, W., TOMPKINSON, B., TOMPKINSON, D., STRUNZ, G. W., YATAGAI, M. 1982. Effects of some naturally occurring chemicals and some extracts of non-host plants on feeding of spruce budworm (*Choristoneuria fumiferana*). Life Sciences and Agricultural Experimental Station, University of Maine Technical Bulletin 107. p.114.
96. CALHOUN, L. A., FINDLAY, J. A., MILLER, J. D., WHITNEY, N. J. 1992. Metabolites toxic to spruce budworm from balsam fir needle endophytes. Mycol. Res. 96: 281–286.
97. FINDLAY, J. A., LI, G., PENNER, P. E. 1995. Novel diterpenoid insect toxins from a conifer endophyte. J. Nat. Prod. 58: 197–200.
98. TODD, D. 1988. The effects of host genotype, growth rate and needle age on the distribution of a mutualistic, endophytic fungus in Douglas fir plantations. Can. J. Forestry Res. 18: 601–605.
99. PORTER, D. 1986. Mycoses of marine organisms: and overview of pathogenic fungi. In: The Biology of Marine Fungi, (S.T. Moss, ed.), Cambridge University Press, New York, NY, pp. 141–153.
100. YEH, Y. A., OIAH, E., WENDEL, J. J., SLEDGE, G. W., JR., WEBER, G. 1994. Synergistic action of taxol with tiazofurin and methotrexate in human breast cancer cells: schedule dependence. Life Sci. 54: 431–435.
101. ROWINSKY, E. K., CAZENAVE, L. A., DONEHOWER, R. C. 1990. Taxol: a novel investigational antimicrotubule agent. J. Natl. Cancer Inst. 82: 1247–1263.
102. WHEELER, N. C., JECH, K., MASTERS, S., BROBST, S. W., ALVARADO, A. B., HOOVER, A. J., SNADER, K. M. 1992. Effects of genetic, epigenetic, and environmental factors on taxol content in *Taxus brevifolia* and related species. J. Nat. Prod. 55: 432–440.
103. HAN, K. H., FLEMING, K., LOPER, M., CHILTON, W. S., MOCEK, U., GORDON, M. P., FLOSS, H. G. 1994. Genetic-transformation of mature Taxus - an approach to genetically control the in-vitro production of an anticancer drug, taxol. Plant Sci. 95: 187–196.
104. STIERLE, A., STROBEL, G., STIERLE, D. 1993. Taxol and taxane production by *Taxomyces andreanae*, an endophytic fungus of Pacific yew. Science 260: 214–216.
105. MANDER, L. N. 1992. The chemistry of gibberellins: an overview. Chem. Rev. 92: 573–612.
106. LANG, A. 1970. Gibberellins: structure and metabolism. Ann. Rev. Plant Physiol. 21: 37–570.
107. BAMBURG, J. R. 1983. Biological and biochemical actions of the trichothecenes mycotoxins. Prog. Mol. Subcell. Biol. 8: 41–110.
108. FEINBERG, B., MCLAUGHLIN, C. S. 1989. Biochemical mechanism of action of trichothecene mycotoxins. In: Trichothecene Mycotoxicosis: Pathophysiologic Effects, Vol. 1 (V. R. Beasley, ed.), CRC Press, Boca Raton, FL, pp. 27–35.
109. CUTLER, H. G., JARVIS, B. B. 1985. Preliminary observations on the effects of macrocyclic trichothecenes on plant growth. Exper. Environ. Bot. 25: 115–128.
110. PROCTOR, R. H., HOHN, T. M., MCCORMICK, S. P. 1995. Reduced virulence of *Gibberalla zeae* caused by disruption of a trichothecene toxin biosynthetic gene. Mol Plant-Microbe Interact. 8: 593–601.
111. COSSETTE, F., MILLER, J. D. 1995. Phytotoxic effect of deoxynivalenol and gibberella ear rot resistance of corn. Nat. Toxins 3: 383–388.

112. SNIJDERS, C. H. A., KRECHTING, C. F. 1992. Inhibition of translocation and fungal colonization in fusarium head blight resistant wheat. Can. J. Bot. 70: 1570–1576.
113. MILLER, J. D. 1989. Effects of Fusarium graminearum metabolites on wheat cells. In : Phytotoxins and Plant Pathogenesis, (A. Graniti, R. Durbin, and A. Ballio, eds.), NATO ASI Series Vol. H27, Springer-Verlag, Berlin, pp. 449–452.
114. MILLER, J. D., ARNISON, P. G. 1986. Degradation of deoxynivalenol by suspension cultures of the fusarium head blight resistant wheat cultivar Frontana. Can. J. Plant Pathol. 8: 147–150.
115. SEWALD, N., LEPSCHY VON GLEISSENTHALL, J., SCHUSTER, M., MULLER, G., APLIN, R.T. 1992. Structure elucidation of a plant metabolite of 4-desoxynivalenol. Tetrahedron: Asymmetry 3: 953–960.
116. FIELDER, D. A., COLLINS, F. W., BLACKWELL, B. A., BENSIMIN, C., APSIMON, J. W. 1994. Isolation and characterization of 4-acetyl-benzoxazolin-2-one (4-ABOA), a new benzoxazolinone from *Zea mays*. Tetrahedron Letters 35: 521–524.
117. JARVIS, B. B., MIDIWO, J. O., TUTHILL, D., BEAN, G. A. 1981. Interaction between the antibiotic trichothecenes and the higher plant *Baccharis megapotamica*. Science 214: 460–462.
118. JARVIS, B. B., MIDIWO, J. O., BEAN, G. A., ABOUL-NASR, M. B., BARROS, C. S. 1988. The mystery of trichothecene antibiotics in *Baccharis* species. J. Nat. Prod. 51: 736–744.
119. JARVIS, B. B., MOKHTARI-REJALI, N., SCHENKEL, E. P., BARROS, C. S., MATZENBACHER, N. I. 1991. Trichothecene mycotoxins from Brazilian *Baccharis* species. Phytochemistry 30: 789–797.
120. KUTI, J. O., JARVIS, B. B., MOKHTARI, N., BEAN, G. 1990. Allelochemical regulation of reproduction and seed germination of two Brazilian *Baccharis* species by phytotoxic trichothecenes. J. Chem. Ecol. 16: 3441–3453.
121. LAMBOY, W. F. 1984. Evolution of flowering plants by fungus-to-host horizontal gene transfer. Evol. Theory 7: 45–54.
122. JARVIS, B. B. 1991. Macrocyclic trichothecenes. In: Mycotoxins and Phytoalexins, (R. P. Sharma and D. K. Salunkhe, eds.), CRC Press, Boca Raton, FL, pp. 361–421.
123. CHAPMAN, D. J., RAGAN, M. A. 1980. Evolution of biochemical pathways: evidence from comparative biochemistry. Ann. Rev. Plant Physiol. 31: 639–678.
124. PIROZYNSKI, K.A., 1981. Interactions between fungi and plants through the ages. Can. J. Bot. 59: 1824–1827.
125. HOHN, T. M., OHLROGGE, J. B. 1991. Expression of a fungal sesquiterpene cyclase in transgenic tobacco. Plant Physiol. 97: 460–462.
126. ECKES, P., DONN, G., WENGENMAYER, F. 1987. Genetic engineering with plants. Angew. Chem. Int. Ed. Engl. 26: 382–402.
127. SCHELL, J. S. 1987. Transgenic plants as tools to study the molecular organization of plant genes. Science 237: 1176–1183.
128. PIETRA, F. 1995. Structurally similar natural products in phylogenetically distant marine organisms, and a comparison with terrestrial species. Chem Soc. Rev. 24: 65–71.
129. SMITH, M. W., FENG, D.-F., DOOLITTLE, R. F. 1992. Evolution by acquisition: the case for horizontal gene transfers. Trends Biochem. Sci. 17: 489–493.
130. AMÁBILE-CUEVAS, C. F., CHICUREL, M. E. 1993. Horizontal gene transfer. Am. Scientist 81: 332–341.
131. HERON, C. 1992. The networks of botanical creation. New Scientist 133: 40–44.
132. ROSENTHAL, G.A., BERENBAUM, M.R. (eds.) Herbivores Their Interactions with Secondary Plant Metabolites, 1991, Second Edition, Vols. 1 and 2, Academic Press, New York, NY. p. 468, p. 493.
133. NAHRSTEDT, A. 1989. The significance of secondary metabolites for interactions between plants and insects. Planta Medica 55: 333–338.
134. BOPPRÉ, M. 1990. Lepidoptera and pyrrolizidine alkaloids - exemplification of complexity in chemical ecology. J. Chem. Ecol. 16: 165–185.

135. EISNER, T., MEINWALD, J. 1987. Alkaloid-derived pheromones and sexual selection in Lepidoptera. In: Pheromone Biochemistry, (G. D. Prestwich and G. I. Bloomquist, eds.), Academic Press, Orlando, FL, pp. 251–269.
136. EHRLICH, P. R., RAVEN, P. H. 1964. Butterflies and plants: a study in coevolution. Evolution 18: 586–608.
137. BERENBAUM, M. R. 1991. Coumarins. In: Herbivores Their Interactions with Secondary Plant Metabolites, (G. A. Rosenthal and M. R. Berenbaum, eds.), Second Edition, Vol. 1, Academic Press, New York, NY, pp. 221–249.
138. IVIE, G. W., BULL, D. L., BEIER, R. C., PRYOR, N. W., OERTLI, E. H. 1983. Metabolic detoxification: mechanism of insect resistance to plant psoralens. Science 221: 374–376.
139. BERENBAUM, M. 1983. Coumarins and caterpillars - a case for coevolution. Evolution 37: 163–179.
140. JANZEN, D. H. 1980. What is coevolution? Evolution, 34: 611–612.
141. DESJARDINS, A. E., PLATTNER, R. D., SPENCER, G. F. 1988. Inhibition of trichothecene toxin biosynthesis by naturally occurring shikimate aromatics. Phytochemistry 27: 767–771.
142. DESJARDINS, A. E., SPENCER, G. F., PLATTNER, R. D. 1989. Tolerance and metabolism of furanocoumarins by the phytopathogenic fungus *Gibberella pulicaris* (*Fusarium sambucinum*). Phytochemistry 28: 2963–2969.
143. BU'LOCK, J. D. 1980. Mycotoxins as secondary metabolites. In: The Biosynthesis of Mycotoxins: A Study in Secondary Metabolism, (P. S. Steyn, ed.), Academic Press, New York, NY, pp. 1–16.
144. BRAYTON, K. P., CHEN, Z., ZHOU, G., NAGY, P. L., GAVALAS, A., TERENT, J. M., DEAVAEN, L. L., DIXON, J. E., ZALKIN, H. 1994. Two genes for *de novo* purine nucleotide synthesis on human chromosome 4 are closely linked and divergently transcribed. J. Biol. Chem. 269: 5313–5321.
145. ZORIO, D. A. R., CHENG, N. N., BLUMENTHAL, T., SPIETH, J. 1994. Operons as a common form of chromosomal organization in *C. elegans*. Nature 372: 270–272.
146. MARTIN, J. F. 1992. Clusters of genes for the biosynthesis of antibiotics: regulatory genes and overproduction of pharmaceuticals. J. Ind. Microbiol. 9: 73–90.
147. HOHN, T. M., MCCORMICK, S. P., DESJARDINS, A. E. 1993. Evidence for a gene cluster involving trichothecene-pathway biosynthetic genes in *Fusarium sporotrichioides*. Curr. Genet. 24: 291–295.
148. DESJARDINS, A. E. HOHN, T. M., MCCORMICK, S. P. 1993. Trichothecene biosynthesis in *Fusarium* species: chemistry, genetics, and significance. Microbiol. Rev. 57: 595–604.
149. TURNER, G. 1992. Genes for the biosynthesis of β-lactam compounds in microorganisms. In: Secondary Metabolites: Their Function and Evolution, (Davies, J., ed.), John Wiley & Sons, Chichester, England, pp. 113–128.
150. STONE, M. J., WILLIAMS, D. H. 1992. On the evolution of functional secondary metabolites (natural products). Mol. Microbiol. 6: 29–34.
151. WESSON, R. 1991. Beyond Natural Selection, The MIT Press, Cambridge, MA, p. 353.
152. CANE, D. E. 1994. Polyketide biosynthesis: molecular recognition or genetic programming? Science 263: 338–340.
153. CORTES, J., WIESMANN, K. E. H., ROBERTS, G. A., BROWN, M. J. B., STAUNTON, J., LEADLAY, P. F. 1995. Repositioning of a domain in a modular polyketide synthase to promote specific chain cleavage. Science 268: 1487–1489.
154. VINING, L. C. 1992. Secondary metabolism, inventive evolution and biochemical diversity - a review. Gene 115: 135–140.
155. DEMAIN, A. I. 1995. Why do microorganisms produce antimicrobials? In: Fifty Years of Antimicrobials: Past Perspectives and Future Trends, (P. A. Hunter, G. K. Darby, and N. J. Russell, eds.), Cambridge University Press, New York, pp. 205–239.
156. KAUFFMAN, S. A. 1993. The Origins of Order, Oxford University Press, New York, NY, p. 709.

Chapter Twelve

AN EXPLANATION OF SECONDARY PRODUCT "REDUNDANCY"

Richard D. Firn[1] and Clive G. Jones[2]

[1] Department of Biology
University of York
York YO1 5DD, United Kingdom
[2] Institute of Ecosystem Studies
Millbrook, New York

Natural Products in Microbes and Plants 295
The Screening Hypothesis of Natural Products Diversity 298
Biological Activity—A Variable Dependent on Dose and Assay System ... 299
    "The Dose Maketh the Poison" 299
    What Kind of Biological Activity? 300
Are Potent Biologically Active Compounds Easily Found in Extracts of
    Microbes or Plants? 301
Are Biologically Active Compounds Easily Found in Collections of
    Synthetic Chemicals? 302
Why is Potent Biological Activity a Rare Property for a Chemical to
    Possess? 303
Biosynthetic Routes to Generate Chemical Diversity 304
The Constraints of Exploiting the Biological Activity of Irreversible
    Inhibitors. 305
How Could Secondary Product Pathways Have Evolved if Biological
    Activity Were a Rare Property? 306
    Some Simple Scenarios 307
Summary 309

NATURAL PRODUCTS IN MICROBES AND PLANTS

Natural products are elaborated by many classes of organisms but they are made in especially large numbers by some types of bacteria, fungi and plants. It

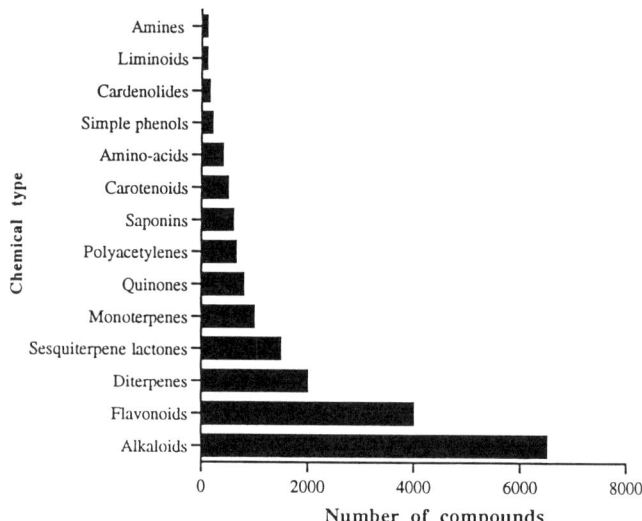

Figure 1. The main classes of natural products found in higher plants. (Derived from Harborne[1]).

is on these secondary chemicals that this paper will focus. In excess of 18,000 natural products have been characterised,[1] and estimates of the total number in plants alone exceeds 500,000.[2] The majority of natural products described in plants are made by a relatively small number of key pathways (Fig. 1), and the structural diversity is therefore largely a consequence of diversity of product within these few broad groups.

The actual number of substances made by any individual species is unknown because no complete analysis for a single species has been reported. Most analyses have been very selective, either in terms of the class of substance or by the type of biological activity being sought. However, attempts to quantify a minor component of a plant extract necessitates a considerable purification before sensitive methods of analysis can be used. Anyone needing to be reminded of the chemical diversity of a plant extract need only study the capillary GC traces of extracts of plant oils and note the numerous small peaks that crowd the baseline of the chromatogram (Fig. 2).

Although this book focuses primarily on phytochemicals, we will address natural products from both plants and microbes. The reason for the inclusion of a broader range of organisms is that natural products most likely evolved in organisms billions of years before land plants and the selective pressures that directed their evolution must have operated for a much longer time in microbes than in plants (Fig. 3).

Surprisingly, there seems to have been little cross fertilization of ideas between those interested in natural products in microbes and those interested in

AN EXPLANATION OF SECONDARY PRODUCT "REDUNDANCY"

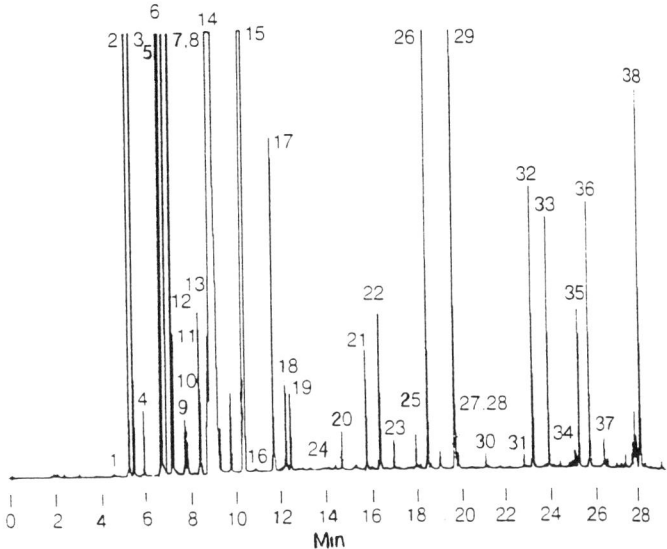

Figure 2. The complexity of extracts of natural products, as illustrated by a GC of the substances in an extract of Lemon Oil (0.5g lemon oil on a 100m x 0.25mm SPB-1 capillary column. By permission of Supelco).

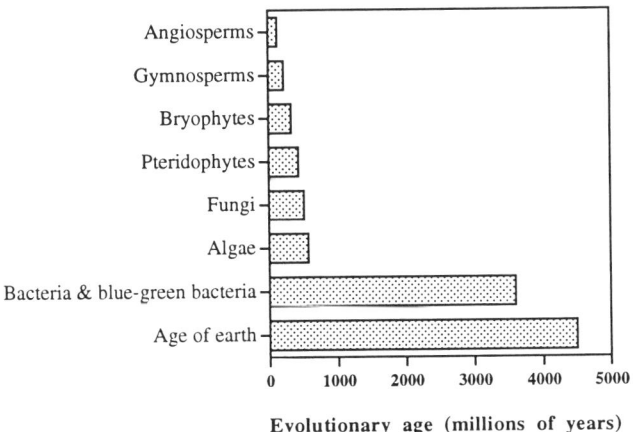

Figure 3. The approximate age of plant and microbial taxa and of the earth. (Derived from information from Raven et al[3]).

such compounds in plants. Most microbiologists have focused their discussions on the pharmaceutical exploitation of natural products as antibiotics, or as other therapeutic agents, with only passing interest, but much dispute, over the role of these compounds in the organisms from which they were extracted.[3] In contrast, the great majority of those interested in secondary plant products have been particularly interested in their ecological and evolutionary roles, even though the pharmaceutical value of these substances is well recognized. Chemists have aided both groups but usually have been more interested in the chemical structures than in the roles these compounds play.

Given the different emphases of the research on microbial and plant chemicals, it is not surprising that the scientists in two groups rarely discuss topics of mutual interest. However, it is still remarkable that people working on microbial natural products cannot agree on what role such compounds play, or indeed whether they have any role in micro-organisms,[4] while most people working on plant natural products hold the view that each and every chemical has, or has had, an important role in evolution of the organism.[5,6]

THE SCREENING HYPOTHESIS OF NATURAL PRODUCTS DIVERSITY

The Screening Hypothesis is an attempt to provide a model for the evolution of secondary metabolism, applicable to both microbes and plants, that may explain the observed chemical diversity.[7] The hypothesis is based on some fundamental constraints that are a consequence of the way in which chemical structure relates to the biological activity of a molecule. Two observations are central to the hypothesis: (a) high potency, specific biological activity is a rare property for any chemical to possess, because such activity is limited by the stringent requirements for high affinity binding to receptor sites. The main evidence to support this statement, which will be considered in more detail later, comes from the thousands of screening programmes that have been conducted by agrochemical companies seeking insecticides, fungicides, herbicides and molluscicides and the huge screening efforts made for biological activities of interest to the pharmaceutical industry. (b) plants contain a considerable diversity of compounds but few are highly biologically active; whereas arguments based solely on cost-defence benefits predict that most plants should contain a moderate diversity of highly active compounds.

The first argument (rarity of high potency, specific biological activity) suggests that organisms that exploit the biological activity of a substance must have evolved mechanisms to generate chemical diversity, inevitably synthesising many chemicals that have no useful biological activity alongside the very occasional one that possesses high biological activity. If natural selection, operating solely on a cost-defence benefit basis, eliminated all but the biologically active chemicals, insufficient chemical diversity would remain to generate further chemical diversity, and such organisms would be at a selective disadvantage once "target" adaptation occurred. Recognising this limitation, the Screening Hypothesis proposes that mechanisms must exist to encourage the generation and retention of chemicals, irrespective of their activity, and that plants and microbes will therefore have a large number of low potency compounds and rather few highly active substances.

The purpose of this paper is to consider in greater depth the evidence that high potency biological activity is a rare property. The reasons for focusing on this one aspect of the Screening Hypothesis are two-fold Firstly, if valid, this argument undermines many alternative hypotheses about the role of natural products. Secondly, if high potency biological activity is a rare property, models for natural products evolution must explain how the massive redundancy that is required is generated and retained.

BIOLOGICAL ACTIVITY—A VARIABLE DEPENDENT ON DOSE AND ASSAY SYSTEM

Is biological activity a rare property for a chemical to possess? A meaningful answer requires a definition of the term biological activity.

"The Dose Maketh the Poison"

At a time when natural products were mainly of interest only to herbalists, the "father" of modern toxicology, Paracelsus (1493–1541) astutely observed that whether one thought of a substance as a poison or not depended entirely on the dose that was administered. It is therefore obvious that a discussion of the biological activity of any compound is meaningless unless some comparative standard is adopted with regards to the dose. The problem is especially acute when the biological activity is evaluated on the basis of an adverse effect on an organism, because a sufficient dose of many substances will have an adverse effect on virtually any organism.[8] For the purposes of this discussion it seems sensible that the only realistic basis for judging the biological activity of a substance is whether the compound would have a significant biological effect at a dose that a recipient organism would receive. A biological activity that is found when the compound is applied to test organisms at unrealistic concentrations cannot sensibly be regarded as meaningful.

The importance of these considerations can be illustrated by considering model dose-response curves for three hypothetical substances C1, C2 and C3 (Fig. 4). It is apparent that all three substances can be regarded as showing biological activity in that all saturate the response when applied at high concen-

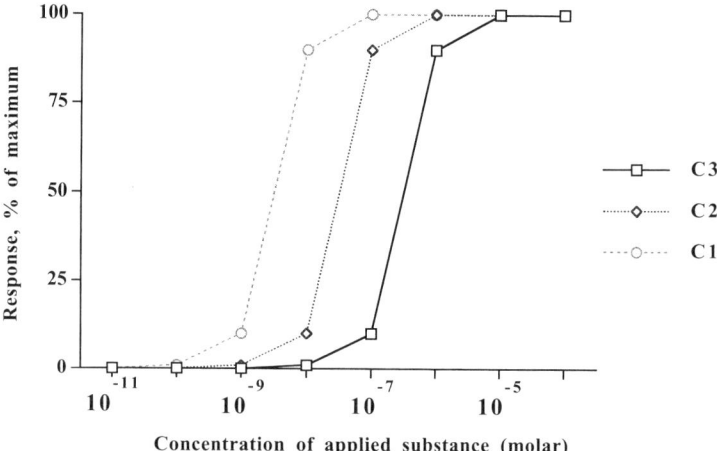

Figure 4. "The dose maketh the poison". The bioassay dose-response curves for three hypothetical substances C1, C2 and C3 illustrate that whether or not a substance is considered to be biologically active depends on the dose being considered. For naturally occurring compounds, the only relevant dose is that to which a target organism will be normally be exposed. If the actual concentrations in the environment of C1, C2 and C3 were 10^{-8}M, then only compound C1 could be considered to possess a biological activity that was meaningful.

trations ($>10^{-5}$M). However, suppose that each substance occurs in a plant at a concentration of 10^{-8}M. In that case, only compound C1 can be considered to possess meaningful biological activity. Compounds C2 and C3 show no significant activity at their endogenous concentrations and would therefore confer no selective advantage to the plant.

What Kind of Biological Activity?

It is obvious that the wider the range of organisms for which biological activity of a compound is assessed, the greater the chance that some effect will be found. However, when discussing the evolution of natural products, it must always be borne in mind that the selective pressures that operate on any individual within a population at any one time will be quite specific and only a limited range of opportunities will be available where chemical interactions can be beneficially used. Thus, the only types of biological activity that should be considered when discussing evolutionary arguments concerning natural products are those that are realistic. It is apparent that a real problem exists in reaching agreement about what is realistic. However, it will be useful to establish some broad guidelines. For instance, in the case of natural products derived from a soil micro-organism, it would be reasonable to argue that specific biological activity (when tested at "realistic" does as explained above) found only in a mammalian test system is more likely to be fortuitous than meaningful. Likewise a substance extracted from a terrestrial plant is more likely to possess fortuitous biological activity against a marine organism than meaningful activity. It is of course possible to argue that some fundamental function involving highly conserved proteins found in most organisms provides a target that is accessible even in non-target organisms. However, such assertions demand that the real target organism is identified and the conserved function demonstrated in that organism. The principle that should be accepted is that the possession of any type of biological activity should not in itself be sufficient reason to argue that the substance has been selected by evolution because it possesses that type of biological activity.

ARE POTENT BIOLOGICALLY ACTIVE COMPOUNDS EASILY FOUND IN EXTRACTS OF MICROBES OR PLANTS ?

The fact that microbially-derived antibiotics found by screening make up most of a $16.5 billion *per annum* antimicrobial pharmaceutical sales[9] explains why so many large antibiotic screening programmes have been conducted during the last half century. However, while the commercial and humanitarian success of the huge screening effort is obvious, the magnitude of this effort in relation to the number of antibiotics discovered suggests that such compounds are not found easily. Essentially groups who report the success rate of antimicrobial screening agree that finding naturally occurring antimicrobials in biologically derived samples is a rare event. For example: (i) 400,000 microbial cultures were assayed over 10 a year period and only three utilisable compounds were found;[10,11] (ii) 21,830 isolates screened in one year gave 2 possible compounds;[12] (iii) 10,000 microorganisms gave one clinically effective agent.[13] Jones and Firn[7] provide examples of the low frequency of detecting highly biologically active compounds in plant extracts that illustrate the same point. Thus, whatever the source of the natural products and whatever the organism used in the screening program, the percentage of compounds with high potency, specific biological activity is very low.

Those sceptical of such evidence argue that commercial constraints applying to many screening programs result in biased data which underestimate the true number of biologically active compounds because: (i) the high selectivity required for a substance to be of commercial use means that many active compounds will be rejected at an early stage because of lack of specificity; (ii) some active compounds are unstable and are either lost during isolation or are unsuitable for use; (iii) some compounds are too difficult to synthesise or extract from cultures and are never developed; (iv) some compounds have already been isolated (and possibly patented before); (v) the type of biological activity found is not meaningful in terms of the particular usage sought; (vi) the screening process is inappropriate; (vii) some forms of biological activity are dependent on the presence of other compounds (synergists) which are lost during the purification processes before the substance is bioassayed.

Each of the above reasons could be valid in some circumstances but even taken together they only partially account for the fact that screening natural products has produced relatively few biologically active compounds. Thus, evidence is available to support the contention that potent, specific biological activity is hard to find. More helpful in this debate would be evidence that contradicts this view. To the best of our knowledge such evidence does not exist, primarily because screening of most compounds found in plants or microbes against all or most organisms with which they interact have not been carried out. Our studies on bracken fern,[14] a plant renowned for its chemical diversity[15] showed that only one or two chemicals could be considered to have meaningful biological activity.

ARE BIOLOGICALLY ACTIVE COMPOUNDS EASILY FOUND IN COLLECTIONS OF SYNTHETIC CHEMICALS?

Screening for many forms of biological activity have been undertaken on a large scale with synthetic compounds, and the evidence from such work confirms the view that few structures possess significant biological activity even when assayed at high concentrations. Unfortunately few screening trials produce published results because of the commercial sensitivity of the information gained (and also because negative results are rarely accepted for publication). Hence there is a natural bias to the reporting of biological activity. However, those involved in screening trials acknowledge that it is hard to find specific, high potency biological activity, and considerable effort has been expended in developing automated *in vitro* screens because such screens can assess biological activity in over 20,000 compounds or extracts per month.[9] There would obviously be no need for such high capacity screening methods if the proportion of chemicals that possessed biological activity were higher than it seems to be. Screening programmes using synthetic compounds show that the ratio of "active" to "inactive" compounds is not obviously different from that found in collections of naturally occurring natural products. For example: (i) 100,000 different organophosphorus compounds have been synthesised and screened for insecticidal activity from which about 100 have been marketed;[16] (ii) the success rate in finding a commercial agrochemical fell from 1 in 1800 in 1956, to 1 in 7500 in 1970 to 1 in 15,000 in 1978;[17] (iii) in 1977 a survey of 36 companies suggested that 41,000 compounds were being screened during that year for biological activity that could be exploited in the agrochemical market,[18] which illustrates the effort that has to be expended to compensate for the low proportion of compounds showing biological activity; (iv) in one company, between 3000 and 7500 substances had to be screened on average in order to find one useful agrochemical.[19]

Once such massive screening efforts produce a lead compound, large numbers of chemicals structurally related to the lead compound are made and tested. This interest in defining the structural requirements required for biological activity has resulted in a wealth of information about the relative activity of structurally related compounds.[16] Structure-activity studies show that small changes to the structure of the most active compound of any series can greatly reduce biological activity. It is rare for a wide range of related chemical structures to possess similar biological activity of equal potency. Examples to support these assertions can be found in numerous structure-activity relationship studies (insecticides,[20] fungicides,[21] auxins[22] and phenoxy- herbicides[23]).

WHY IS POTENT BIOLOGICAL ACTIVITY A RARE PROPERTY FOR A CHEMICAL TO POSSESS?

The developing knowledge of the way in which ligands interact with binding sites suggests that in order for a molecule to possess potent, specific biological activity[24] it must interact reversibly with a target protein (biological activity resulting from irreversible interactions will be discussed later). The two key features discussed earlier, high potency and high specificity, indicate that the target protein and the chemical must interact in a very specific way. This interaction demands that the chemical possesses exactly the right three dimensional structure such that its charge structure is attracted to a matching compatible charge structure on certain amino acids of the protein. If the chemical has exactly the right structure, there will be a high affinity binding of the chemical to the receptor site. Such high affinity binding insures that the chemical will bind to the protein even when the chemical is present in solution at very low concentrations. Structural analogues of a high potency chemical that have a less than optimal charge distribution will have a lower affinity for the protein and will only interact significantly if they are present at much higher concentrations. The binding affinity of the most biologically active molecule in a series is usually orders of magnitude greater than that of the least active member of chemically related compounds[25] (Fig. 5). These studies of ligand-receptor interactions support the view that high potency biological activity results from the fact that a binding site on a target protein shows great discrimination for particular structural features of the ligand. Hence, at a molecular level, this provides an explanation why potent, specific biological activity is a property that only a few chemicals possess.

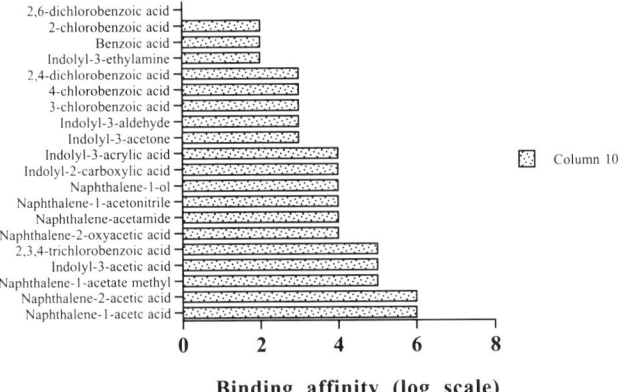

Figure 5. The strict structural requirements necessary to show high potency biological activity is usually matched with a similar structure-activity relationship for ligand-receptor interactions. As an example, the binding affinity (K_d) of a number of compounds which show auxin-like activity in plants, or which are structurally related to compounds possessing such activity, has been shown to be correlated with their biological activity.[25] Similar results have been reported with respect to many biologically active ligands.

BIOSYNTHETIC ROUTES TO GENERATE CHEMICAL DIVERSITY

If there is a low probability that any new natural product being made by an organism will benefit that organism as a result of its biological activity (for the reasons outlined above), some predictions can be made about features that may be expected in biosynthetic pathways generating such compounds. For instance, features that promote the generation of chemical diversity might be advantageous. Mechanisms that allow the generation of chemical diversity but which reduce the cost of redundancy would also be expected. In other words, it can be predicted that the rules that apply to primary metabolism need not be universal rules applicable to all metabolism.

Protein synthesis is an excellent example of a system that is capable of generating massive chemical diversity by joining a mixture of monomers (amino acids) in a huge variety of ways. This is demonstrated by animals which possess an immune system. In response to an antigenic challenge, animals produce a large number of antibodies in order to compensate for the fact that the majority will not have the appropriate molecular structure to enable them to bind the antigen. The immune system generates biological diversity, screens for the appropriate biological activity, and then amplifies the production of the most effective molecules. The immune system tolerates massive redundancy of individual antibodies. In theory, polypeptide synthesis is an efficient way to generate chemical diversity in any organism. However, there are severe limitations for plants and microbes seeking to use this route to the production of biologically active compounds. Although numerous different proteins can be generated, the chemistry of the linking peptide bonds is a constant, repeating point of vulnerability and all target organisms will possess proteases capable of attacking such bonds.

The biosynthesis of most natural products is by sequential elaboration or transformation of structures, using linear or branched pathways (although in some cases the linking of monomers is part of the backbone elaboration). In such biosynthetic routes, new chemical diversity is dependent on maintaining all the compounds in the pathway because the loss of the capacity to make any precursor will result in a loss of all subsequent products. This is a severe constraint.

THE CONSTRAINTS OF EXPLOITING THE BIOLOGICAL ACTIVITY OF IRREVERSIBLE INHIBITORS

The discussion of biologically active compounds so far has been based on the assumption that the chemicals act by reversibly binding to specific target molecules in the cell. However, inhibitors are known which interact covalently with proteins and other substances in the cell, and it is necessary to consider the implications of such an irreversible mode of action.

The most obvious implication of covalent binding is that the action of the compound will be irreversible. Once an organism has received a harmful dose of an irreversible inhibitor which reacts with a protein for example, the organism will be only be able to recover if the target protein is resynthesised. Consequently, the susceptibility of the organism to such covalent inhibitors will depend to some extent on the half life of the protein.

Another significant property of irreversible inhibitors is that they possess their particular mode of action because they possess specific functional chemical groups.[26] In order for such compounds to form covalent bonds with their targets, they must be chemically reactive or capable of becoming chemically reactive when exposed to particular conditions (for example photochemically reactive compounds require photo-activation by light). Synthetic compounds which are photochemically active have been made many times as an aid to identifying the target protein for a ligand.[27] After exposure of the target protein to the radiolabelled photochemically reactive ligand and a period of exposure to UV light, the protein become radiolabeled and can then be isolated and identified. Information on the properties of these synthetic photochemically reactive compounds can be used to gain an insight into the use of similar compounds in nature. In general, such compounds only provide highly specific labelling of the target protein if the ligand binds to the target protein at low ligand concentrations. If the ligand binds only weakly to its target, high concentrations of the ligand must be used in order for it to become associated. If the ligand is present at high concentrations, it becomes associated with many proteins and "non-specific binding" occurs as it reacts covalently with many proteins - the specificity of the ligand-receptor interaction is masked. Thus, an organism making a chemically or photochemically reactive compound will be faced with the problem that if the compound is not potent, it can potentially have adverse effects as a result of "non-specific" interactions with non-target protein. Such compounds, will be most concentrated in the organism that make them and any "non-specific" action will irreversibly inhibit vital processes by chance, and thus, the production of low potency chemically reactive compounds will be selected against. Clearly, compartmentation of the production and storage of such compounds can reduce, but not eliminate, the risk to the organism making them. Reversible inhibitors also will act randomly with other chemicals but such interactions will only have an effect for a short duration, and the process influenced will recover when the inhibitor disassociates. The specificity against target organisms might also be expected to be a problem unless the photochemical or chemically reactive compound can act at very low concentrations. The fact that rather few such chemically or photochemically active natural products are known may be a consequence of these constraints. The furanocoumarins, however, are an example of such compounds that have been found in plants.[28]

It is likely that the structure-activity requirements for biological activity of an irreversible inhibitor will be more associated with the chemically (or photochemically) active substituent group than with the overall structure of the molecule. Thus, a series of structurally similar compounds might show similar biological activity if they all possess the same active group simply because this dominant feature governs activity. Consequently, the proportion of compounds showing biological activity should be higher than in a group of compounds which act by reversible binding only to their target. It is expected that chemically or photochemically active compounds will possess a wider range of biological activity (especially if tested at high concentrations) than compounds that bind reversibly because irreversible inhibitors possess the ability to react chemically to some extent with many proteins (and some non-proteins).

HOW COULD NATURAL PRODUCT PATHWAYS HAVE EVOLVED IF BIOLOGICAL ACTIVITY WERE A RARE PROPERTY?

Before considering this question it is necessary to summarize existing ideas about the way in which biochemical pathways may have evolved. It is widely accepted that the mutation of a gene coding for an enzyme will result in various outcomes:[29] (i) the enzyme activity will be unaltered (the most likely scenario);

(ii) the kinetic properties of the enzyme will be changed, usually in a detrimental way; (iii) the enzyme will change its substrate specificity; (iv) sites necessary for the allosteric control of the enzyme will be altered.

It is not widely believed that a mutation of the gene coding for an enzyme will change the type of transformation carried out. Thus most biochemical inventiveness[30] will arise from existing catalytic functions being applied to new substrates rather than new types of transformations being applied to the original substrate. Such biochemical inventiveness requires that multiple copies of a gene must exist if the gene being mutated codes for an enzyme that is involved in primary metabolism, otherwise a loss of an essential part of primary metabolism would result.

Some Simple Scenarios

Combining current ideas about the evolution of biochemical pathways with the fact that there is a low probability of any new chemical possessing a useful biological activity at the concentration at which it is made, one can construct some simple scenarios to illustrate the constraints that might have operated during the evolution of natural product pathways. Consider a "primitive" microbe, possessing only a primary metabolism growing in competition with several similar organisms. Individuals will obviously be competing with each other. A mutation of a duplicated gene may allow the mutated enzyme to make a new product (X) from an existing primary metabolite (Fig. 6).

Because of the low probability that X will have any useful biological activity, the mutant will, in nearly all cases, be selected against, with the selection intensity being dependent on the costs incurred due to the loss of some primary product that results when X is produced. However, let us assume that the rare event occurs, and that X possesses biological activity at a concentration at which it is made. There will be a high probability that the biological activity will be detrimental to the producer, because X will necessarily be similar in structure to compounds taking part in primary metabolism. The opportunity exists for X to act as a substrate analogue or to interfere with allosteric control mechanisms within the producer, especially as X will be in higher concentrations within the producer than in any other organism. Let us assume, however, that X acts as an antibiotic against competitors when made in amounts that do not negatively effect the producer. The mutant will thrive at the expense of its competitors whose growth will be inhibited by the antibiotic. However, the ability to grow in the presence of X must exist in the producing organism, and such an ability will also be selected for in competing organisms. It is possible that a mutation will occur in the genes of the target organisms that will enable them to metabolise, sequester, excrete or tolerate X. Consequently the producing organism will become subject to competition from any X-resistant mutants that will inevitably arise. This new competition will apply selection pressure on the producer of X.

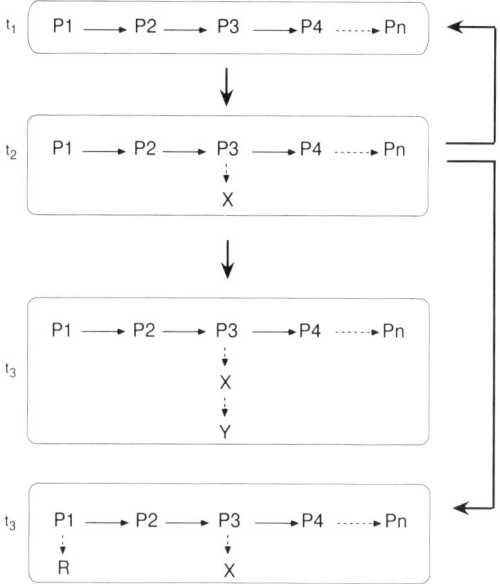

Figure 6. A simplified, possible scenario for the initial evolution of a secondary product pathway. At time t_1, a simple linear primary biosynthetic pathway P1-Pn exists. At t_2, a new biosynthetic capacity arises from a mutation of a duplicated gene and X is produced, the production of which benefits the organism. At t_3 a compound Y could arise from a new mutation allowing an elaboration of X, or R could arise directly from primary metabolism (P1 -> R). Once X has evolved, at any time the loss of the biosynthetic capacity leading to X could occur. Thus after t_2, three possible scenarios could occur.

Once a competitor arises that can grow in the presence of X, the producer of X will be in competition with that species and also with strains of the its own species that still exist that do not produce X or with strains that have lost the ability to make X. Eventually a new mutant capable of producing a new secondary product might evolve. Let us suppose that this new natural product, Y, arises from X. The probability of Y having any biological activity is low for the reasons outlined previously. Furthermore, because Y is derived from X, it is likely that any biological activity of Y will be similar to that shown by X, to which it is structurally related. If X is no longer an effective antibiotic, there is a high probability that Y, which is made from X, will also be ineffective. The chance of X and Y both possessing potent biological activity is low, and the chance of X and Y having potent, different types of appropriate biological activity must be extremely low. Unless X was very potent, i.e. was made in only small amounts at low cost, surviving strains that no longer produce X will compete successfully with the strains making X.

It is apparent that at each step in a sequential natural product pathway, the probability that the new compound will have an appropriate biological activity (i.e. it possesses the right structural features to possess high potency, specific biological activity) is low, and this problem is compounded as the pathway extends. It is hard to avoid the conclusion that it would be extremely difficult to evolve long sequential pathways in this manner. Furthermore, as the biosynthetic chain lengthens, the increasing number of enzymes involved means that mutations that result in loss of parts of that sequence will become more frequent, and it would become more difficult to sustain the pathway after the terminal product has lost its beneficial role.

Another possible scenario that can be considered is that once the producer organism making X is suffering from competition from an organism that can cope with X, gene duplication of the gene coding for the enzyme making X could be used to enhance the rate of synthesis of X by the producer. This strategy compensates for loss of potency by increasing production. However, it suffers from the fact that defence costs will rise. Unless the producer gains a net fitness benefit in the presence of the competitor, the increased costs will reduce fitness. Furthermore, the competing organism can predictably rely on similar gene duplication strategies for counter measures.

In the scenario just described, the new chemical Y, produced after X lost its effectiveness, was postulated to arise from X. An alternative evolutionary scheme could be envisaged where a new compound R takes over the role of X, where R is derived from a primary metabolites (Fig 6). This scenario has the advantage that R has a greater chance than Y of possessing a different mode of action to X. However, this scenario suffers from the disadvantage that the total chemical diversity available is limited to the kind of substances that can be generated by a single transformation of a primary metabolite. This is a severe limitation and when combined with the fact that many such compounds have a high chance of being detrimental to the producer, it suggests that such a scenario is implausible.

These simple scenarios illustrate the constraints that operate as a result of the rarity of high potency biologically active compounds. Yet massive chemical diversity has been produced within many organisms. Hence some extra elements would seem to be required in constructing these scenarios. As we see it, the challenge is to propose new scenarios which more sensibly explain natural product evolution.

In an initial attempt to create a credible scenario for the evolution of natural product pathways, we proposed that certain properties must be possessed by secondary metabolism that increase the chances for the generation and retention of natural product diversity.[7] These properties were branched pathways, matrix transformations, combining of pathways, modes of pathway regulation, and relatively low substrate specificity of some enzymes. How these properties enhance the generation and retention of natural products has been outlined[7] and will be considered in more detail later (Firn and Jones, in preparation). The ideas are an attempt to devise a set of broad rules for natural product metabolism based on the thesis elaborated in this paper i.e. that potent, specific biological activity is a rare property.

SUMMARY

The Screening Hypothesis is a model for the evolution of natural product diversity that is based on evidence accumulated over several decades which has shown that specific, high-potency biological activity is a rare property for a molecule to possess. Evidence to support this observation comes from studies conducted at the organism, organ, tissue, cell and molecular levels.

The proportion of naturally occurring molecules showing biological activity will depend on the concentrations at which they are assayed. It is argued that the meaningful comparative discussion of biological activity of any endogenous compounds can only be made with reference to their concentration at the site of action. When biological activity is only shown by a substance when assayed at concentrations above those that occur in natural situations, it may be of pharmacological interest, but such activity should not be assumed to be meaningful in terms of the role(s) that the substance might play in the organism that produces it.

The Screening Hypothesis is based on concepts which place severe constraints on the evolutionary framework governing the generation, retention and use of natural products. It suggests that these constraints operate at a high level and that mechanisms to generate and retain chemical diversity must have arisen at an early stage in the evolution of life. Indeed, it seems likely that natural products may have been produced as a consequence of "biochemical inventiveness" which might have been inevitable when primary metabolism was evolving. "Secondary metabolism" might well have evolved concurrently with primary metabolism, and primitive organisms may have evolved their "secondary metabolism" to exploit the inevitability of the generation of some chemical diversity.

It is apparent that a debate as to whether any natural product is "redundant" becomes a semantic argument. If one defines redundant as no longer playing a role (i.e. the biological activity of a compound, which was once of value to the producer, is no longer exploited), the Screening Hypothesis suggests that the many natural products never had a role resulting from inherent biological activity. Hence they cannot be considered redundant in this sense of the word. Likewise if one defines redundant as being superfluous or over-copious,[31] the Screening Hypothesis provides arguments against the view that the majority of natural products evolved as a result of superfluous production. Hence most natural products cannot be considered redundant in this sense. The Screening Hypothesis suggests that the term "redundant" is only appropriate in a few circumstances. The Hypothesis suggests that the synthesis of many compounds which bring no short term benefit to the producer is a necessary part of the overall mechanisms employed by plants and microbes to produce the occasional chemical which does possess useful biological activity. Most natural products are only "redundant" in the way that most antibodies are "redundant". The production of the majority of these substances often results in no short-term benefit but short-term costs are compensated for by longer term benefits. The majority of natural products and antibodies are not redundant in any commonly accepted use of the word but are a necessary consequence of the need to generate chemical diversity.

ACKNOWLEDGMENTS

CGJ thanks the Mary Flager Cary Charitable Trust and the John Simon Guggenheim Memorial Foundation for financial support. Contribution to the programme of the Institute of Ecosystem Studies.

REFERENCES

1. HARBORNE, J.B. 1988. Introduction to Ecological Biochemistry. Academic Press, London, p. 242.
2. MENDELSOHN, R., BALICK, M. 1995. The value of undiscovered pharmaceuticals in tropical forests. Economic Botany 49: 223–228.
3. RAVEN, P.H., EVERT, R.F., EICHHORN, S.E. 1992. Biology of Plants. Worth Publishers, New York, p. 791.
4. CHADWICK, D.J., WHELAN, J. 1992. Secondary Metabolites: Their Function and Evolution. John Wiley, Chichester, p. 299.
5. FRAENKEL, G. 1959. The reason d'etre of secondary plant substances. Science 129: 1466–1470.
6. SWAIN, T. 1975. Flavonoids. In: The Flavonoids, (J.B. Harborne, T.J. Maby, H. Maby. eds.), Chapman and Hall, London, p. 1204.
7. JONES, C.G., FIRN, R.D. 1991. On the evolution of secondary plant chemical diversity. Philo. Trans. Royal Soc. 333: 273–280.
8. RODRICKS, J.V. 1992. Calculated Risks. Cambridge University Press, Cambridge, p. 256.
9. NISBET, L.J. 1992. Useful functions of microbial natural products. In: Secondary Metabolites: Their Fucntionand Evolution. (D. J. Chadwick, J. Whelan, eds.), John Wiley, Chichester, p. 299.
10. FLEMING, I.D., NISBET, L.J., BREWER, S.J. 1982. Target directed antimicrobial screens. In: Bioactive Mirobial products: Search and Discovery. (J.D. Bu'lock, N.J. Nisbet, D.J. Winstanley, eds.), Academic Press, London, pp 107–130.
11. NELSON, T.C. 1961. Proceeding of the National Academy of Research Council Supplement. 1961. p. 891
12. WOODRUFF, H.B., HERNANDEZ, S., STAPLEY, E.D. 1979. Evolution of antibiotic screening programme. Hindustan Antibiotic Bull. 21: 71–84.
13. WOODRUFF, H.B., MACDANIEL, A.E. 1958. Antibiotic approach in strategy of chemotherapy. Soc. General Microbiology Syposium 8: 29–48.
14. JONES, C.G., FIRN, R.D. 1979. Resistance of *Pteridium aquilinum* to attack by non-adapted insects. Biochem. Syst. Ecol. 7: 95–101.
15. JONES, C.G. 1983. Phytochemical variation, colonization and insect communities: the case of bracken fern. In: Variable Plants and Herbivores in Natural and Managed Systems. (R.F. Denno, M.S. McClure, eds.), Academic Press, New York, pp. 513–558.
16. HASSALL, K.A. 1990. The Biochemistry and Uses of Pesticides. MacMillan, London, p. 511.
17. BRAUNHOLTZ, J.T. 1981. Crop protection: the role of the chemical industry in an uncertain future. In: Crop Protection Chemicals: Directions of Future Development. (L. Fowden, I.J. Graham-Bryce, eds.), Royal Society, London, pp. 19–33.
18. MENN, J.J., HENDRICK, C.A. 1981. Crop protection: the role of the chemical industry in an uncertain future. In: Crop Protection Chemicals: Directions of Future Development. (L. Fowden, I.J. Graham-Bryce, eds.), Royal Society, London, pp. 57–71.
19. EDSON, E.F. 1973. Crop protection. Philo. Trans. Royal Soc.B267: 93–102.
20. HENRICK, C.A. 1982. Juvenile hormone analogs: structure-activity relationships. In: Insecticide Mode of Action. (J.R.Coats, ed.), Academic Press, New York, pp. 315–402.

21. WOODCOCK, D. 1977. Structure-activity relationships. In: Systemic Fungicides (R.W. Marsh, ed.), Longman, London, pp. 32–84.
22. WAIN, R.L., FAWCETT, C.H. 1969. Chemical plant growth control. In: Plant Physiology - A Treatise. Volume VA. (F.C. Steward, ed.), Academic Press, New York, pp. 231–298.
23. VELDSTRA, H. 1944. Researches on plant growth substances, IV and V. Relation between chemical structure and biological activity. Enzymologia, II: 97–163.
24. BRANDON, C., TOOZE, J. 1991. Introduction to Protein Structure. Garland, New York, p. 302.
25. RAY, P.M., DOHRMANN, P.M., HERTEL, R. 1977. Specificity of auxin-binding sites on maize coleoptile membranes as possible receptor sites for auxin action. Plant Phys., 60: 585–591.
26. BAKER, B.R 1967. Design of active-site-directed irreversible enzyme inhibitors. Wiley, New York, p. 325.
27. WOLD, F. 1977. Affinity labelling - an overview. In: Methods on Enzymology, Vol. 46 (W.B. Jakoby, M. Wilchek, eds.), Academic Press, New York, pp. 3–14.
28. BERENBAUM, M. 1990. Evolution of specialization in insect-umbellifer associations. Ann. Rev. Entomol. 35: 319–343.
29. JENSEN, R.A. 1976. Enzyme recruitment in evolution of new function. Ann. Rev. Microbiol. 30: 409–425.
30. ZÄHNER, H., DRAUTZ, H., WEBER, W. 1982. Novel approaches to metabolites screening. In: Bioactive Microbial Products: Search and Discovery. (J.D. Bu'lock, N.J. Nisbet, D.J. Winstanley, eds.), Academic Press, London, pp. 51–70.
31. GEDDIE, W. 1959. Chambers's Twentieth Century Dictionary. Chambers, Edinburgh. p. 1396.

INDEX

Abies, 183–184, 190, 194, 199, 207
Abies grandis, 184–186, 188–189, 191, 193
Acalymma vittatum, 144
4-Acetylbenzoxazolin-2-one (4-ABOA), 284
Activated oxygen, 132, 135–136
Alkaloids, 60
  biological activity, 92, 97–103, 107
    invertebrates, 100–102
    vertebrates, 97–99, 106
  environmental variation, 104–105, 126
  ergot, 82–83, 86–88, 90–92, 95, 99, 102, 107, 279
    biological activity, 92
    biosynthesis, 90, 92–94, 105, 107–108
    clavines, 82, 92
    ergopeptides, 82, 91
    ergopeptines, 90–92, 103–104
    ergovaline, 86, 91–92, 95–96, 100, 102–105, 107
    lysergic acid, 82, 90–92, 103
    peptine derivatives, 82
  indole diterpene, 88, 93, 95; *see also* lolitrems
  localization, 95
  lolines, 87, 95–97, 99, 101–105; *see also* pyrrolizidine
    n-acetyl-loline, 86–87, 90, 102–104
    n-formyl-loline, 86–87, 90, 102–104
  lolitrems, 86–87, 93–95, 99, 103, 106–107, 109; *see also* indole diterpenes
    biosynthesis, 94, 107
    lolitrem B, 93–96, 101, 105–106, 279–280
    paxilline, 93–95, 101, 105–106
  morphine, 271
  nicotine, 126, 140
  pyrrolizidine, 87, 282; *see also* lolines

Alkaloids (*cont.*)
  pyrrolopyrazine, 88, 92
    biosynthesis, 93
    peramine, 82, 86–87, 92–93, 95–96, 99–102, 104–105, 107
  response to wounding, 137
  steroidal glycoalkaloids (SGA)
    solanidine, 123
    tomatidine, 123–124
    tomatine, 124, 126
  tetrahydrocannabinol (THC), 271
  translocation, 97
Allelopathic, 102
Alliaria petiolata, 67–68, 74–75
Anacardic acid, 128–129
Analogue synergism, 13
Anticancer, 280
Antifeedant, 125, 139, 156–159, 161–165, 167–168, 172–175, 278
Antifungal, 102, 123, 133, 135, 239, 253, 283
Antiherbivore, 38–49, 97, 99, 108–109, 123, 127, 139, 144, 156, 232–233, 240–241, 245–246, 273, 278, 280
Antimammalian activity, 47–49, 97
Antimicrobial, 3, 123, 125, 127, 132, 139, 144, 254, 279, 301
Antinematode, 101
Antinutritive, 232–233, 239, 241, 245
  defense proteins, 233, 239–240
Antioxidant, 240
Antipathogen, 139
Argentine stem weevil, 100, 103
Arabidopsis, 131, 134, 254, 258–259
Ascomycetes, 274
Aspen, 26–51
  abiotic interactions, 49
    UV radiation, 49
  bird interactions, 47

313

Aspen (cont.)
  clonal variation, 27, 35, 37, 41–47, 49, 51
  insect interactions, 38–47
  pathogen interactions, 37–38
  mammal interactions, 47–50
Aspergillus species, 277–279
Associative learning, 72–73, 75
Azadirachta indica, 156, 159, 162, 171–172
Azadirachtin, 159–164, 171–174

Baccharis species, 281–282
Balance of stimulants/deterrents: see Mixtures
Barbarea vulgaris, 67–68, 71, 75
Bark beetles, 180–183, 185, 187–189, 192–195, 206
1,4-Benzenoid diradical, 269–270
Benzoic acid derivatives, 257
Big poplar sphinx moth: see *Pachysphinx modesta*
Binding
  receptors, 303
  site competition, 13–14
  specificity, 16, 270–272, 298, 303–306
Bioactivation, 14
Bioassay-driven fractionation, 159, 165, 171–172
Bioassays, 2–3, 9, 17, 62, 66, 68, 95, 102, 162, 165, 173, 175, 302
Biological activity, 2, 7, 9–12, 156, 159–160, 162–163, 165, 168, 171–174, 279, 284–285, 296, 298–310
  additive effects, 12–13, 15
  broad spectrum, 9, 51, 190, 254–255, 259, 284
  mode of action, 9, 12–13
  multiplicity, 9
  nonadditive effects, 12
  synergism, 4, 13, 17
  UV-independent effects, 11
  UV-mediated effects, 10
Boronyl acetate, 182, 198–199, 206–207
Botanical insecticide, 156–158, 165, 174
Brassicaceae: see Cruciferae
Broad spectrum activity, 9, 51, 190, 254–255, 259, 284

Cabbage butterflies: see *Pieris* species
Canker: see *Hypoxylon mammatum*
Carbohydrates, 200–202, 206
  galactose, 201–203, 206
  inducers, 244

Carbon–nutrient balance hypothesis, 33–34, 36, 43, 70
Carcinogens, 277–279, 282, 284
Cardenolides, 60–62, 64, 67, 71
  digitoxigenin derivatives, 64, 74–75
  erychroside, 62–63, 74–75
  erycordin, 62–63
  erysimoside, 62–63, 74–75
  feeding deterrents, 63–64, 74
  oviposition deterrents, 62–63, 74
  strophanthidin derivatives, 62–63, 67, 73–74
Catalases, 256–257
Catechol, 37–39, 41, 122–123, 125, 140, 232, 238
Ceratocystis, 191, 192
Ceratocystiopsis ranaculosis, 193–197
Chemical defense, 57–58, 60–67, 181, 187–194, 197–200, 208–209, 273, 286; see also Defense, Deterrence
Chemoreception, 61–62, 67–68, 71–72, 74–75
Chitinases, 133, 135, 140, 254–256
Chlorogenic acid, 66, 72–75, 232, 238, 240
  deterrent activity, 66, 72–75
Choristoneura conflictana, 38, 41
Choristoneura fumiferana, 198, 200, 204
Choristoneura occidentalis, 200–204, 209
Clavicipitaceae, 82–83
  Balansieae, 82–83
    Claviceps species, 83, 92, 108
    Claviceps purpurea, 279
Coevolution, 203, 283
Colletotrichum circinans, 122
Colletotrichum graminicola, 130
Colletotrichum lagenarium, 134, 140, 144
Colorado potato beetle, 240
Common sense scenario, 2, 4, 16
Complexity theory, 265–266, 268–269, 273, 286
Condensed tannins, 27, 30–31, 33–35, 38, 40–43, 46–47, 51, 144, 183, 187, 195, 201, 209
Conifers, 181–183, 185–186, 189–190, 192, 194–195, 197, 204, 273, 280
Coniferyl alcohol, 132
Coniferyl aldehyde, 132
Coniferyl benzoate, 27, 30, 47, 51
Copaifera, 180, 190–191, 198–199
Cost defense, 298; see also Optimal defense hypothesis
Cotton, 128, 130, 144
Cruciferae, 57–76

Cucumber: *see Cucumis sativa*
Cucumis sativa, 127, 134–135, 139–141, 144, 255
Cucurbita, 128
Cucurbitacins, 60, 71, 136, 139, 144
   cucurbitacin B, 127
   cucurbitacin E, 65, 73, 75, 127
   cucurbitacin I, 65, 73–75
   environmental variation, 127
   feeding deterrents, 64–65
   ovipositron deterrents, 65, 67
Cyanogenic glycosides, 124, 136, 217, 220–226
   biosynthesis, 220–223, 226
   distribution, 218–220, 224
   gynocardin, 218, 226
   intermediates, 221, 223
   linamarin, 218, 226
   localization, 222–226
   lotaustralin, 218–224, 226
   sequestration, 218, 220–221, 226
Cyanogenesis, 217–219, 225
Cyanide bomb, 218
Cystein-rich proteins, 254
Cytotoxic, 267, 284

Defense
   constitutive, 122–129, 145
   enzymatic, 246
   induced, 32, 70, 122, 125, 129–139, 145
   pathogen induced, 238
   wound induced, 233, 236, 241–246
Dendroctonus, 188–189, 191–193, 209
Desensitization, 174
Deterrence, 62–67, 75, 194, 205
   feeding, 58, 62–66, 68, 71–72, 74–75, 100–101, 103–104, 157
   natural enemy recruitment, 58
   oviposition, 58, 62–63, 65–67, 69, 71–72, 74, 100, 157
Deuteromycetes, 274
Disease resistance, 255–256, 258–260, 281
Dose response, 299–301
Dynemicin A, 269–270

Eastern spruce budworm: *see Choristoneura fumiferana*
Enediyne, 269–270
Endophytic fungi, 279
   Acremonium species, 83–85, 88, 93, 95–98, 101–102, 105–106, 279
   Epichloë species, 82–86, 88, 102, 109
Epilachna varivestis, 161–162, 164, 167

Erysimum cherianthoides, 60, 62–64, 70–71, 74
Essential oils: *see* Terpenoids
Esterases, 38–40, 172
European corn borer: *see Ostrinia nubilalis*

Fatty acid derivatives
   anandamide, 271
   okadaic acid, 274–275
   synthase genes, 286
Feeding attractant, 60–61, 65, 67–69, 72, 127; *see also* Stimulants
Festuca arundinacea, 83, 88, 92–93, 97, 104–105; *see also* Tall fescue
Fescue toxicosis, 98–99, 103, 107
Firs: *see Abies*
Flavonoids, 273
   cyanidin, 266–267
   isoflavonoids, 130, 144
   luteolin, 266–267
   lutiolindin, 130
   pterocarpans, 130
Forest tent caterpillars: *see Malacosoma disstria*
Fungal endophytes, 190
Fungal natural products, 272, 274–277
   aflatoxin, 276–279, 285
   deoxynivalenol (DON), 276, 281
   FK506, 272
   fumonisin B, 276
   induction, 275
   patulin, 276–277
   trichothecenes, 276–277, 281–283, 285
   zearalenone, 276
Fungal pathogens, 182, 184, 187, 190–196, 203–204
Fungitoxic: *see* Antifungal
Furanocoumarins, 283
   angelicin, 6, 9, 12–13, 18, 283
   begapten, 5, 12–14, 18, 126
   biosynthesis, 6, 8
   enzyme specificity, 7, 15
   detoxification, 15, 283
   cytochrome P450 monooxygenases, 4, 13–15
   inhibition, 15
   distribution, 4–7, 18
   diversity, 3, 18
   evolutionary diversification, 15
   imperatorin, 5
   isopimpinellin, 12
   metabolism, 14
   psoralen, 6, 9, 12, 14, 283

Furanocoumarins (*cont.*)
   sphondin, 5, 13, 126
   toxicity, 283–284
   types
      angular, 5, 12, 18
      linear, 5, 9
   xanthotoxin, 5, 7, 12–14, 126, 283
Fusarium graminearum, 281
Fusarium sambucinum, 123–124
Fusarium solani, 124, 131

Gibberellins, 276, 281–282
Glucanases, 133, 254
β-1,3-Glucans, 133, 135
β-Glucosidases, 38, 40, 45, 217–218, 223–224
Glucosinolates, 59, 63, 66, 68, 70–72, 74, 124
   deterrence, 67, 69, 72
   glucobrassicin, 61, 67
   hydrolysis, 59
   induction, 70
   sensitivity to, 61
   sinigrin, 61, 67
   stimulants, 60–61, 65, 67–69, 72
   structure, 59
   toxicity, 60
Glycosidases, 136, 138
Glucoslytransferase, 222–223
Gossypol, 144
Grass/endophyte associations
   environmental influence, 82, 84, 86, 104, 108
   host-fungal genome influence, 82, 84, 86, 105, 108
   manipulation, 106–109
Growth inhibition, 190, 194, 240
Gypsy moth: *see Lymantria dispar*

Habituation, 73–74, 159, 174
Heliconius, 226
Horizontal gene transfer, 267, 282
Host-recognition cues, 17
Hydrogen cyanide (HCN), 218–219, 222–223, 225
Hydrogen peroxide, 132, 135, 256–257
Hydroxynitril lyases, 217 218, 223–224
6-Hydroxy-2-cyclohexenone (6-HCH), 38–39, 41
Hydroxyproline-rich glycoproteins, 132, 135
Hymenaea, 180, 190–191
Hypersensitive response (HR), 129–130, 132–133, 145, 238, 258
Hypoxylon mammatum, 37

Iberis amara, 60, 64–65, 71, 74
Ips, 189, 209
Immunosuppressants, 272
Immunophilins, 272
Indoles
   camalexin, 131
Inhibition, 189, 191, 193, 196–199, 205
Insecticide, 157, 175; *see also* Antifeedant, Antiherbivore, Biological activity, Botanical insecticide
Isothiocyanates: *see* Mustard oils

Jasmonic acid, 135, 138, 141–142, 244
Jones–Firn hypothesis, 1–2, 4, 7, 15–16, 18; *see also* Screening hypothesis

Laccases, 233
β-Lactam, 285
Large aspen tortrix: *see Choristoneura conflictana*
Larval growth inhibitors, 157–158, 165, 167–169, 171–173, 175
Lectins, 266
Lepidoptera, 218, 226
Lignans, 182, 187
Lignin, 128, 131, 134, 136, 182, 187
Limonoids, 159–161, 163, 165–166, 168–169, 171–175
   azadirachtin, 159–164, 171–174
   azadiradione, 162, 164
   cedrelanolide I, 169–170
   3,12-deacetyl-toosendanin, 165–167
   3,12-dideacetyl-toosendanin, 165–167
   gedunin, 162–164, 166–167
   hirtin, 169–171
   khivorin, 167
   marrangin, 161–162
   meliacins, 162, 164
   nimbin, 163–164, 173
   nymanin C, 169–170
   ochinolide, 163–164
   piscidinol C, 169–170
   salannin, 163–164, 173
   sandoricin, 169–170
   sendanin, 163–166, 171
   3-tigloylazadiractol, 162–163, 173
   toosendanin, 165–167, 172
   turaflorins, 169–170
Linamarase, 224–226
Lipopolysaccharides, 266–267
Lipoxygenase (LOX), 135, 138–139, 141–142

Lolium perenne, 87–88, 92, 96–98, 279–280; *see also* Ryegrass
Lotus corniculatus, 220–221
Lymantria dispar, 35, 42–44, 46

Malacosoma disstria, 35, 38, 42
Manduca sexta, 126
Meliaceae, 159–160, 166, 168–169, 171, 175
  Melia species, 161
  Melia toosendan, 165–166, 172
Methyljasmonate, 138, 141–142, 242–244, 246
Mexican bean beetle: *see Epilachna varivestis*
Mixtures, 3–4, 12–13, 16–17, 66, 68–72, 74, 156, 168, 171, 173–174, 179–209, 273, 302
  defensive characteristics, 204–206
  nutrient interactions, 200–204, 206
  terpenoid, 179–209
Molecular parsimony, 7
Multifunctional: *see* Multiplicity
Multiple defense, 134, 144
Multiplicity, 9, 51, 59, 76, 143, 145, 246, 256, 283–284
Multitrophic interactions, 83, 85–86, 101, 138–139, 145; *see also* Tritrophic
Mustard oils, 59–60
Mustard oil glycosides: *see* Glucosinolates
Mutagenicity, 278–284
Mycotoxins, 274, 277
  aflatoxins, 276–279, 285
Myzus melonis, 144
Myzus persicae, 173

Nasturtium: *see Tropaeolum majus*
Natural products
  biosynthesis, 6, 8, 220–223, 226, 268–269, 274, 277, 281–282, 284–286, 304–309
  classes, 296
  diversity, 7, 15, 19, 181–182, 296, 298–299, 304, 309–310
  ecological roles, 298
  environmental effects, 206–209, 275–276, 286
  evolution, 17–18, 298, 300, 306–309
  genetic control, 182, 203–205, 274, 286
  localization, 274, 276
  origins, 268, 281–282, 286
Natural Selection, 17–18
Neem, 156–157, 161, 172–175
Neem tree: *see Azadirachta indica*
Nitrogen, 126, 200, 209, 276

Oats, 124, 130
Octadecanoid wound signaling pathway, 233, 244–246
Oleoresins, 181–188, 194
  constitutive, 183–185, 188–189, 194
  effects on beetles, 188–194
  mixture effects, 189–194
  variation, 185–187
  wound induced, 183–195, 204, 206
Oligogalacturonides, 143
  galacturonic acid, 143
Onion, 122, 125
Ophiostoma, 191–192, 194–197
Optimal defense hypothesis, 36, 137
Ostrinia nubilalis, 167–169, 171, 174
Oxidation, 231, 233, 237–239, 242
  catechol oxidase, 232–233
  monophenolase, 232–233, 235

P450 dependent oxidase, 223
Pachysphinx modesta, 43–45
Papilio species, 40–41
Parsnips: *see Pastinaca sativa*
Pastinaca sativa, 4–6, 12, 14, 16–17
Pathogen defense, 232, 237–239, 246
Pathogenesis related proteins (PR), 130, 133–135, 140, 146, 254, 256
Penicillin, 276
Peptides
  prosystemin, 138, 142–143
  systemin, 138, 142–143
Peroxidase (POD), 125, 132–136, 138, 140, 233, 238–239, 242
Phenolics, 137, 182–183, 187, 195–197, 200–201, 204, 209, 273
  4-allylanisole, 182, 190–191, 195
  σ-diphenolics, 231, 233–235, 237, 240
  monophenolics, 231, 233–234
Phenolic glycosides, 27, 31, 33–35, 38, 40–46, 48–49, 51
  biological activity, 27, 38
  biosynthesis, 28
  salicylates, 27, 30, 38, 48–49, 51
    salicin, 27–29, 32, 37–39, 47
    salicortin, 27–29, 32, 37–45, 51
    tremulacin, 27–29, 32, 38, 42–45, 51
    tremuloidin, 27–29, 32, 38
  seasonal variation, 31–32
  nutrient effects, 33
Phenoloxidase (PO), 133–134, 136, 138, 142
Phenylpropanoids, 130, 138, 142, 182, 187, 190
  biosynthesis, 237

Phytoalexins, 130–131, 143–144, 283
Phytophotodermatitis, 5, 7
Phytophthora infestans, 133–135
Phytophthora parasitica, 254–255, 258
Pierid butterflies: *see Pieris* species
Pieris species
  Pieris napi, 59–76
  Pieris rapae, 59–76
Pines: see *Pinus*
Pinus, 183, 188–191, 193–194, 199
  Pinus contorta, 184–185, 188, 191
Plant defense, 254, 260
  inducible, 256; *see also* Systemic acquired resistance
Plant nutrition, 70
  nitrogen levels, 70–71
Pleuroplaconema, 206
Polyketide synthetases (PKS), 276–277, 279, 285–286
Polyketides, 276, 285
Polypeptides, 271, 305
  endorphins, 271
  enkephalins, 271
Polyphenol oxidases (PPO), 125, 132–134, 138, 231–246
  ditribution, 235–236
  functions, 233, 236–240, 246
  induction, 236, 241–246
  localization, 236–237
  specificity, 234–235
  structure, 234–235
Populus species, 136
Populus tremuloides, 26; *see also* aspen
Potato, 123, 126, 128, 135, 235
Primary nutrients, 209
Proteinase inhibitors (PI), 138–139, 142–143, 233, 240–246
Prosystemin, 241–244
Protocatechuic acid, 122–123, 125, 232
Pseudomonas syringae, 255, 258

σ-Quinones, 231, 233–234, 238–240, 246

Reactive oxygen species (ROS), 256–257
Redundancy, 19, 75–76, 86, 122, 135, 137, 145–146, 221, 226, 246, 304, 310
Resistance, 169, 173–174
Resource allocation theory, 58, 75, 137–138, 142; *see also* Carbon–nutrient balance hypothesis

Resource availability, 33, 36; *see also* Carbon–nutrient balance hypothesis
Rocaglamide, 169–171
Roridins, 281–282
Ryegrass, 84, 87, 93, 96–97, 100, 103, 105, 107; *see also Lolium perenne*
Ryegrass staggers, 98, 103, 107, 279

Salicaceae, 26, 28, 30, 40, 47
Salicylaldehyde, 45, 48
Salicylate, 135
Salicylic acid (SA), 29–30, 140, 142, 244–245, 256–257, 260
Salix species, 47
Saponins, 139
  avenacins, 124
Scolytidae: *see* Bark beetles
Screening, 2–3, 301–303; *see also* Bioassays
Screening hypothesis, 298–299, 309–310
Secondary metabolism, 1
  diversity, 7, 15, 19
  evolutionary origin, 17–18
Self-assemble, 268–269, 274, 284–286
Self-organizing: *see* Self-assemble
Sensitivity, 59, 68–69, 72–74
Sesquiterpenes, 130, 181–182, 187, 190, 198, 201, 204, 206, 276
Shikimic acid pathway, 27, 36, 51, 190
Signal, 139–143, 146, 233, 241, 244–245, 256–259, 266, 272
Silica, 128
Sipermite: *see Tetranychus urticae*
Solanaceae, 130
Sorghum, 130
Spodoptera litura, 159, 163–165, 167–169, 171–174
Stilbenes, 182, 187, 195
Stimulants
  feeding, 60–61, 66–68, 71–72
  oviposition, 60–62, 65–67, 69, 71–72
Structural–activity relations, 160–168, 175, 286, 303–306
  insect growth regulatory (IGR), 161–162, 173
  quantitative structure–activity relations (QSAR), 161–162
Sugar cane, 128
Suicide substrates, 14–15
Symbiosis, 82–83, 85, 99, 109, 267, 279
  agonism, 82
  mutualism, 82, 85, 109, 280–281
  pleiotropic, 82, 84, 86

Synergistic effects, 4, 12–13, 15, 17, 101, 165–
    166, 171–172, 194, 205, 239, 255, 302
Systemic acquired resistance (SAR), 129,
    134–135, 137, 141, 144–145, 253–260
  constitutive, 253–256, 259–260
  genes, 254–255, 260
    nah G, 256–258
    PR-1, 254, 260
    PR-5, 255, 260
    signals, 256–259
  induction, 253–254
    chemical, 255–259
    pathogen, 254, 258–259
    wounding, 254
Systemic resistance: *see* Systemic acquired resistance
Systemin, 241–246

Tall fescue, 83–84, 87, 90, 96–98, 101–103, 105,
    107; *see also Festuca arundinacea*
Taxanes, 280–281
Taxol, 280
Taxus species, 280
Terpenoids, 142, 179–209; *see also* Limonoids
  chemical defenses, 181, 187–194, 197–200,
    208–209
  coevolutionary interactions, 203
  diterpenes, 181–182, 187–188, 198
  diversity, 181–182
  environmental influences, 206–209
  genetic control, 182, 203–205
  mixtures, 179–209
  monoterpenes, 126, 181–182, 184–194,
    200–202, 206–207
  nutrient dynamics, 200–204
  primary nutrients and, 200–203
  seasonal variability, 207
  sesquiterpenes, 130, 181–182, 187, 190, 198,
    201, 204, 206, 276

Terpenoids (*cont.*)
  tetracyclic, 127
  triterpenoids, 156, 159
    limonoids, 159–161, 163, 165–166,
    168–169, 171–175
Tetranychus articae, 144
Thioglycosides: *see* Glucosinolates
Tiger swallowtails: *see Papilio* species
Tobacco, 128, 133–135, 137, 140, 144, 254–
    256
Tobacco mosaic virus (TMV), 141, 144, 254,
    256, 258
Tobacco necrosis virus (TNV), 144
Tomato, 124, 128, 138, 142–143, 235, 238,
    240–246
Toxicity, 60, 192–194, 199, 202, 205, 278–279,
    284
Traumatin, 141–142
Tremorigenic, 95, 99, 102, 106, 279
Trichomes, 128–129, 235, 239
Trichothecenes, 276–277, 281–283, 285
Tritrophic interactions, 45–46, 101, 104,
    180–181, 273; *see also* Multitrophic
    interactions
Tropaeolum majus, 66, 72–73

Umbelliferae, 283

Vasoconstriction, 102–103
Verticillium albo-atrum, 144

Western spruce budworm: *see Choristoneura
    occidentalis*
Wheat germ diet, 72–73
Wounding, 136–138, 142–143

Zygaena species, 219, 226
Zygaena trifolii, 218–225
Zygaenidae, 218–220